THEORETICAL AND APPLIED MECHANICS

THEORETICAL AND APPLIED MECHANICS

Volume 37

Proceedings of the 37th Japan National Congress for Applied Mechanics, 1987

Edited by Japan National Committee for Theoretical and Applied Mechanics   Science Council of Japan

UNIVERSITY OF TOKYO PRESS

© UNIVERSITY OF TOKYO PRESS, 1989
ISBN 4–13–066104–3
ISBN 0–86008–441–8
Printed in Japan
Second Printing, 1989

All rights reserved. No part of this publication may be reproduced or transmitted in any form or by any means, electronic or mechanical, including photocopy, recoding, or any information storage and retrieval system, without permission in writing from the publisher.

PREFACE

The 37th Congress on Applied Mechanics (NCTAM-37) was held at the Science Council of Japan in Tokyo on December 16, 17 and 18, 1987, under the joint sponsorship of the National Committee for Theoretical and Applied Mechanics of the Science Council of Japan (NCTAM), and the following ten related societies: the Japan Society for Aeronautical and Space Sciences, the Japan Society of Applied Physics, the Architectural Institute of Japan, the Japan Society of Civil Engineers, the Japanese Society of Irrigation, Drainage and Reclamation Engineering, the Mathematical Society of Japan, the Japan Society of Mechanical Engineers, the Mining and Metallurgical Institute of Japan, the Society of Naval Architects of Japan, and the Physical Society of Japan. The Architectural Institute of Japan and the Japan Society of Applied Physics were responsible for the Congress this year, Professor H. Maruo acting as chairman of the Organizing Committee.

The Congress was composed of symposia and technical sessions. Sectional lectures were presented at the symposia on the following subjects: (A) Applied Mechanics of New Materials, (B) Vortices and Waves, (C) Supercomputer and Computational Mechanics and (D) Frontirs of Applied Mechanics.

This 37th volume of NCTAM Proceedings contains thirty-seven contributed papers selected by the Editorial Committee. Partial financial support was given by the Grant-in-Aid for Publication of Scientific Research Result from the Ministry of Education, Science and Culture of Japan, for the publication of this volume.

February 1989

                              Isao IMAI
                                  Chairman, Publication Committee
                                  NCTAM-37 Proceedings

                              Hidenori HASIMOTO
                                  Chairman, Editorial Committee
                                  NCTAM-37 Proceedings

Publication Committee
Chairman
Isao IMAI, Professor Emeritus, University of Tokyo

Masuo HUKUHARA, Professor Emeritus, University of Tokyo
Ryuma KAWAMURA, Professor Emeritus, University of Tokyo
Masataka NISIDA, Science University of Tokyo
Toshie OKUMURA, Professor Emeritus, University of Tokyo
Ko SUZUKI, Professor Emeritus, University of Tokyo
Itiro TANI, Professor Emeritus, University of Tokyo
Yoshikatsu TSUBOI, Professor Emeritus, University of Tokyo
Teruyoshi UDOGUCHI, Professor Emeritus, University of Tokyo
Masao YOSHIKI, Professor Emeritus, University of Tokyo

Editorial Committee
Chairman
Hidenori HASIMOTO, Professor Emeritus, University of Tokyo

Yasuhiko HANGAI, Professor, University of Tokyo
Manabu ITO, Professor, University of Tokyo
Tadahiko KAWAI, Professor Emeritus, University of Tokyo
Kozo KAWATA, Professor Emeritus, University of Tokyo
Hajime MARUO, Professor, Yokohama National University
Fumio NISHINO, Professor, University of Tokyo
Hiroyuki OKAMURA, Professor, University of Tokyo
Koichi OSHIMA, Professor, Institute of Space and Astronautical Science
Masaru SAKATA, Professor, Tokyo Iustitute of Technology
Jumpei SHIOIRI, Professor Emeritus, University of Tokyo
Hideo TAKAMI, Professor, University of Tokyo
Hiroshi TSUJI, Professor Emeritus, University of Tokyo
Yoshiyuki YAMAMOTO, Professor Emeritus, University of Tokyo

CONTENTS

Preface

I. Report on IUTAM Symposium

* Fundamental Aspects of Vortex Motion
.... H. Hasimoto 3

* Introductory Remarks on 1987 IUTAM Tokyo Symposium on Nonlinear Water Waves
.... K. Horikawa 9

II. Shock Waves

Reflection of Plane Shock Waves Over a Dust Layer
.... T. Adachi, S. Kobayashi, and T. Suzuki 23

Shock Wave/Boundary Layer Interactions Induced by a Flat-Faced Fin
.... N. Saida 31

Performance of a Slotted Wall Test Section for Aerodynamic Testing in a Shock Tube
.... Y. Yamaguchi 39

III. Jet and Spiral Flow

Wall Jet of Ideal Fluid Flowing along Deflected Surfaces
.... M. Hatano 51

A Bounded Jet Flow Considering the Initial Turbulence
.... M. Nakashima, T. Nozaki, K. Hatta, and T. Mori 63

A Study of Short Fibre Transportation Using a Spiral Flow
.... Y. Matsumae, X.M. Cheng, M. Takei, and K. Horii 73

IV. Numerical Dynamics

Breakdown of the Karman Vortex Street Due to Natural Convection: Numerical Simulation
.... K. Noto and R. Matsumoto 83

An Implicit, High-Order, Accurate Upwind Scheme for Unsteady Euler Equations
.... K. Matsuno 95

Numerical Simulation of Flow Ejected from a Nozzle by Electric Force
.... Y. Takeda, D. Takahashi, K. Ishii, and H. Takami 105

---

* Invited lecture papers.

Effect of Ground on Wake Roll-Up behind a Lifting Surface
.... K. Zhu and H. Takami   115

Macroscopic Dynamic Simulations of the Impact of Flexible Beams
.... K. Yamamoto   125

Study of Response Phenomena to the Change of External Field in Vortex-Like Phase for XY-Type Spin Glass State by Computer Simulation
.... Y. Natsume, K. Fujimoto, and T. Yoshihara   137

V. Mechanics of New Materials

* Mechanics and Design of Composite Materials and Structures
.... I. Kimpara   149

* Optimum Design of Composite Structures
.... Y. Hirano   159

Calculation of the Two-Dimensional Effective Thermal Conductivity of Media with Regularly Dispersed Parallel Cylinders
.... N. Watari and N. Oshima   171

Anisotropic Theory of Rapidly Deforming Granular Assembly
.... M. Nakagawa   183

Application of the Method of Continuous Distribution of Dislocations to an Interface Crack Problem
.... H. Kasano, K. Shimoyama, and H. Matsumoto   195

Scattering of Elastic Waves in a Particulate Composite with Interfacial Layers (Phase Velocity and Attenuation)
.... Y. Shindo, S.K. Datta, and H.M. Ledbetter   207

Whole-Field Strain Measurements by Moiré and their Application to Composite Materials
.... K. Kunoo, H. Ohira, K. Ono, and N. Sato   219

VI. Mechanics in New Environments

* Space Structures and Applied Mechanics
.... K. Miura   227

* Magnetomechanical Behavior of Solids
.... K. Miya and T. Takagi   235

VII. Waves and Vibrations

Stability of a Rotor Containing Viscous Fluid
.... S. Morishita and K. Okuzono   253

Natural Vibration Analysis of a Five-Span Continuous Rigid-Frame Bridge with V-Shaped Legs
.... T. Hayashikawa and N. Watanabe   263

Dynamic Response Analysis of Specimen-Load Cell Systems in an Impact Loading Based upon Wave Propagation Theory
.... M. Itabashi and K. Kawata   273

Self-Excited Oscillation of a Closed-Engine-Governor Loop: Equilibrium
  Instability and Limit Cycle
.... Y. KAWAZOE 281
Free Vibration Analysis of a Distributed Flexural Vibrational System by
  the Tranfer Influence Coefficient Method
.... T. KONDOU, A. SUEOKA, D.H. MOON, H. TAMURA, and T. KAWAMURA 289
Characteristics of Acoustic Radiation from Plate Girder
  Railway Bridge
.... T. SUGIYAMA, Y. FUKASAWA, and Y. GOMI 305
Chaos in Autonomous Elastic Systems
.... K. SUMINO, H. SOGABE, and Y. WATANABE 315

VIII. Elasticity, Buckling, and Fatigue

Expression of Three-Dimensional Deviatoric Tensor through
  Complex Number ω
.... M. SATAKE 329
Stress Analysis of an Elastic Half-Space Having an Axisymmetric Cavity
.... H. HASEGAWA and K. KONDOU 339
Inplane Strength Analysis of Tapered Thin-Walled Steel Beam-Columns
  by Dynamic Relaxation Method under Displacement-Control Loading
.... I. MIKAMI, Y. MIURA, S. TANAKA, and Y. SHINNAI 347
Stability Analysis of Circular Cylindrical Shells Subjected to
  Tangential Follower Force
.... K. MITSUI and K. SUMINO 361
Buckled Configuration of Laterally Supported Columns in the
  Inelastic Range
.... J. SUZUYA 373
Buckling of an Annular Sector Plate Subjected to In-Plane Moment
.... K. TAKAHASHI, Y. KONISHI, Y. NATSUAKI, and M. HIRAKAWA 381
Development of Rail/Wheel High-Speed Contact Fatigue Testing
  Machine and Test Results
.... M. ISHIDA, Y. SATOH, Y. SATO, and S. MATSUYAMA 389

Index of Authors 397

I
REPORT ON IUTAM SYMPOSIUM

Fundamental Aspects of Vortex Motion

——IUTAM Symposium——

Hidenori HASIMOTO

Department of Mechanical Engineering, Hosei University, Tokyo

This is a short report on the IUTAM Symposium 'Fundamental Aspects of Vortex Motion' held in Tokyo from 31 August to 4 September 1987. It gives the background and the motivation of the Symposium and a brief sketch of the program and the lectures which appear in the Proceedings of the Symposium as a special issue of *Fluid Dynamics Research 3* (1988).

I. INTRODUCTION

The International Symposium on Fundamental Aspects of Vortex Motion was held in the Congress Hall of the Science Council of Japan from 31 August to 4 September 1987. This Symposium was planned by our National Committee of Theoretical and Applied Mechanics and was proposed by the chairman of the Committee, Professor Imai, to the International Union of Theoretical and Applied Mechanics (IUTAM).

The Symposium was motivated by discoveries of various phenomena in fluid flows, which could probably be understood on the basis of the fundamental properties of vortex motion. These phenomena were revealed by the advent of high-speed computers and new experimental techniques and have been explored by many researchers whose work has been recognized internationally.

The proposal was approved by the General Assembly of IUTAM in August 1984, and I was asked to chair the International Scientific Committee whose other members were H. Aref (San Diego, U.S.A.), J.J. Keller (Baden, Switzerland), T. Maxworthy (Los Angeles, U.S.A.), D.W. Moore (London, U.K.), R. Moreau (Grenoble, France), E.-A. Müller (Göttingen, West Germany), P.G. Saffman (Pasadena, U.S.A.) and L. van Wijngaarden (Enschede, Netherlands).

It is a pleasure to note that the IUTAM Symposium on Nonlinear Water Waves was also held before this symposium at the University of Tokyo, co-chaired by Professor H. Maruo and Professor K. Horikawa. A detailed report is given by Horikawa in this volume.

The local organizing committee for our symposium, chaired by Professor Imai, began its work in close contact with the scientific committee. We were also supported by a number of organizations:

Architectural Institute of Japan, Japan Society of Applied Physics, Japan Society of Civil Engineers, Japan Society of Fluid Mechanics, Japanese Society of Irrigation Drainage and Reclamation Engineering, Meteorological Society of Japan, Oceanographical Society of Japan, Physical Society of Japan and Science Council of Japan.

Since the Proceedings of the Symposium were published as Volume 3 (1988), 1–440, of

Fluid Dynamics Research and in book form as *Vortex Motion* (North-Holland), and an excellent review by H. Aref and T. Kambe appeared in the *Journal of Fluid Mechanics* (1988), 190, 571–595, I shall confine the present report to a short sketch.

II. THE BACKGROUND AND A SHORT SKETCH OF THE SYMPOSIUM

Vortex motions, extreme examples of which are big eddies in the Naruto Straits or the eye of a typhoon, govern all kinds of universal phenomena. Tip vortices of wings and back vortices behind buildings, bridge girders, automobiles and ships are typical common examples.

Starting with quantum vortices in liquid helium and the flow of protoplasma in organic cells to the atmosphere and the ocean of the earth, and then to the great black hole, our cosmos is full of vortices. It is well known that we cannot determine the fluid phenomena without information on the distribution of vortices.

In spite of their universality and importance, the motions of vortices present many unsolved problems, and their study is one of the difficult branches of fluid mechanics.

The advent of high-speed computers and the recent remarkable progress in experimental techniques are revealing many phenomena which probably could be understood only by the exploration of fundamental properties of vortex motion.

In fundamental research on vortex motion, Japan played a leading role; the names of T. Terada, K. Terazawa and S. Fujiwara were prominent in the early years, and many important contributions have been and are being produced in Japan. This is the background to the opening of the Symposium in this country and the reason for the number of participants from Japan, i.e., 63.

The total number of participants from abroad was 47: 11 from the U.S.A., 9 from West Germany, 6 from France, 5 from England, 5 from The Netherlands, 3 from Australia, 2 from the People's Republic of China, and 1 each from Canada, Israel, New Zealand, Switzerland, Norway, and Taiwan.

Particular attention was paid to the following topics: 1. Formation and modeling, 2. Numerical methods and numerical simulation, 3. Measurement and visualization, 4. Mechanism of concentration, entanglement and self-interaction, 5. Formation of cell structure and large-scale structure, 6. Instability, chaos and statistics of vortices, 7. Vortex shedding and vortex motion around solid bodies, 8. Vortex-induced vibrations and waves, and 9. Vortex motion in a rotating, stratified or reacting fluid.

The same topics are also reflected in seven general lectures:

H. Hasimoto on Elementary aspects of vortex motion, P.G. Saffman on The stability of vortex arrays to two- and three-dimensional disturbances, H.K. Moffatt on Generalized vortex rings with and without swirl, J.J. Keller, W. Egli and R. Althaus on Vortex breakdown as a fundamental element of vortex dynamics, E.-A. Müller and F. Obermeier on Vortex sound, T. Maxworthy on Waves on vortex cores, H. Aref, J.B. Kadtke, I. Zawadzki, L.J. Campbell and B. Eckhardt on Point vortex dynamics: recent results and open problems.

It was a source of great pleasure that many proposals for papers were made. Except for a few which were outside our subject scope, they were all excellent, and we were obliged to divide the conference into two sessions, oral (25 min.) and visual. The latter were a new

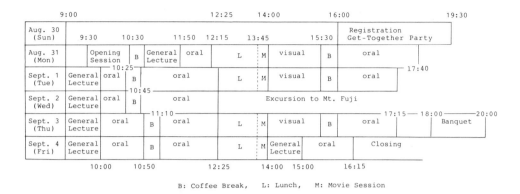

Fig. 1. Time schesoul.

attempt to arouse intimate contact between the audience and the authors, making efficient use of various media of communications, i.e., posters, videos, computer simulations, exhibitions of experiments, and films. They were preceded by short essentical introductory speeches (6 min.) and discussions organized by the skillful chairmen, Hama, Tsinobar and Zabusky.

We still remember the atmosphere of excitement in the well-equipped auditoria of the Congress Hall, where oral and movie sessions were also held.

Starting with the opening address by Professor Imai, the technical program was busy until the closing evening of September 4, except for the afternoon of September 2 which was devoted to an excursion to Mt. Fuji. The symposium was preceded by a get-together party on the evening of August 30. A banquet was held at the Kasumigaseki Building on Thursday evening, and some participants went on a post-Congress tour to Nuruto Channel to see the big eddies.

The number of papers selected from among more than 70 proposals was 64; these were divided into 35 oral and 29 visual presentations. For the proceedings they were classified into eight categories: two-dimensional vortices (10), vortex rings and three-dimensional vortices (17 papers), reconnection of vortices (4), vortex breakdown (2), stability and turbulence (10), vortex and sound (7), vortices in compressible fluid (4) and vortices in statified and rotating fluid (6), though they are interrelated too closely to be classified definitively.

These categories are represented and correlated by the seven general lectures. In the opening lecture, various elements of vortices are introduced in terms of the transient flow past a point force singularty. After a broad overview of the Symposium, the present status of idealization such as point vortices or vortex filaments was discussed with reference to several examples, i.e., the twisting solitary wave on a filament and then regular or chaotic motions of point vortices on the basis of Hamiltonian formalism.

The latter was surveyed comprehensively by Araf et al., who reported further remarkable progress in determining steady moving configurations and numbers of point vortices, and in discovering so-called chattering, i.e., spikey change in scattering time of two vortex pairs according to change in their impact parameters. Eight open questions were also presented.

If cross-sections of vortices are finite, various new phenomena are revealed, as shown clearly by Saffman's survey. Starting with bifurcations into many branches of the form of steadily rotating vortices with or without external strain, stability analyses are extended to the pair, the single row and the Karman street of vortices including three-dimensional perturbations. The onset of singularity in the boundary curve or the amalgamation of vortices is heralding the repeated filamentation or reconnection of vortices discussed by many participants.

The topological approach of Moffatt will be useful in choosing general essential features without grappling with complicated details. Using analogies of approach to magnet static equilibria in a viscous perfectly conducting fluid, an entire class of axisymmetric vortex rings with swirling velocity could be presented.

Maxworthy's discussions include waves of varicose, helicoidal and fluted form and their interactions, including recent experiments on liquid He 4. Various points of controversy in the non-linear regime are pointed out, related to soliton propagation and vortex breakdown.

The vortex breakdown in tubes of varying cross-sections is discussed comprehensively by Keller et al., who confined their study to the axisymmetric transition, which was perfectly explained by an extension of Benjamin's hydraulic theory.

Müller and Obermeier's stress on the vortex motion as the only source of aerodynamic sound in low Mach number flow was impressive. Various sources such as elliptic vortex rings, colliding circular rings, vortex-body, etc., are taken as examples of agreement between theory and experiment. Scattering of sound by vortex motion was also included.

Remarkable progress was reported and discussed with enthusiasm. An example is intensive studies concentrating on the mechanism of reconnection processes, the presence of viscosity being of course necessary.

Numerical studies by Kida and Taoka as well as Melander and Zabusky reveal the process of fingering and bridges after antiparallel alignment, which could be explained partially by Takaki and Hussain's analytical treatment with the emission of a vortex ring in mind. Related work was presented by Oshima et al. on elliptic rings as well as by Kiya and Ishii on the interaction of vortex line and ring.

These results are also tied to the question of conservation of helicity, as was pointed by Moffatt. The appearance of helicity of different sign in grid turbulence at different places reported by Tsinobar is also related to this problem and is waiting for clarification in the near future, e.g., on, the occasion of the IUTAM Symposium on Topological Fluid Mechanics to be held in Cambridge in 1989.

Acknowledgment

We should like to express our gratitude for the generous financial support from the Commemorative Association for The Japan World Exposition (1970), The Mitsubishi Foundation, the Shimadzu Science Foundation, the Turbulence Fund, Asahi Kogyosha Co., Ltd., Ebara Corporation, Fuji Heavy Industries, Ltd., Fuyō Data Processing & Systems Development, Ltd., Graphtec Corp., Hino Motors, Ltd., Hitachi, Ltd., Honda R & D (Wako Center), Kajima Corporation, Kanomax Japan, Inc., Maezawa Industries, Inc., Mazda Motor Corporation, Mitsubishi Motor Corporation, Nissan Diesel Motor Co., Ltd., Nissan Motor Co., Ltd., Ricoh Company, Ltd., Tokyo Electric Power Company, Inc., and Toyota Motor Corporation, in addition to our sponsor, the International Union of Theoretical and Applied Mechanics.

Our thanks are also due to the 36 members of the local Organizing Committee listed below, especially to its executive members; without their devotion the success of the Symposium would not have been achieved.

Local Organizing Committee:

Professors and Doctors I. Imai (Chairman), H. Hasimoto (Vice-Chairman), T. Asai, H. Fujita, K. Gotoh, M. Hino, K. Horikawa, K. Horiuti, K. Ishii, T. Ishihara, T. Kakutani, T. Kambe (Secretary), Y. Kaneda, K. Kawada, M. Kawaguti, R. Kawamura, S. Kida, K. Kusukawa, S. Kuwabara, K. Kuwahara, H. Maruo, T. Matsui, Y. Matunobu, S. Mizuki, J. Mizushima, H. Ohashi, M. Ohji, Y. Oshima, O. Sano, H. Sato, R. Takaki, H. Takami, S. Taneda, I. Tani, T. Tatsumi, Y. Yamamoto, A. Yoshizawa.

Introductory Remarks on 1987 IUTAM Tokyo Symposium on Nonlinear Water Waves

Kiyoshi HORIKAWA

*Department of Civil Engineering, The University of Tokyo, Tokyo**

The IUTAM Symposium on Nonlinear Water Waves was held in August 1987 at the University of Tokyo. The background, aim, and program of the symposium were given first, and then the contents of the three keynote lectures were briefly described. Other papers presented at the 10 oral sessions and the two poster sessions were classified into two categories, the characteristics of nonlinear water waves and the nonlinear interaction between water waves and bodies, and reviewed shortly one by one. Based on the above treatments, a state-of-the-art summary of nonlinear water wave research was made.

I. INTRODUCTION

I-1. Background of the Symposium

The aim of this paper is to introduce the present research activities on nonlinear water waves by reviewing the papers presented at the International Union of Theoretical and Applied Mechanics (IUTAM) Tokyo Symposium on Nonlinear Water Waves in August 1987.

Nonlinear water waves have been treated since the 19th century as one of the most important subjects in the field of fluid mechanics in relation to ocean waves and tidal current. In recent years, numerous marine structures have been constructed in order to exploit and develop the nearshore as well as offshore zones for human activities. Ocean waves are the most important external forces for such marine structures. Therefore the characteristics and behavior of ocean waves have been intensively studied by researchers in related engineering fields.

The subjects treated in the symposium were, for example, 1) to clarify the characteristics of nonlinear water waves in shallow water including breaking waves; and 2) to evaluate the wave forces, especially the impact forces of breaking waves, acting on marine structures. In addition to the above, it should be mentioned that the nonlinearity of ocean waves induces drifting forces acting on a floating body, and hence the nonlinear interaction between a floating body and the drifting force should be clarified to ensure the stability of ships and floating marine structures. The tsunamis generated by the submarine earthquake in the Sea of Japan in 1983 displayed remarkable nonlinear behavior such as the transformation of solitons in the nearshore area. Additionally, researchers in the field of coastal engineering are interested in the effect of nonlinearity of ocean waves on the nearshore current velocity and the coastal sediment transport processes.

As stated previously, water wave study has been one of the important fields of fluid

---

* Present address: Department of Foundation Engineering, Saitama University, Urawa.

mechanics. However the governing equations of wave motion include nonlinear terms, and this fact makes solving such problems difficult in general. Therefore analysis of water waves progressing without deformation has been made since the 19th century. Among these treatments, the trocoidal wave theory in deep water introduced by Gerstner is well known as an exact solution of waves with vorticity. Another solution of water waves in deep water was obtained by Stokes (1847) as an assymptotic expansion of physical quantities. However, in practical cases mainly concern shallow water waves, and therefore numerous expressions corresponding to the above deep water waves have been presented by many researchers.

Another branch of study of water waves is investigation of a solitary wave. The existence of solitary waves was discovered by Russell (1844) and theoretical treatment of this wave type was made by Boussinesq (1871) and Rayleigh (1876). The solitary wave is a nonperiodic, translatory wave progressing in very shallow water without any deformation. Korteweg and de Vries (1985) introduced the so-called K-dV equation to express the nonlinear water waves in such a shallow region. The permanent type progressive waves based on the K-dV equation is called cnoidal waves. Since 1950, the governing equations of nonlinear wave motions in some fields of physics have been investigated and realized to be quite similar to the K-dV equation. Zabusky-Kruskal (1965) solved the K-dV equation numerically by using the newly developed electronic computer and discovered the particular characters of the so-called "soliton." The soliton appearing in very shallow water has attracted great interest from scientists and engineers since 1970.

The Stokes wave theory and the cnoidal wave theory give approximate solutions of water waves based on the perturbation method of the nonlinear equation; however, the numerical calculations treat the exact nonlinear equations by using electronic computers. The marker and cell (MAC) method and the boundary integral method are typical examples of such treatment, and have been established as rigorous methods.

I-2. Objective of the Symposium

Although nonlinear water waves have attracted interest from researchers in both natural and applied sciences, there have been very few chances to exchange views among various disciplines. Even in the engineering field, coastal engineers in civil engineering have had almost no chance to discuss with naval architects. Therefore the aim of the symposium was to rectify in part this situation, and to promote future cooperation among the related fields.

I-3. Outline of the Symposium

The symposium was held at the Sanjo Conference Hall of the University of Tokyo from August 25 to 28, 1987, under the auspices of the University of Tokyo and the Yokohama National University. The cooperating scientific organizations were the Science Council of Japan, Japan Society of Civil Engineers, Society of Naval Architects of Japan, and 12 other societies. The sponsors were IUTAM, the Commemorative Association for the Japan World Exposition, Inoue Foundation for Science, Japan Society for the Promotion of Science, and the Kajima Foundation.

The symposium was organized under the following IUTAM guidelines:
1) Participation was restricted to invited scientists.
2) The number of invited scientists was decided by the Scientific Committee.

3) It was the duty of the Scientific Committee to locate and select active scientists in specified fields (often young people).
4) Young scientists from developing and from small countries were given a fair chance to participate.

Therefore the selection of participants was made by the Local Organizing Committee by referring to the scientists nominated by the Scientific Committee members and by considering the balance of specialties as well as of geographical distribution.

The organization of the symposium consisted of the Scientific Committee (9 members), Local Organizing Committee (19 members), and Exective Committee (14 members). Professor H. Maruo of the Yokohama National University (naval architecture) and the present writer (civil engineering) were co-chairmen of the symposium. The symposium program was made up of three keynote lectures, 35 papers for oral presentation, and 16 papers for poster sessions, among which 54 papers in total were presented at the symposium while five speakers cancelled. The total number of participants was 110 from 16 different countries.

I-4. Proceedings of the Symposium

Extended abstracts were distributed to the participants in order to encourage discussion at the symposium. All of the full papers submitted to the Local Organizing Committee by the authors were reviewed carefully by myself with the coordinative assistance of Dr. Hisashi Kajitani, Professor of Naval Architecture, The University of Tokyo. As a co-editor of the proceedings, I would like to express sincere thanks for the valuable contribution of Professor Kajitani to the overall editing work. The proceedings was published in 1988 through Springer-Verlag.

II. STATE-OF-THE-ART ON NONLINEAR WATER WAVE RESEARCH

II-1. Introduction

It seems fairly difficult to describe the research trends in nonlinear water wave research by reviewing the papers presented at the symposium. The subjects treated in these papers can be classified into two categories: the characteristics of nonlinear water waves, and the nonlinear interaction between water waves and a body. In the following, an outline of these subjects will be made separately.

II-2. Topics of Keynote Lectures

Three well-known professors were invited to give lectures from three different fields: Professor C.C. Mei, MIT; Professor D.H. Peregrine, Bristol University; and Professor D.M. Faltinsen, the Norwegian Institute of Technology. The title of the first keynote lecture was "Nonlinear effects in water wave diffraction" made by Professor Mei, in which various aspects of weakly nonlinear water waves with a narrow band spectrum were treated. The subjects introduced in this paper were: 1) slow sway of a floating body; 2) trapping of long waves on a wide ridge in short swells; 3) penetration of long internal waves behind a breakwater in short surface waves; 4) radiation of long waves over periodic bars and drag scattering; and 5) forward scattering and the nonlinear parabolic approximation. Professor Mei demonstrated that the treatment of nonlinear water waves is a powerful tool to explain the detailed mechanism of complex phenomena which cannot be treated precisely by the linear wave theory.

The next keynote lecture was "Recent developments in the modelling of unsteady and breaking water waves" by Professor Peregrine. A water wave increases its height and its asymmetry in water surface profile due to water depth decrease and finally breaks under certain conditions. The breaking process is quite interesting in the physical sense and is also very important from the engineering viewpoint. Although a number of studies have been made to clarify the physical criterion of wave breaking analytically or experimentally, our knowledge of wave breaking is still far from complete. In order to analyze the phenomenon, there are two approaches. The first approach, represented by the MAC method, is to analyze the governing equations of fluid dynamics in the whole fluid field. The other approach is to treat the problem by the boundary integral methods such as the Green function method. In the latter treatment the unknown variables on the relevant boundary are only calculated. Therefore this treatment is superior to the former from the viewpoint of numerical computation time. Longuet-Higgins and Cokelet (1976) were the first to apply the boundary integral method to analyze wave breaking in deep water. This method was extended further by Peregrine et al. (1980) to treat wave breaking in shallow water, by which they successfully calculated the processes of breaking such as overhanging of surface profile and attaching the overhung surface with that of the preceding wave. However, as the detailed structure of breaking waves is still unclear, further effort is needed to understand the real phenomenon and to model the wave breaking appropriately.

The last keynote lecture was "Second-order nonlinear interactions between waves and low frequency body motion" by Professor Faltinsen. In this paper, recent studies on the behavior and stability of a ship as well as a marine structure in a high sea were reviewed. The perturbation method was applied to obtain the second-order solution with respect to the incident wave height. That solution is useful to analyze the drift force acting on a floating body and the resulting solw drift motion of the body. In this paper a two-dimensional problem was selected to explain the procedures and the numerical results. This procedure can be extended rather easily to three-dimensional problems. The numerical results of drift forces acting on a circular cylinder with its axis in the mean free surface indicate that 1) without a uniform current, the effect of viscous roll damping on drift forces appears remarkably near the roll resonance frequency, and 2) the existing uniform current induces a strong influence on drift forces.

II-3. Characteristics of Nonlinear Water Waves

Papers related to the characteristics of nonlinear water waves were presented at eight sessions. Below, S and P represent oral and poster sessions, respectively, and the figures indicate the session number.

S-1: Evolution of nonlinear water waves (two papers)
S-2: Nonlinear shallow water waves (five papers)
S-3: Nonlinear water waves in a finite region (five papers)
S-4: Breaking water waves (four papers)
S-5: Nonlinear water waves (four papers)
S-6: Wave-current interaction (three papers)
S-10: Second-order nonlinear water waves (two papers)
P-1: Nonlinear water waves (seven papers)

S-1: This session consisted of only two papers. The first paper analyzed the wave packet on a sloping bottom in shallow water based on the nonlinear Schrodinger (NSL) equation and verified the adaptability of the calculated results by comparing with the laboratory data. In this paper the bottom slope was fixed at 1/50 and the transformation of the wave packet on the sloping bottom was classified into three types: 1) symmetric type; 2) forward-leaning type; and 3) forward-leaning and fission type. The second paper studied the wind-generated gravity-capillary waves. The motivation of this study came from the application of remote sensing techniques, such as synthetic aperture radar (SAR), that can provide information on surface waves with the wavelength on the order of 4–40 cm. This kind of waves will be modulated strongly in shallow water by the existence of nonuniform current and varying bottom topography. In this paper the initial evolution of the waves was investigated by means of a dynamic model in which the effect of wind input, viscous dissipation, and three-wave interactions were included. The results of investigation showed that the nonlinear three-wave interaction induces an important effect on wave evolution comparable to that of wind.

S-2: This session consisted of five papers treating the behavior of shallow water waves, solitons, generated under various conditions. The first paper was an analytical treatment of nonlinear forced water waves in a shallow channel on the basis of the Kadomtsev-Petviashvilli (KP) equation. According to the results of that investigation, the steady-state input for nonlinear water waves differs remarkably from that for linear water waves. The second paper was a numerical study on the water surface fluctuation generated by three-dimensional disturbances moving in a channel based on the generalized Boussinesq (GB) model. It is interesting to note that the forerunner is a series of solitons, but the trailing wave pattern is conspicuously three-dimensional. The third paper described a numerical result on the soliton reflected at a sloping beach. The Boussinessq (B) equation was used as the governing equation of this phenomenon, and the edge-layer theory developed by the authors was applied to calculate the incident and reflected waves separately. The result indicated that the reflected wave evolves into one soliton accompanied by an oscillating tail. At the early stage, the waveform appears to evolve into a series of solitons. Later, only one sharp soliton emerges, leaving the oscillating tail behind. The peak height of the surviving soliton is slightly less than that of the incident one. In the next paper the higher order approximation of the KdV equation was introduced and applied to the interaction problem between two solitary waves of different size. The result indicated that the larger (smaller) solitary wave increases (decreases) its height and that a third new solitary wave appears. In the last paper the distribution of energy spectrum of strongly nonlinear swell in shallow water was derived theoretically. The basic assumption was that the shallow water waves are the gathering of solitons which satisfy the KdV equation. Then the energy distribution function was determined by the maximum probability condition, and was compared with field data recorded off the Ogata Coast facing the Sea of Japan and off Caldera Port in Costa Rica facing the Pacific Ocean.

S-3: In this session several features of nonlinear water waves in a limited region were treated. The first paper was a theoretical as well as experimental study on nonlinear sloshing waves generated in a tank. Comparison between experimental results and numerical results based on the NSL equation indicated that 1) the agreement is good qualitatively, but not good quantitatively, and 2) unsteadiness of the wave field always appears in the numerical calculations, but does so in the experimental results only at large forcing amplitude. Based on the above results the authors suggested the necessity of introducing the dissipation term into the governing NSL equation. The second paper discussed the nonlinear waves excited near resonance in a circular basin by storms or earthquakes. In this treatment the author introduced linear damping due to friction within a boundary layer and calculated coefficients in a Fourier-Bessel representation of the fully nonlinear water surface boundary. In the third paper two methods were presented to analyze numerically the two-dimensional nonlinear wave motion using the Lagrangian description. The first one is based on velocity formulation and the second on pressure formulation. The above two methods were applied to analyze finite-amplitude wave behavior, such as sloshing of water in a rectangular tank, and the propagation and reflection of a solitary wave. As far as the waves with moderate amplitudes are concerned, solutions do not significantly depend on the computation method. The fourth paper was related to the nonlinear problem of determining the position of a totally submerged body and to the free surface nonlinearity. The last paper was a theoretical treatment of nonlinear standing waves.

S-4: This session included four papers treating breaking wave characteristics from various approaches. The first three papers observed the steady breaker precisely and formulated the wave breaking phenomenon. In the first paper a submerged two-dimensional foil was set at a water depth below the free surface with an attack angle of 2 degrees to a steady flow in a water-circulating flume. Based on the measuring data of mean velocity and turbulent intensity above the foil, the breaking mechanism was assumed in such a way that the excess flow energy stored at the wave crest due to the increment of free-surface elevation is dissipated through the turbulence production. The stated mechanism of wave breaking was treated as an instability problem of inviscid fluid to introduce the breaking criterion. In the second paper a foil set along the free surface of steady flow was considered to model the spilling breaker. In the third paper, the front part of a spilling breaker was assumed to be a stagnant eddy riding on the front of an underlying gravity wave. The equilibrium of the steady spilling breaker thus defined is the result of balance between the hydrostatic pressures due to the weight of the eddy and the friction on its underlying surface. Following the above concept, nonlinear unsteady equations were used to study the onset of breaking, breaking growth, and history. In the last paper, a modelling and laboratory investigation on the combined action of progressive waves on a sloping beach with the onshore current was reported. This study was based on the concept that the nonlinear interaction between breaking waves and cross-shore currents exists in field. According to that investigation, it was concluded that the wave height and the wave set-up in the breaker zone receive weak influence of the mean flow, although the vertical distribution of water particle velocity in the surf zone shows remarkable difference depending upon the mean flow velocity.

S-5: In this session various aspects of nonlinear water waves were discussed. In the first paper, cross waves generated by a wavemaker were treated theoretically as well as experimentally. A method based on the modified NLS equation which contained a viscous effect was presented. It was confirmed that the method of solution stated in this paper considerably reduced the amount of computation in comparison with the usual perturbation method. The next paper treated the nonlinear evolution equation in which the terms of self-excitation, dispersion, and dissipation are included in order to analyze the features of soliton-like pulses. The initial value problem, steady pulse solution, pulse interaction, and soliton lattice were discussed. The third paper presented a method to analyze the transformation of nonlinear water waves owing to an arbitrary bottom profile using the conformal mapping method. As an example, nonlinear wave evolution over a stepped sea bottom was numerically calculated. In the last paper it was pointed out that the application of the Hamiltonian method is valuable in order to evaluate appropriately the numerical calculation results.

S-6: In this session, the interaction between surface waves and current, and interfacial waves were disucssed. The first paper showed the modelling of the amplification of current effects on short waves by a long wave field, the phenomena of which can be noticed in the analysis of images collected by the SEASAT Synthetic Aperture Radar. This model was applied to images of the English Channel and the Georgia Strait. The second paper demonstrated that the wind velocity shear has an important effect on the stability of nonlinear surface water wave trains. According to the theoretical treatment based on the NLS equation, the instability of Stokes wave trains will be amplified in the case of small surface velocity shear, while they are suppressed in the case of large shear. The above results well explain qualitatively the experimental evidence observed in wind-wave flumes. The last paper presented the results of numerical calculations of interfacial solitary waves with large amplitude. By changing the ratio between the densities of the upper and lower layers as well as the dimensionless vorticity of the upper layer, the interfacial solitary wave profiles were calculated and compared each other.

S-10: Second-order nonlinear water waves were discussed in this session. In the first paper the authors stressed the important effects of second-order components in nonlinear water waves on such practical problems as run-up of irregular waves on a sloping beach and a response of a moved vessel. From such a viewpoint, reproduction of second-order waves in the laboratory and analysis of prototype wave data were discussed. In the second paper the deformation of the two-dimensional low-frequency wave spectrum in the shallow water region was treated under the recognition of the importance of the second-order low frequency waves in the various phenomena appearing in the shallow water region.

P-1: Numerous topics related to nonlinear water waves were discussed in this poster session. The first paper described experimental results on the nonlinear resonant interaction between two waves with different periods generated by two wavemakers set in a ship model basin with 90 degrees. As a result of two-wave interaction the third wave appears. The spatial distribution of the above wave amplitude was compared with the numerical solution based on the Zakharov equation. The agreement is fairly good when the wave steepness of incident waves is less than a certain value. The second paper treated the transformation of irregular waves propagating on a uniformly sloping beach. The wave profile was measured at six sites, while the computation of wave profiles was made starting from the data recorded at the point nearest to the wavemaker on the basis of the B equation. Agreement between the recorded data and the calculated ones seemed fairly good. The third paper gave the numerical results of one-dimensional and weakly two-dimensional waves in varying channels based on the KP equation. The phenomena of solitary wave scattering by a field of bars and solitary wave evolution in a diverging channel were treated by the one-dimensional model. Shallow water wave focusing caused by shoals was analyzed by the two-dimensional model and compared with laboratory data on the formation of Mach-stem due to grazing incidence on a vertical wall. The next paper described a theoretical and experimental treatment of precursor solitary waves generated by moving disturbances. The GB equation and the inhomogenous Korteweg-de Vries (IKdV) equation were applied for the numerical calculation of the problem, and compared with the data from laboratory investigations. Based on the above comparison the characteristic difference of the governing equations was discussed. The subsequent three papers related to breaking waves. The first paper discussed the transformation of a near-breaking solitary wave propagating on a uniform water depth. The authors calculated the long-term evolution of the instability of a solitary wave by using the time-stepping method. The results showed that there are two cases, breaking and nonbreaking, depending upon the magnitude of the parameter which represents the degree of initial perturbation. The second paper was a numerical study of wave breaking on a beach using the MAC method in which the Navier-Stokes equations of motion and of continuity were used as governing equations. In the third paper newly developed techniques to observe the surface profile and water particle velocities of breaking waves and the results obtained were presented.

Numerous results related to nonlinear water waves were presented in these sessions. Instead of summarizing them, I would like to comment on them below.

Studies on nonlinear water waves were initiated in the 19th century. One important achievement has been the formulation of the KdV equation (1895), which was neglected for many years due to the difficulty in obtaining solutions. In the 1960s it was recognized that the governing equation of nonlinear oscillation in the numerous physics fields is similar to the KdV equation. Some years later the method of numerical analysis was widely applied to analyze nonlinear equations where the KdV equation was a typical example. Then Zabusky and Kruskal (1965) discovered the soliton, a finding that aroused keen interest in the characteristic behavior of solitons as water waves. A number of subsequent theoretical studies have been carried out on related subjects, mainly using numerical analysis. However, in engineering disciplines the study of nonlinear long waves in shallow water was initiated in the 1980s in relation to the behaviors of swells and tsunamis. Reflecting the general trend, the treatment of the soliton was one of the central subjects at the symposium. In addition to the KdV equation, several others such as the GB, KP, and NLS equations were utilized as the governing equation of investigated phenomena. More efforts are needed to elucidate the adaptability or limitations of these equations.

The mechanism of wave breaking is particularly interesting. At the present stage appropriate empirical formulas are practically used to evaluate such breaking criteria as breaker height and breaker depth. The main reason for this is the essential difficulty in solving the phenomenon analytically. The most legitimate method to approach the problem is to solve the governing equations of fluid motion with suitable boundary conditions. However this approach is extremely difficult even for numerical analysis. In order to avoid this difficulty, Longuet-Higgins and Cokelet applied the boundary integral method to calculate the transformation of the deep water wave profile up to the breaking in their 1976 paper. Since then their method has been applied to the breaking process of shallow water waves. In addition, various experimental trials have been done to formulate the wave breaking process using stationarized models. These efforts seem promising to clarify the structure of breaking waves in the near future.

II-4. Nonlinear Interaction between Water Waves and Bodies

Papers related to the nonlinear interaction between water waves and bodies were presented in four sessions.

S-7: Second-order wave-body interaction (three papers)
S-8: Slow-drift wave force (four papers)

S-9: Body-induced nonlinear water waves (three papers)

P-2: Nonlinear wave-body interaction

S-7: Higher-order wave-body interaction has been a subject of interest for many years in the field of naval architecture and has been recognized as important in recent years in the design of floating marine structures. The first paper treated the nonlinear hydrodynamic forces induced on a half-immersed circular cylinder by two-dimensional forced oscillation. The semi-Lagrangian time-stepping method using the singularity distribution concept on boundary surfaces was employed in the computation of a fully nonlinear potential flow model. In some cases the numerical computation was broken down due to generation and breaking of the jet appeared in a limited region near the interface between the water free surface and body surface. However, as the jet is confined within a thin layer, the thin fluid film of jet was removed from the fluid region in order to continue the computation. Agreement between the calculated result and experimental data is satisfactory. It was also pointed out that the fluctuation of hydrodynamic coefficients corresponding to the heaving motion is mainly caused by the free surface nonlinearity. The next paper introduced the method for calculating the nonlinear diffraction loads acting on three-dimensional bodies of arbitrary shape up to the second order of the steady state perturbation solution. The method presented was applied to the cases of a circular cylinder and a rectangular cylinder, and the results were compared with the unsteady solutions based on the fully nonlinear equations. The agreement is fairly good in the case of a periodic incident wavetrain. However, in the case of a modulated incident wavetrain, the computation of the fully nonlinear equation cannot be continued for sufficiently long duration due to the numerical instability compared with the steady-state solution. The last paper introduced the calculation methods of nonlinear wave-body interaction effects. The employed potential theory is based on a quasi-third-order regular perturbation expansion in terms of small parameter and Green's function integral-equation method. Based on the results the author discussed the contribution of the third-order forces on corrected added-mass and damping coefficients of a ship.

S-8: In this session slow-drift wave forces induced by the nonlinear interaction of waves and a floating body or a semisubmerged body were discussed. The first paper dealt with the second-order wave forces acting on a horizontal circular cylinder in irregular waves. According to the comparison between calculated and experimental data, good agreement can be observed in the time series of free surface profiles and of wave forces. However, in the Fourier components of horizontal and vertical wave forces, discrepancies were observed in the higher frequency components of horizontal wave forces and the lower frequency components of vertical wave forces. The second paper treated the moored submerged marine structure. Slow-drift forces acting on a submerged body are different from the forces on a floating body in such a way that the slow-drift oscillation caused by the second-order potential is predominant in comparison to that expressed by the product of the first-order potential. In order to analyze such phenomena, the authors applied the simple but consistent approximate method. The third paper dealt with the slowly varying wave drift forces acting on a moored ship. In the last paper the second-order fluid forces acting on a moored ship induced by irregular waves was analyzed precisely and the results were compared with those obtained by a simplified method. As an example, a floating circular cylinder hinged at the sea bottom was analyzed. The moment of drift forces around the hinge and the drifting motion of the hinged cylinder were calculated. The calculated results were compared with the result obtained by Newman under regular waves. The agreement between both results is in general fairly good. However, when the peak frequency of the incident wave spectrum lies in the frequency range where the second-order waves make an important contribution, Newman's method tends to underestimate both the drift moment and motion.

S-9: In this session body-induced nonlinear water waves were discussed. The first paper dealt with nonlinear waves generated by a ship moving at low Froude number. The bow and stern wave fronts and the wave resistance coefficient were calculated and compared with previous results. The comparison demonstrated the importance of nonlinear wave-body interaction. The second paper discussed the characteristics of stationary waves superposed to the flow around a body in a uniform stream. The third paper treated the nonlinear interaction between long waves and obstacles such as a floating body, submerged body, and a step on the bottom.

P-2: The title of this session was nonlinear wave-body interaction. However there were some exceptions, which should be classified into the session on the characteristics of nonlinear water waves, or a special session. The first paper was related to nonlinear instability observed in an open channel flow or a jet flow. Here the authors treated the jet flow as one-dimensional nonlinear unsteady flow and found that the infinite velocity modulation at the jet outlet develops due to the nonlinear effect of flow motion. The resulting harmonically modulated shape of the jet flow is in good agreement with that obtained experimentally. The next paper was an analytical and experimental investigation on a long-internal wave of finite amplitude. The result shows that the solitary wave profile obtained theoretically agrees quite well with the experimental one, but differs remarkably from the corresponding KdV and the Benjamin-Ono theoretical curves. The next paper treated the phenomenon of transmission and reflection of a soliton caused by the existence of a two-dimensional rectangular cylinder floating on the free surface. In order to solve this problem the authors applied the so-called matched-asymptotic method. That is to say, the B equation and the Laplace equation were utilized as governing equations in the outer and inner regions. In the third paper the nonlinear interaction between the waves and a uniform current in the region with a constant water depth was calculated by using the perturbation method up to the fourth-order solution. Based on the theoretical result the nonlinear interaction of waves, regular or irregular, and a vertical wall was discussed. The fifth paper analyzed the surface waves generated by the vertical oscillation of a submerged body with large amplitude and the fluid forces acting on the body. Even though the linearized boundary condition was used for the computation, the water surface profile obtained theoretically agreed with the experimental one. The sixth paper presented a new approach to nonlinear water waves generated by a body moving steadily at a free surface. Based on this method the deformation of the free surface induced by a submerged disturbance or a slender shape ship was demonstrated. In the seventh paper the evaluation of breaking wave forces acting on a semisubmergible cylinder with a lowerhull by using the boundary element method for wave profile in the vicinity of the body and the Morison formula for wave forces. The wave force acting on a fixed column was also investigated. It was indicated that the existence of swell strongly affects the maximum value of impact forces. The next paper evaluated the large slow drift of a ship in slightly modulated seas, and the final paper dealt with the statistics of slow-drift oscillations with nonlinear restoring forces.

As stated above a number of aspects of wave-body nonlinear interaction were extensively discussed. This subject has been one of the central interests in naval architecture. At the initial stage of research related to this subject it was treated as a linear problem, although in recent years the nonlinearity on the free surface and the body surface has been introduced in the analysis. An important result of recent studies is the reevaluation of the wave resistance forces. Another aspect of engineering interest is slow-drift forces acting on a floating body, which are induced as the result of the nonlinear interaction between waves and floating bodies. Even though the magnitude of drift forces is relatively small in comparison to the wave forces acting on a body, it is possible that the slow-drift motion induces the resonant oscillation of a floating body with a long natural frequency resulting in instability of the body. Reflecting this, these subjects were extensively treated.

III. CONCLUSIONS

——On the Achievements of the Symposium——

I believe that the symposium accomplished its aim of creating a venue for discussion. The fields in which the participants are interested are so broad that their approaches are different. This kind of recognition is a result of the symposium, and is a gateway to mutual understanding and cooperation in research activities. I hope the proceedings of the symposium will accelerate the development of our knowledge on the behavior of nonlinear water waves.

The symposium was successfully held as the result of cooperation of members of the Theoretical and Applied Mechanics Committee in the Science Council of Japan, represented by Chairman Isao Imai. As a co-chairman of the Local Organizing Committee, I would like to express my sincere thanks to all the members of the Local Organizing Committee and the Executive Committee of the IUTAM Symposium on Nonlinear Water Waves.

II
SHOCK WAVES

Reflection of Plane Shock Waves Over a Dust Layer

——The Effects of Permeability——

Takashi ADACHI, Susumu KOBAYASHI, and Tateyuki SUZUKI

Department of Mechanical Engineering, Faculty of Engineering, Saitama Institute of Technology, Okabe, Saitama

The aim of the present paper is to describe reflections of plane shock waves from a dust layer and to discuss the effect of permeability of the dust layer on the shock reflection through reference experiments using a multiguttered bed, which is an idealized two-dimensional model of the dust layer. We used pseudostationary flows in a shock tube. The working gas was air. "Amberlite," an ion-exchange resin, was used as a model dust. It is chemically inert in our experimental conditions and its particles are nearly spherical. The experimental parameters were incident shock Mach number M_i and reflecting wedge angle θ_w. Wave configurations were examined through shadowgraphs for the case of $1.10 \leq M_i \leq 1.60$, and $20° \leq \theta_w \leq 50°$. One of the results is that the domain of Mach reflection over a dust layer on a rigid wedge on the (Mach number, incident angle)-plane is smaller than that over a dust-free, smooth, rigid wedge owing to the porosity of the dust layer. Another important result is that the reference experiments using a guttered bed in place of a dust layer made clear the structure of the dispersion of reflected shock waves over a dust layer, which was observed in our previous report (see Fig. 2 in the paper by Suzuki and Adachi[12]), that is, it is revealed that leading of dispersed reflected shock is generated on the surface of the dust layer and subsequent shock, compression, and rarefaction waves are generated at the dust particles inside the layer and on the solid surface under the dust layer.

I. INTRODUCTION

The interaction of a spherical blast wave front with a planar surface can be reasonably treated as an oblique shock reflection over a wedge in a shock tube, providing that the spherical wave front has developed enough to resemble its two-dimensional equivalence.[1] The equivalent wedge angle is geometrically related to the incident shock angle made between the blast wave front and the planar surface. Two types of oblique shock reflections have been observed (Fig. 1) for relatively weak incident shocks. Many experimental and theoretical studies on this subject have been reported for the reflection over a smooth rigid wedge.[2-7] Since in real problems there are some obstacles or dusts on the planar surface, the phenomenon of shock reflection becomes very complex. It is important to study the effects of surface roughness[8] and dust on the shock reflection. Increasing practical interest in studying shock reflections from such a dust layer on the surface requires investigation of the effects of a dust layer on shock reflections experimentally, because theoretical investigation is very difficult due to high nonlinearity.[9]

We have reported experiments on shock reflections over a dusty wedge.[10-12] We found that: 1) the critical incident angle from regular to Mach reflections over a dust layer is

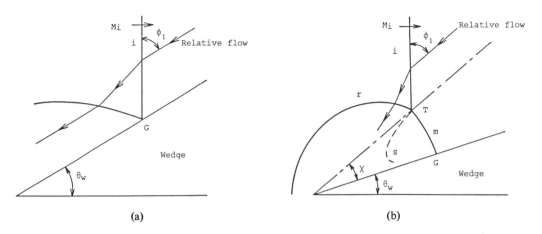

Fig. 1. Schematic diagrams of types of oblique shock wave reflections from smooth wedge in a shock tube. (a) Regular reflection; (b) Mach reflection. i: incident shock; r: reflected shock; m: Mach stem; s: slip line; G: reflecting point; T: triple point; θ_w: wedge angle; ϕ_1: incident angle; χ: triple-point trajectory angle.

smaller than that over the dust-free rigid surface; and 2) broad reflected shock waves were observed. We conjectured there that both 1) and 2) were due to the permeability of the dust layer.

The aim of the present paper is to examine systematically the effects of porosity of the dust layer on the shock reflection for the case of relatively weak incident shocks by reference experiments and to describe more precisely our understanding of shock reflections from a dust layer than in our previous paper.[12]

In the reference experiments multiguttered beds were placed on a rigid wall in place of dust for the reasons: a dust layer is made of round particles. The layer is characterized by surface roughness and porosity in addition to the physical properties of dust particles and the thickness of the layer. It has spaces or pores among the particles through which air can flow and pressure waves can propagate. As the spaces or pores are practically three dimensional, analysis of air flows in the pores and shock interactions with the dust particles which compose the pore wall is very complicated. Regarding a meandering pore as a straight gutter, we made a multiguttered bed as a two-dimensional model of a dust layer. The cross section of the multiguttered bed and its size are shown in Fig. 2 and Table 1,

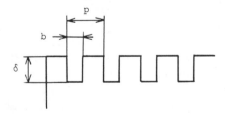

Fig. 2. Sectional view of guttered beds.
δ: Depth of an equivalent pore; b: equivalent width of a pore; p: pitch of the gutter.

Table 1. Configuration of multiguttered bed.

	Model A	Model B1	Model B2	Model B3
Pitch of gutters (mm)	0.35	0.35	0.45	0.35
Depth of a gutter (mm)	0.15	1.0	1.0	2.0
Width of a gutter (mm)	0.15	0.15	0.15	0.15
Porosity	0.43	0.43	0.33	0.43

respectively. The width b of a gutter would correspond to a diameter of a pore in a dust layer. The depth δ of a gutter should be equivalent to the length of a pore from the initial surface of the dust layer to its bottom on the rigid wall. The pitch p of gutters is regarded as the surface roughness. The porosity of the multiguttered bed is calculated as b/p.

Structures or wave configurations of reflected shocks from a dust layer have been examined through shadowgraphs. Air was the working gas. "Amberlite" was used as the model dust, which is chemically inert in our experimental conditions, and whose particles are nearly spherical.

II. EXPERIMENTAL APPARATUS AND METHOD

We used pseudostationary flows in a shock tube with a wedge, as most other investigators studying shock reflections have done. The shock tube here was placed with its longitudinal axis tilted on a horizontal floor. The angle between the longitudinal axis of the shock tube and the floor is nearly equal to the wedge angle so that: 1) the model dust on the wedge surface would be prevented from slipping down; and 2) the direction of gravity would be normal to the dust layer. The driver section of the shock tube is 1.2 m long. The driven section is 3.9 m long and has a rectangular cross section (30 mm wide and 65 mm high). Its last section has optical viewing windows (62 mm high and 94 mm long). A wedge 30 mm wide and 55 mm long was used, with a trough of depth 1 mm and/or 2 mm on its surfaces.

We used air as a working gas and "Amberlite IRC-50" as a model dust, which is an ion-exchange resin (Tokyo Organic Chemistry Ltd., Tokyo). Its physical properties are shown in Table 2. Before every experiment the model dust was spread in the trough of the wedge and cut level with the rest of the wedge surface. Reflected shock waves were visualized through shadowgraphs using a flash lamp with pulsewidth of 700 ns.

In a reference experiment one of the four kinds of multiguttered beds (shown in Table 1) was placed on the wedge surface. Visualization was also made by the shadowgraph method.

The experimental parameters were incident shock Mach number M_i and reflecting wedge angle θ_w, that is, $1.10 \leq M_i \leq 1.60$, and $20° \leq \theta_w \leq 50°$. The initial gas in front of the incident shock was at room temperature and atmospheric pressure.

Table 2. Physical properties of model dust layer.

Approx. mean diameter of particle	0.3 mm
Approx. mean mass of a particle	0.002 mg
Approx. material density	980 kg/m^3
Approx. porosity	0.5

III. RESULTS AND DISCUSSION

III-1. Structure of Reflected Shock

Typical shadowgraphs illustrating reflected shocks from a dust layer on a wedge are shown in Fig. 3. Incident shock, reflected shock, Mach stem, and slip line can be seen in the neighborhood of the triple point in Fig. 3 (a), which shows that the reflection pattern is the Mach reflection type. Particles are not observed in the flow field behind the reflected shock in these experimental conditions. The effect of dust dispersion need not be taken into consideration in the following.

The reflected shock waves in Fig. 3 (b) is different from the reflected waves of Fig. 1 (a) and (b), which are typical kinds of regular and Mach reflection on a smooth, dust-free wedge. It has a dark triangle zone near the reflecting point. We cannot explain its structure from this shadowgraph only. We will leave its discussion to the reference experiments.

Both reflected shock waves near the reflecting points in Fig. 3 (a) and (b) are more dispersed than those from a dust-free wedge. The first leading reflected shock seems to be generated when the incident shock is reflected from the initial surface of the dust layer. The following shock, compression, and rarefaction waves seem to be generated at the particles inside the dust layer and on the solid surface under the layer.

In order to make this reflection mechanism clear, systematic experimental results of shock reflection over a multiguttered bed, which is one of the simplest two-dimensional models of a dust layer to examine the effect of porosity on the reflection, are shown in Fig. 4. The initial conditions are the same as those in Fig. 3. Figure 5 also shows shadowgraphs of the reflected shock wave over another multiguttered bed, whose gutter is deeper than that one used in Fig. 4. From Figs. 4 (b) and 5 (b) we can explain the structure of the dark triangle region in the regular reflection shown in Fig. 3 (b). The nearest edge to the incident shock of the triangle is confirmed to be the first leading reflected shock wave generated at the surface. The left edge is the second shock reflected at the bottom of the gutter. It seems that the distance between the first and the second depends on the depth of the layer (or the depth of a gutter), that is, the equivalent depth of pores is larger, and the distance between the first and the second reflected shock becomes larger provided that

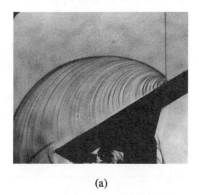
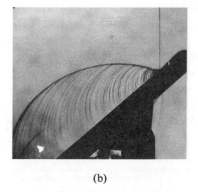

(a)   (b)

Fig. 3. Typical shadowgraphs of reflected shocks from a dust layer, of which thickness δ is 1 mm. Incident shock propagates up the dusty wedge from left to right.
(a) $M_i = 1.4$, $\theta_w = 30°$; (b) $M_i = 1.4$, $\theta_w = 40°$.

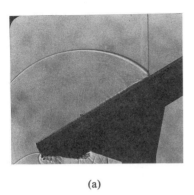
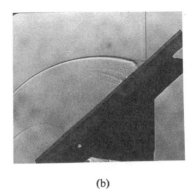

(a) (b)

Fig. 4. Typical shadowgraphs of reflected shocks from multiguttered bed model B1.
(a) $M_i = 1.4$, $\theta_w = 30°$; (b) $M_i = 1.4$, $\theta_w = 40°$.

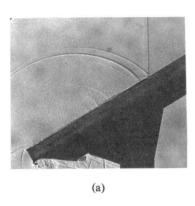
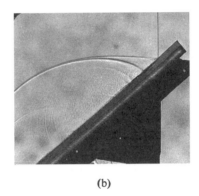

(a) (b)

Fig. 5. Typical shadowgraphs of reflected shocks from multiguttered bed model B3.
(a) $M_i = 1.4$, $\theta_w = 30°$; (b) $M_i = 1.4$, $\theta_w = 40°$.

the initial conditions are the same. This distance (the length of the edge of the triangle on the surface of the bed) is identified by easy estimation by calculating the time required for the incident shock to reach the bottom of the gutter and to return to the surface of the multiguttered bed. In case of Mach reflection, the same structure is observed except that the second wave comes up with the first as time proceeds (see Figs. 4 (a) and 5 (a)). The discussion above shows that the dispersion of reflected shock wave always appears where a reflecting wall is porous and placed on a rigid wall.

III-2. Transition from Regular to Mach Reflection

The reflection pattern was divided into regular and Mach type from shadowgraphs of reflected shock waves from a dust-free and dusty wedge, and plotted on the (M_i, ϕ_1)-planes in Fig. 6 (a) and (b), respectively, where ϕ_1 is the incident angle defined in Fig. 1 and equal to $90° - \theta_w$ for regular reflection. It is seen that the domain of the Mach reflection over a dusty wedge on the (M_i, ϕ_1)-plane is smaller than that over a dust-free wedge. The pattern

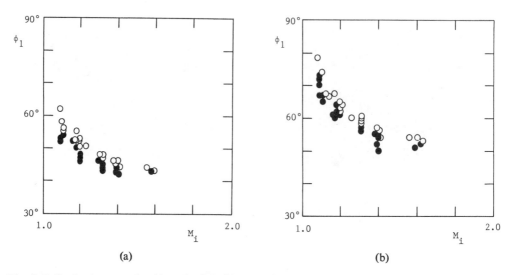

Fig. 6. Reflection patterns plotted on the (M_i, ϕ_1)-plane. (a) Over smooth rigid surface; (b) over dusty surface ($\delta = 1$ mm). The open symbol denotes Mach reflection, and the filled one regular reflection.

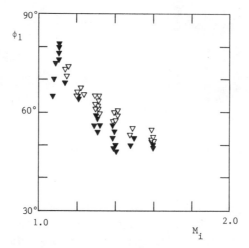

Fig. 7. Reflection patterns over multiguttered bed, model B1. The open symbol denotes Mach reflection, and the filled one regular reflection.

of shock reflection from the multiguttered bed, model B1, are illustrated on the (M_i, ϕ_1)-plane in Fig. 7. Comparing Fig. 6 (b) with Fig. 7 it is seen that the transition boundary over the dusty wedge is nearly the same as that over the multiguttered bed, model B1, which has the most similar porosity and depth to the dust layer of four multiguttered beds. It is confirmed that the multiguttered bed worked well to determine the transition boundary of a dust layer, when the porosity and depth of the bed are nearly equal to those of the layer. Transition boundaries from regular to Mach reflection obtained from experiments using various types of bed are summerized in Fig. 8. Comparing curves (7) with (8), (3)

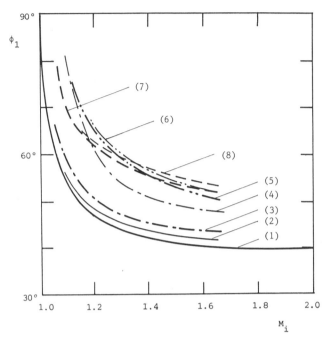

Fig. 8. Transition boundaries from regular to Mach reflection. Curve (1) shows "detached criterion,"[13] (2) shows the boundary obtained from experiments over smooth rigid wedge, (3) that over multiguttered bed model A, (4) over multiguttered bed model B2, (5) over multiguttered bed model B1, (6) over multiguttered bed model B3, (7) over model dust layer ($\delta=1$mm), and (8) over dust layer ($\delta=2$mm).

with (5), and (5) with (6) in Fig. 8, it is seen that the domain of regular reflection becomes larger as the depth increases. From curves (4) and (5), it is seen that the domain becomes larger as the porosity increases. Since the transition from regular to Mach reflection occurs when the flow behind the reflected shock wave cannot keep parallel to the wedge surface (detached criterion), the critical incident angle increases when the shock is reflected on a permeable bed such as a dust layer or guttered bed which acts as an air sink in the vicinity of the reflecting point.

IV. CONCLUSIONS

Shock reflections from a dust layer were studied experimentally under pseudostationary flows within a shock tube. The parameters were the incident shock Mach number and the wedge angle. Reference experiments were done over a multiguttered bed in order to investigate the effects of permeability and depth of the dust layer. The following conclusions were reached:

1) The critical incident angle from regular to Mach reflection over a dusty surface is larger than that over the dust-free, rigid wedge for the Mach number range we observed, $1.10 \leq M_i \leq 1.60$. The effect of permeability is significant.

2) Dispersed reflected shock wave was observed over the dust layer. From the reference experiments, it is seen that the first leading shock wave is generated at the initial surface

of the dust layer, while the following waves are generated at the dust particles inside the dust layer and on the solid surface under the layer.

3) The multiguttered bed worked effectively as a two-dimensional model of a dust layer in order to examine the effects of its porosity and depth on the shock reflections.

Acknowledgments

The authors would like to express their appreciation to Professor A. Sakurai of Tokyo Denki University, Professor K. Takayama of Tohoku University, and Professor N. Oshima of Nihon University for their helpful discussions.

REFERENCES

1) von Neumann, J., Oblique reflection of shocks. *John von Neumann Collected Works, Vol.6*, Pergamon Press (1963), pp. 238–299.
2) Dewey, J.M., Walker, G.D., Lock, G.D., and Scotten, L.N., The effects of radius of curvature and initial angle on transition from regular to Mach reflection of weak shocks reflected from curved surfaces. *Proc. 14th Int. Symp. Shock Tubes & Waves* (1983), pp. 144–149.
3) Henderson, L.F. and Woolmington, J.P., Mach reflection in the diffraction of weak blast waves. *Proc. 14th Int. Symp. Shock Tubes & Waves* (1983), pp. 160–165.
4) Takayama, K. and Ben-Dor, G., A reconsideration of the hysterisis phenomenon in the regular $\rightleftarrows$ Mach reflection transition in truly nonstationary flows. *Proc. 14th Int. Symp. Shock Tubes & Waves* (1983), pp. 135–143.
5) Dewey, J.M. and McMillin, D.J., Observation and analysis of the Mach reflection of weak uniform plane shock waves. Part 1. Observation. *J. Fluid Mech.*, Vol. **152** (1985), pp. 49–66.
6) Glaz, H.M., Colella, P., Glass, I.I., and Deshabault, R.L., A detailed numerical, graphical, and experimental study of oblique shock wave reflections. *UTIAS Rep.*, No. 285 (1986), pp. 1–380.
7) Ben-Dor, G., A reconsideration of the three-shock theory for a pseudo-steady Mach reflection. *J. Fluid Mech.*, Vol. **181** (1987), pp. 467–484.
8) Ben-Dor, G., Mazor, G., Takayama, K., and Igra, O., Influence of surface roughness on the transition from regular to Mach reflection in pseudo-steady flows. *J. Fluid Mech.*, Vol. **176** (1987), pp. 333–356.
9) Dewey, J.M., *Conclusions and Recommendations of the Fourth International Mach Reflection Symposium* (1984), p. 4.
10) Suzuki, T. and Adachi, T., The reflection of a shock wave over a wedge with dusty surface, presented at the Fourth International Mach Reflection Symposium; *Trans. Japan. Soc. Aero. Space Sci.*, Vol. **28** (1985), pp. 132–139.
11) Suzuki, T. and Adachi, T., Structure of the Mach-type reflection on a dust layer. *Theoret. Appl. Mech.*, Vol. **34**, University of Tokyo Press (1986), pp. 73–80.
12) Suzuki, T. and Adachi, T., Comparison of shock reflections from a dust layer with those from a smooth surface. *Theoret. Appl. Mech.*, Vol. **35**, University of Tokyo Press (1987), pp. 345–352.
13) Ikui, T. and Matsui, K., *Mechanics of Shock Waves*, Korona Press, Tokyo (1983), pp. 126–134 (in Japanese).

Shock Wave/Boundary Layer Interactions Induced by a Flat-Faced Fin

Nobumi SAIDA

Department of Mechanical Engineering, Aoyama Gakuin University, Tokyo

This paper presents an experimental study of shock wave/turbulent boundary layer interactions induced by a flat-faced fin placed on the floor of a wind tunnel. The experiments were carried out in an 8 × 10 cm² supersonic wind tunnel mainly at free stream Mach number 2.48. The effects of fin height were studied by varying the height to thickness ratio (H/D) from 0.6 to 10. Surface static pressure measurements, oil flow studies, and Schlieren photographs of the flow field were made. The results show that the flow characteristics along the centerline attain asymptotic values even for fin height less than 2 times the asymptotic triple point height.

I. INTRODUCTION

In supersonic flow past a blunt fin mounted on a flat plate, a bow shock is formed ahead of the fin. This detached shock causes the boundary layer to separate from the surface, resulting a separated flow region. The shock wave emanating from the separated flow region impinges on the bow shock, and results in a highly complex, three-dimensional, inviscid-viscid interaction flow field. This problem has been extensively studied experimentally by many investigators. Particularly for a circular cylinder or a hemicylindrical leading edge fin.[1-17] From these studies it was found[12] that when fin height H is on the order of 2-3 H_{tp} (the asymptotic triple point height) the root shock structure is isolated from the free end by a two-dimensional central region and the dominant parameter controlling the centerline flow field is the leading edge diameter D. On the other hand, it is known that the leading edge shape affects the flow field. For a flat-faced fin the few existing experiments[18,19] and numerical simulations[20] show that the centerline upstream influence increases significantly compared with the hemicylindrically blunted case. However, as far as is known, no systematic studies have been made of the effects of flat-faced fin height.

The purpose of the present experiment was to clarify the effects of fin height of a flat-faced fin through tests using different height fins. Streamwise surface pressures of the flat plate were measured, and the shock wave shape in the plane of symmetry and the separation line were also measured.

II. EXPERIMENTAL APPARATUS AND PROCEDURE

The experiments were conducted in a 8 × 10 cm² supersonic blowdown wind tunnel. The supersonic nozzle used in the present experiments was unsymmetric and was fitted beneath the ceiling of the tunnel. The floor was a flat plate extending upstream from the throat to the test section with the model mounted on it. The model geometry and the coordinate system used in the present experiments are shown in Fig. 1. The models were a

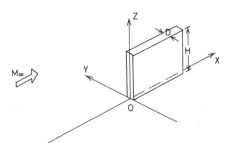

Fig. 1. Sketch of model and coordinate system.

Table 1. Test conditions.

M_∞	$Re_\infty(1/m)$	δ (cm)	Model (H/D)
2.48	3.9×10^7	0.56	0.6, 1.2, 2.4, 4.0 6.0, 8.0, 10
1.98	3.7×10^7	0.46	0.6, 1.2, 4.0, 8.0

slab of thickness $D = 6$ mm and height H of 3.6, 7.2, 14.4, 24, 36, 48, and 60 mm corresponding to H/D of 0.6, 1.2, 2.4, 4.0, 6.0, 8.0, and 10.0.

In this study, almost all tests were conducted at a free stream Mach number of 2.48. The undisturbed boundary layer growing on the test surface was fully turbulent. Additional tests at $M_\infty = 1.98$ were also made for typical models. The incoming flow conditions and the models used are given in Table 1. Measurements of surface pressure and flow visualization were made for all models. For measurements of the surface pressure of the flat plate, pressure holes 0.6 mm in diameter were drilled at intervals of 10 mm in seven rows ($y = -36, -24, -12, 0, 6, 18, 30$ mm). By sliding the model in the x-direction detailed pressure measurements were made. Along the leading edge of the fin, pressure holes 0.6 mm in diameter were drilled at intervals of 2 mm. All surface pressure measurements were made with a conventional mercury manometer. To visualize the flow, an oil film technique was employed to determine the separation and reattachment lines of the surface. The oil was a silicon oil mixed with titanium dioxide and oleic acid. Schlieren photographs were taken to visualize shock waves and shear layers in the plane of symmetry.

III. RESULTS AND DISCUSSION

III-1. Flow in the Plane of Symmetry

Schlieren photographs of the flow field upstream from the fin leading edge were taken for each case and the measured shock shapes are compared in Fig. 2. As can be seen, a lambda type shock is formed for a $H/D = 2.4$ fin and with the increase of H/D the shock shape tends toward an asymptotic shape. For $H/D = 8.0$ and 10.0 fins the difference is no longer discernible. The location of the shock triple point H_{tp}/D is in the range from 2.7 to 2.9, corresponding to a fin of $H/D = 2.4$ to 10. The separated shock angle β is insensitive to fin height and depends only on the Mach number. Even for the smallest case of the $H/D = 0.6$ fin, the measured value is nearly equal to the hemicylindrical blunt fin.

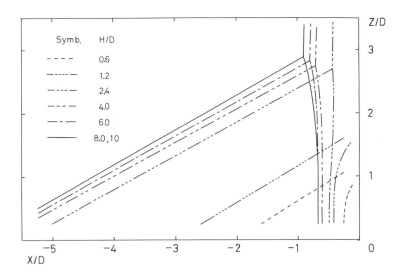

Fig. 2. Shock wave shapes in the plane of symmetry.

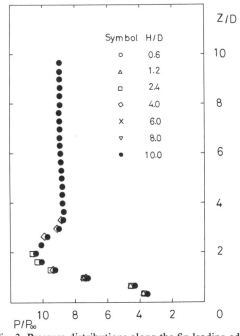

Fig. 3. Pressure distributions along the fin leading edge.

Figure 3 shows pressure distributions along the leading edge of flat-faced fins. The ordinate is the distance of a point measured normal to the flat plate z normalized by D, and the abscissa the surface pressure normalized by the free stream static pressure p_∞. Almost all the results in Fig. 3 fall nearly onto the $H/D = 10$ fin curve. This is particularly true

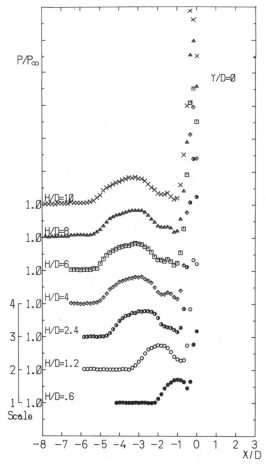

Fig. 4. Pressure distributions on a plate in the plane of symmetry.

when $H > H_{tp}$. When $H < H_{tp}$, the pressure tends to be a little larger below the pressure bump which is formed by impingement of the jet. These results show that the free end effect of the fin is small.

The effect of fin height on upstream pressure distribution along the centerline is presented in Fig. 4. It can be seen that the upstream influence length L_u and the plateau region increase with the increase in H/D. For the $H/D = 0.6$ fin, after reaching the plateau region pressure drops slightly and then increases, but the maximum pressure is less than the asymptotic value. With the increase in fin height the trough pressure decreases and the maximum pressure increases and tends toward an asymptotic curve. It should be noted that for a $H/D = 4.0$ fin an asymptotic state is reached. A similar trend can also be seen at $M_\infty = 1.98$. A numerical simulation performed by Hung[20] for $M_\infty = 4.9$ revealed that the trough region corresponds to the horseshoe vortex. In Fig. 5 the upstream influence distance L_u and the separation distance L_{sep} measured along the centerline of the flat plate from the leading edge to the upstream influence point and the separation point are plotted

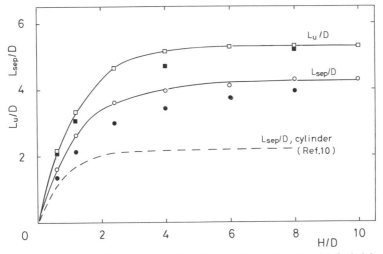

Fig. 5. Variation of separation and upstream influence lengths with fin height.

against fin height H/D. A correlation curve for a circular cylinder obtained by Sedney and Kitchens[10] is also shown for comparison. In this figure the filled symbol is for $M_\infty = 1.98$, and it can be said that the asymptotic result is reached at $H/D \sim 6$. Recent measurements by Dolling[19] for a semi-infinite flat-faced fin at $M_\infty = 4.9$ show that L_u and L_{sep} are about 6 and 5.5 D. These results suggest that for a flat-faced fin L_u and L_{sep} depend on the Mach number.

III-2. Surface Pressure Distribution and Flow Pattern

In Fig. 6 the flat plate pressure distributions for different fin heights are compared by taking the origin at the upstream influence point x_u of each row. For fins of $H/D \geq 2.4$ the pressure curves take a similar shape until the trough region in the spanwise extent of $y/D = 2.0$. The second peak pressure decays and moves downstream with an increase in y/D. A similar trend can also be seen at $M_\infty = 1.98$.

Typical oil flow patterns of a flat plate are shown in Fig. 7, where all the flow patterns are similar. The surface streamlines from upstream and downstream join tangentially and form a separation line. Near the midpoint between the separation line and the fin front face the surface streamlines largely deflect. This may be due to the presence of a high-pressure plateau region. The reverse flow from the high pressure front face region decelerates near the rear part of the plateau region, and off the centerline it turns sharply in the spanwise direction forming a visible line.

In Fig. 8 the shape of the separation line on the flat plate is compared for different fin heights. The origin is taken at the separation point on the centerline. It can be seen that the separation line is similar in shape for $H/D \geq 2.4$. Below $H/D = 2.4$ the spanwise separation distance diminishes largely with the decrease in fin height. From these results we can conclude that, like cylindrical cases, the asymptotic result occurs when fin height H is on the order of 2 times the asymptotic triple point height, but if we consider only the centerline region even for the fin height of $\sim H_{tp}$ shows a nearly asymptotic result.

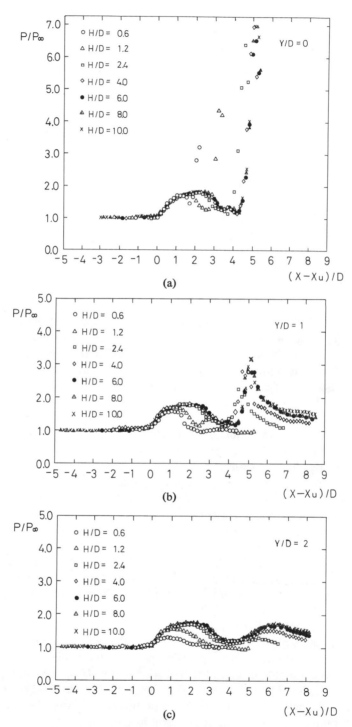

Fig. 6. Streamwise pressure distributions on a plate.
(a) $y/D=0$; (b) $y/D=1$; (c) $y/D=2$.

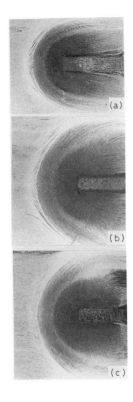

Fig. 7. Oil flow photographs at $M_\infty = 1.98$.
(a) $H/D=1.2$; (b) $H/D=4.0$; (c) $H/D=8.0$.

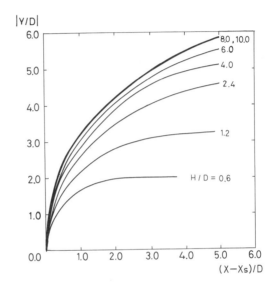

Fig. 8. Variation of separation line with fin height.

IV. CONCLUSIONS

An experimental study has been made of the shock wave/boundary layer interaction induced by a flat-faced fin. The effects of fin height on the upstream influence was studied and the following conclusions were derived.

1) The fin leading edge pressure distribution is independent of fin height.

2) On the centerline an asymptotic result occurs at a fin height about 1.5 times the asymptotic triple point height.

3) The upstream influence and the separation lengths depend weakly on the Mach number.

Acknowledgments

The author would like to thank Messrs. S. Udo and T. Yamauchi for their able assistance in the performance of the experiments.

REFERENCES

1) Korkegi, R.H., Survey of viscous interactions associated with high Mach number flight. *AIAA J.*, Vol. 9, No. 5 (1971), pp. 771–784.
2) Sykes, D.M., The supersonic and low-speed flows past circular cylinders of finite length supported at one end. *J. Fluid Mech.*, Vol. 12, Part 2 (1962), pp. 367–387.
3) Halprin, R.W., Step induced boundary-layer separation phenomena. *AIAA J.*, Vol. 3, No. 2 (1965), pp. 357–359.
4) Westkaemper, J.C., Step induced boundary-layer separation phenomena. *AIAA J.*, Vol. 4, No. 6 (1966), pp. 1147–1148.
5) Voitenko, D.M., Zubkov, A.I., and Panov, Yu. A., Supersonic gas flow past a cylindrical obstacle on a plate. *Izv. ANSSSR MZhG* (Fluid dynamics), Vol. 1, No. 1 (1966), pp. 84–88.
6) Voitenko, D.M., Zubkov, A.I., and Panov, Yu. A., Existence of supersonic zones in three-dimensional separation flows. *Izv. AN SSSR MZhG* (Fluid dynamics), Vol. 2, No. 1 (1967), pp. 13–16.
7) Price, A.E. and Stallings, R.L., Investigation of turbulent separated flows in the vicinity of fin type protuberances at supersonic Mach numbers. *NASA TN* D-3804 (1967).
8) Westkaemper, J.C., Turbulent boundary-layer separation ahead of cylinders. *AIAA J.*, Vol. 6, No. 7 (1968), pp. 1352–1355.
9) Kaufman, L.G. and Korkegi, R.H., Shock impingement caused by boundary layer separation ahead of blunt fins. *AIAA Paper* 73-236 (1973).
10) Sedney, R. and Kitchens, C.W., Jr., Separation ahead of protuberances in supersonic turbulent boundary layers. *AIAA J.*, Vol. 15, No. 4 (1977), pp. 546–552.
11) Dolling, D.S., Cosad, C.D., and Bogdonoff, S.M., An examination of blunt fin-induced shock wave turbulent boundary layer interactions. *AIAA Paper* 79-0068 (1979).
12) Dolling, D.S. and Bogdonoff, S.M., Scaling of interactions of cylinders with supersonic turbulent boundary layers. *AIAA J.*, Vol. 19, No. 5 (1981), pp. 655–657.
13) Dolling, D.S. and Bogdonoff, S.M., Blunt fin-induced shock wave/turbulent boundary-layer interaction. *AIAA J.*, Vol. 20, No. 12 (1982), pp. 1674–1680.
14) Özcan, O. and Holt, M., Supersonic separated flow past a cylindrical obstacle on a flat plate. *AIAA J.*, Vol. 22, No. 5 (1984), pp. 611–617.
15) Hung, C.M. and Kordulla, W., A time-split finite-volume algorithm for three-dimensional flowfield simulation. *AIAA J.*, Vol. 22, No. 11 (1984), pp. 1564–1572.
16) Saida, N. and Hattori, H., Shock wave-turbulent boundary layer interactions induced by a blunt fin. *Trans. Jpn. Soc. Aeronaut. Space Sci.*, Vol. 27, No. 76 (1984), pp. 67–77.
17) Hung, C.M. and Buning, P.G., Simulation of blunt-fin-induced shock wave and turbulent boundary-layer interaction. *J. Fluid Mech.*, Vol. 154 (1985), pp. 163–185.
18) Saida, N., Separation ahead of blunt fins in supersonic turbulent boundary layers. Presented at *IUTAM Symposium on Turbulent Shear Layer/Shock Wave Interaction*, Palaiseau, France, September 1985.
19) Dolling, D.S. and Rodi, P.E., Upstream influence and separation scales in fin-induced shock boundary layer interaction. Private communication.
20) Hung, C.M., Computation of separation ahead of blunt fin in supersonic turbulent flow. Presented at the *International Conference on Fluid Mechanics*, Beijing, July 1987.

Performance of a Slotted Wall Test Section for Aerodynamic Testing in a Shock Tube

Yutaka YAMAGUCHI

Department of Aeronautical Engineering, The National Defense Academy, Yokosuka, Kanagawa

> This paper presents results of a study to confirm the permissible range of angle of attack and the upper and lower limits of the blockage factors for the slotted wall test section designed for the aerodynamic testing of two-dimensional transonic airfoils in a shock tube. The shadowgraph method was employed to test three symmetric airfoils. The tests were carried out at a nominal Mach number range from 0.80 to 0.83. The results show that the present test section produces the steady flow field around an airfoil of a blockage factor up to 0.064 and, will greatly reduce wall interference effects, compared with a solid wall test section.

I. INTRODUCTION

The shock tube can produce flows of subsonic, transonic, and low supersonic Mach number regimes, and can be used as a trisonic intermittent wind tunnel. But shock tubes have not been employed extensively in airfoil testing. In the early days of the evolution of shock tube experiments, only a few researchers performed the testing of airfoils using shock tubes as testing equipment.[1-4]

There were several reasons for such a situation. A major one was that the uniform flow duration time is too short to measure aerodynamic forces except by optical methods. The duration time is on the order of several milliseconds. A second reason was the flow Reynolds number problem. Usually, shock tube flows have low Reynolds numbers because of their low initial pressures of driven sections.

Recently, high Reynolds transonic flow has attracted interest because of a desire to conserve energy and to design more efficient transonic aircraft. Existing conventional, ambient-temperature transonic wind tunnels do not serve those purposes and the aircraft industry has suffered from this lack of high Reynolds transonic testing capability.

In the course of developing high Reynolds number transonic wind tunnels, it was felt that there was a need for a smaller, less costly, high Reynolds number test facility, particularly one in which research and testing of two-dimensional transonic airfoils could be carried out in order to provide fundamental flow data to augment analytical studies of transonic airfoil flows.

Cook et al.[5,6] focussed their attention on the potential of shock tubes, and studied the feasibility of airfoil testing in shock tubes for high Reynolds number transonic flows. To solve unrealistically heavy construction problems in using air as the working gas, and the γ effect on aerodynamic forces acting on airfoils, they proposed the use of a heavier gas mixture of $\gamma = 1.4$ other than air, and showed the superiority of the mixture of Ar and Freon-12 over air theoretically.

Another important problem in transonic wind tunnel testing, including shock tube experiments, is to minimize the wall interference on aerodynamic forces. For conventional wind tunnels, the development of an adaptive wall test section is now underway.[7] The adaptive wall test section is the ideal solution for eliminating wall ineterference. But, for shock tubes, it is very difficult to adapt test section walls to the streamlines of a test condition in such a short uniform flow duration time of several milliseconds.

Cook and his associates also worked on eliminating or at least relaxing wall interference in shock tube experiments, and developed two methods for the purpose. One is to use a contoured wall test section, and the other is a slotted wall test section. They showed that a contoured wall test section is good for investigating flows of known boundary conditions, but, when the angle of attack is changed or another airfoil is installed, the contoured walls should be changed to correspond to each condition. This is not a practical method, and a slotted wall test section has more flexibility in terms of airfoils tested and the range of angle of attack than a contoured one. If a slotted wall test section has a plenum chamber on the order of the test section volume as do conventional transonic wind tunnels, uniform flow is not established in the test section in the duration time of the hot air region. So, they experimentally determined the geometry of the test section with airfoils of blockage factor F_B of 0.06 at a near-zero angle of attack. F_B is defined as a ratio of the airfoil frontal area at zero incidence to the test section cross sectional area.

In a previous report[8] the present author performed a shock tube experiment on the steady shock wave formation on airfoils of $F_B = 0.027$ and 0.04 at zero angle of attack to evaluate the usefulness of an Ar-Freon-12 mixture with $\gamma = 1.43$ and a slotted wall test section, as proposed by Cook. It was found that the mixture showed excellent performance as expected, and the test section worked well. That study also indicated that the test section was not universally valid in terms of blockage factors of airfoils, although it was slightly modified from Cook's. This paper presents the preliminary results on the permissible range of angle of attack and upper limit of the blockage factor of airfoils in the previous slotted wall test section adopting air as the driven gas instead of the Ar-Freon-12 mixture.

II. EXPERIMENTAL SET-UP

II-1. Shock Tube and Test Section Configuration

The shock tube used in the previous paper was modified in the driving mechanism of the diaphragm spear and the section behind the test section including the dump tank. The modified shock tube has a dump with a volume of 1.3 m³, which is about 4 times larger than the previous one, and is sufficient to reduce the final pressure of the shock tube.

The previous driving mechanism was of electrical solenoids, and almost fully occupied the portion 30 cm from the end flange of the driver section, reducing the effective length of the section. The spear of the new mechanism is driven by a pneumatic cylinder mounted on the end flange of the driver section, making the effective length of the section 2.5 m. Mylar film 0.075 mm thick was used as the diaphragm. Shock speed was measured by two pressure transducers (PCB 111A21) placed 265 mm apart in front of the observation window. Figure 1 shows a schematic diagram of the shock tube.

The test section configuration employed for this shock tube is slightly modified one from Cook's proposed one, which is the optimum for airfoils with the blockage factor F_B of

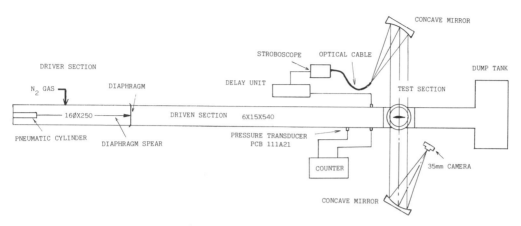

Fig. 1. Schematic view of shock tube (unit: cm).

0.06 at zero angle of attack. Details of this configuration were given previously[8] and the plenum chambers attached to the slotted walls have 1.1% of the volume of the entire test section. The upper and lower walls have four slots, respectively, and those slots are 3 mm wide. The opening ratio of the test section is thus 20%, but the slotted walls may easily be changed, and it is easy to vary the opening ratio. The cross sections of the test and driven sections are 60 mm wide and 150 mm high. The test section has a pair of plastic glass observation windows 150 mm in diameter, which are 25 mm thick and have a hole 5 mm deep for supporting a model at their centers.

The driven gas was air at 15.47 Kpa, and the driver gas was a mixture of air and nitrogen. The nominal test shock Mach number ranged from 0.80 to 0.83.

II-2. Optical Method

Observations of flows around models were performed using the shadowgraph method. The optical arrangement is shown in Fig. 1. The present system consists of a pair of concave Schlieren mirrors 200 mm in diameter and with a 1500-mm focal length. As a light source, a single flash of a strobotron (DSX-1B, Sugawara Lab. Inc.) is guided to the focal point of the collimation mirror with an 60-cm long plastic optical cable (Eska CH2012, Mitsubishi Rayon Ltd.). The outer diameter of the optical cable is 3 mm, and it contains seven optical fibers of 0.5 mm diameter. The trigger signal for the light source is fed independently by another pressure transducer as shown in Fig. 1.

II-3. Airfoil Models

In this study three symmetrical airfoils were used: the NACA 0012 airfoil and the 8% and 12% thick biconvex circular arc airfoils. These airfoils are made of aluminum alloy, and have a 60-mm span. The chord lengths are 75, 50, and 80 mm for the NACA 0012 and the 8%, 12% circular arc airfoils, respectively. A pin 3 mm in diameter was plugged into each end of the span. The location of the pin is the 50% and 47% chord points from the leading edge for the circular arc airfoils and the NACA 0012, respectively. The airfoils were fixed tightly between two glass windows by putting the pins into the holes on the glass.

III. RESULTS AND DISCUSSION

III-1. Circular Arc Airfoils

The biconvex circular arc airfoils were tested to find the extreme values of the blockage factor for this test section. The blockage factors at $\alpha = 0°$, F_B, of the 8% and 12% circular arc airfoils are 0.027 and 0.064, respectively.

A previous report[8] shows that a steady shock wave was not established at $\alpha = 0°$ on the 8% circular arc airfoil within the testing time primarily due to unsteady shock waves reflecting from the upper and lower walls and chambers. Figure 2 shows the variation of the shock profiles of the 8% airfoil with time at $\alpha = 2°$, and that the steady condition was not achieved. For $\alpha = 4°$ it was almost same as $\alpha = 0°$ and $2°$, and it may be said that shock wave development up to $\alpha = 4°$ is essentially same as that of the zero angle of attack case in the present test section. But, as shown in Fig. 3, the shock wave profiles and locations are more stable at 6° than at 2° and 4°. Although there is a problem which will be mentioned later, it can be concluded that a steady shock wave was formed on the 8% biconvex airfoil at $\alpha = 6°$. If the equivalent blockage factor. F_B^* is defined as the airfoil frontal area at nonzero angle of attack to the test section cross sectional area, A_F^*/A_{TS}, the values of F_B^* at α less than 6° are almost equal to that of zero angle of attack. On the other

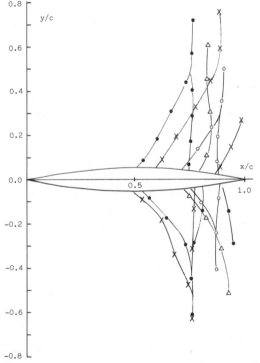

Fig. 2. Shock wave profiles of the 8% biconvex airfoil.
$M_\infty = 0.81$; $R_e = 2.8 \times 10^5$; $F_B = 0.027$; $\alpha = 2°$.
●: $t = 2.21$ ms;
×: $t = 2.41$ ms;
△: $t = 2.61$ ms;
○: $t = 2.81$ ms.

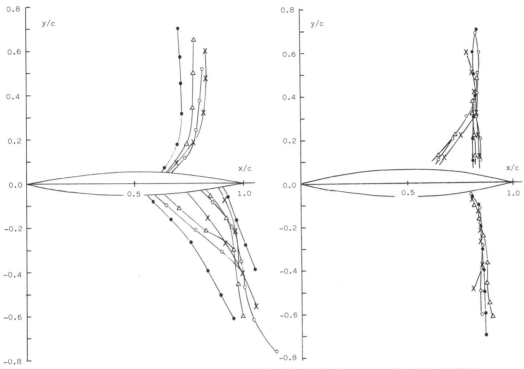

Fig. 3. Shock wave profiles of the 8% biconvex airfoil. $\alpha = 6°$.
●: $t = 2.01$ ms;
×: $t = 2.21$ ms;
△: $t = 3.01$ ms;
○: $t = 3.21$ ms.

Fig. 4. Shock wave profiles of the 12% biconvex airfoil.
$M_\infty = 0.81$; $R_e = 4.8 \times 10^5$; $F_B = 0.064$; $\alpha = 0°$.
●: $t = 2.03$ ms;
○: $t = 2.13$ ms;
×: $t = 2.23$ ms;
△: $t = 2.33$ ms.

hand, F_B^* at 6° is about 0.035. This indicates that a steady transonic flow may develop around an airfoil in the present test section if its $F_B \fallingdotseq 0.035$.

F_B of the 12% biconvex circular arc airfoil is 0.064 and exceeds the optimum value of 0.06 for Cook's original test section. In Fig. 4, shock wave development is shown for the 12% biconvex airfoil at $\alpha = 0°$, and a steady shock is established on the airfoil and is stable within the testing time. This is true up to $\alpha = 6°$. Figure 5 shows the shadowgram of the 12% airfoil at $\alpha = 4°$. In this photograph the airfoil is deformed around the mid-chord point. Such deformation is due to the optical defect of the plastic glass windows, caused by drilling holes for the airfoil support pins. But, when α exceeds 4°, an oblique shock wave appeared at the leading edge as shown in Fig. 5. Shock development at $\alpha = 4°$ is shown in Fig. 6. This is the same problem mentioned implicitly in the 8% airfoil case, when the oblique shock wave was also formed at the leading edge at α exceeding 4°. It is not known why such oblique shocks are formed at the leading edges of biconvex airfoils in the above-mentioned condition, although the following explanation might be offered: as the leading edge of a biconvex airfoil is sharp, the stagnation point stays at the leading edge at low angle of attack range. But, when $\alpha \geq 4°$, the stagnation point will move to the

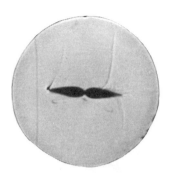

Fig. 5. Shadowgraph of the 12% biconvex airfoil. $M_\infty=0.81$; $R_e=4.8\times 10^5$; $\alpha=4°$.

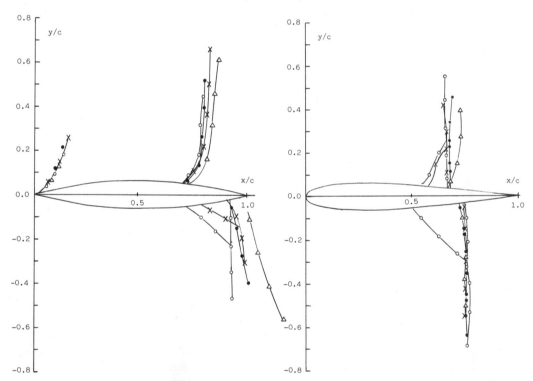

Fig. 6. Shock wave profiles of the 12% biconvex airfoil. $\alpha=4°$.
- ●: $t=1.63$ ms;
- ×: $t=1.83$ ms;
- △: $t=2.03$ ms;
- ○: $t=2.23$ ms.

Fig. 7. Shockwave profiles of the NACA 0012. $M_\infty=0.81$; $R_e=4.7\times 10^5$; $F_B=0.060$; $\alpha=0°$.
- ●: $t=1.63$ ms;
- ×: $t=1.83$ ms;
- ○: $t=2.03$ ms;
- △: $t=2.23$ ms.

lower surface, and the flow velocity around the leading edge becomes very fast. Then, an oblique shock is formed. To the author's knowledge, no optical observations on any biconvex circular arc at $\alpha\geqq 3°$ have been made. Therefore, it should be further examined whether

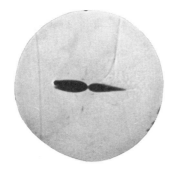

Fig. 8. Shadowgraph of the NACA 0012.
$M_\infty = 0.81$; $R_e = 4.7 \times 10^5$; $\alpha = 8°$.

the above-mentioned phenomenon is peculiar to the present test section or not. But it can be said that steady flow is established around the airfoil with $F_B = 0.064$ up to $\alpha = 6°$ in the present slotted wall test section.

III-2. NACA 0012 Airfoil

Since the present test section configuration is based on Cook's design for airfoils with $F_B = 0.06$, it should also work satisfactorily for such airfoils. To examine this condition, the NACA 0012 airfoil with $F_B = 0.06$ was tested.

Steady shock wave development on the airfoil was almost perfect at $\alpha = 0°$ as expected (Fig. 7). The angle of attack was varied in 2° steps from 0° to 8°. A steady shock wave was established on the airfoil at every angle of attack, and was stable in its shape and location. Figure 8 shows the shadowgraph of the steady state shock wave at $\alpha = 8°$. The deformation of the airfoil on the shadowgram is due to the same reason described in the previous section. No leading edge shock wave was formed, as opposed to the biconvex circular arc airfoils as the flow around the leading edge accelerates more gradually than that of the circular arc airfoils due to the finite leading edge radius of the NACA 0012. The time-averaged locations and profiles of the steady shock waves are plotted in Fig. 9. On this NACA 0012, the shock disappears from the lower surface, and remains only on the upper surface when α becomes higher than 2°.

Figure 10 shows how the shock location on the airfoil surface varies with the angle of attack. The points indicated by the symbol ▲ are taken from a study by Takayama and Itoh[9] in which a NACA 0012 with 5-cm chord was tested in a solid wall test section (6 cm in width, 15 cm in height) using pulsed laser holographic interferometry. ■ denotes Cook's results.[6] As the Mach and Reynolds numbers of the present results are not coincident with those of Takayama and Itoh[9] and Cook,[5] it may not be possible to compare the present results with theirs, directly, but Fig. 10 indicates that the present results qualitatively agree with those of Takayama and Itoh, and also are in good agreement with those of Cook. As for F_B^* at $\alpha = 7°$, the present one is about 0.061 and Ref. 9's 0.041. Therefore, it may be said that, if the cross sectional areas are the same, the present slotted wall test section gives almost the same degree of wall intereference as the solid wall test section does, even when F_B is about 1.5 times of the solid wall section.

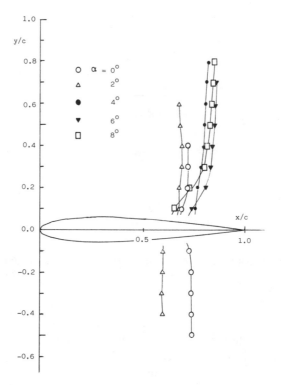

Fig. 9. Steady shock wave profiles of the NACA 0012.

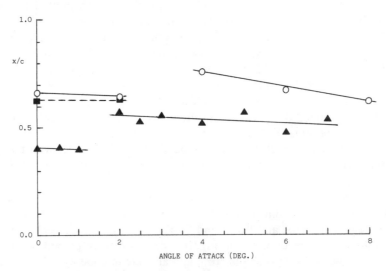

Fig. 10. Shock wave locations on the upper surface of the NACA 0012.
○: Present results; $M_\infty=0.81$; $R_e=4.7\times10^5$; $F_B=0.060$;
■: Results of Cook et al.[5]; $M_\infty=0.83$; $R_e=2\times10^6$; $F_B=0.060$;
▲: Results of Takayama and Itoh[9]; $M=0.76$; $R_e=1.6\times10^6$; $F_B=0.040$.

IV. CONCLUSIONS

The results of this study led to the following conclusions:
1) The angle of attack range and blockage factors of airfoils assuring the steady state flow condition in the present slotted wall test section are $\alpha \geq 6°$ for the airfoil with $F_B = 0.027$ and $F_B \geq 0.035$ for near-zero angle of attack.
2) The airfoil with $F_B = 0.06$ (chord length 75 mm) can be tested up to $\alpha = 8°$ in the present test section.

These results indicate that the present slotted wall test section will permit transonic airfoil testing with a shock tube in a reasonable range of angle of attack. As the present study was performed using the shadowgraph method, further experiments will be needed to study the quantitative agreement between aerodynamic characteristics of airfoils obtained from the shock tube and those of other wind tunnels.

REFERENCES

1) Geiger, F.W. and Mantz, C.W., The shock tube as an instrument for the investigation of transonic and supersonic flow patterns. *Engng. Res. Inst. Univ. Michigan* (1949).
2) Ruetenik, J.R. and Witmer, E.A., Transient aerodynamics of two-dimensional airfoils, Part 2. *WADC Tech. Rep.*, **54-368** (1958).
3) Andrews, P.T. and Ruetenik, J.R., Transient aerodynamics of two-dimensional airfoils, Part 3. *WADC Tech. Rep.*, **54-368** (1959).
4) Ruetenik, J.R. and Herrmann, W., Shock-tube measurements of step-blast loads on a NASA 64A010 airfoil. *ASD-TR*-61-219, *USAF* (1962).
5) Cook, W.J., Presley, L.L., and Chapman, G.T., Shock tube as a device for testing transonic airfoils at high Reynolds numbers. *AIAA J.*, Vol. **17**, No. 7 (1979), pp. 714–721.
6) Cook, W.J., Presley, L.L., and Chapman, G.T., Test section configuration for aerodynamic testing in shock tubes. *Shock Tubes and Waves* (1980), pp. 127–136.
7) Goodyer, M.J. and Wolf, S.W.D., The development of a self-streamlining flexible walled transonic test section. *AIAA J.*, Vol. **20**, No. 2 (1982), pp. 227–234.
8) Yamaguchi, Y., Transonic flows around airfoils in a shock tube slotted wall test section. *Theoret. Appl. Mech.*, Vol. **35**, Univ. of Tokyo Press (1987), pp. 249–257.
9) Takayama, K. and Itoh, K., Unsteady drag over cylinders and airfoils in transonic shock tube flows. *Rep. Inst. High Speed Mech.*, Vol. **51**, No. 377 (1986), pp. 1–41.

III
JET AND SPIRAL FLOW

Wall Jet of Ideal Fluid Flowing along Deflected Surfaces

Masahide HATANO

Department of Aeronautical Engineering, The National Defense Academy, Yokosuka, Kanagawa

The problem of two-dimensional wall jets of an ideal fluid flowing along a deflected surface is investigated by the method of a conventional Kirchhoff-Helmholtz's free streamline analysis. The jet boundaries, the jet width ratio, the flow direction at infinite downstream, the streamlines, and the velocity and pressure distributions are calculated by the numerical method for various deflection angles, wall lengths, and nozzle locations. The results for the jet along an outward-deflected wall surface are compared with the experimental data obtained previously by the present author for the viscous plane wall jet issued from the "small inclined flush slot type Coanda nozzle."

I. INTRODUCTION

The phenomenon of jet flow attachment to a solid surface is well known as the "Coanda effect," and the jet injection nozzle that uses this effect is termed "Coanda nozzle." The present author designed the special "small inclined flush slot type Coanda nozzle," shown in Fig. 1, and used it in his previous experimental investigations on the turbulent wall jet of air in still surroundings.[1-7] The inclination angle of this injection slot $\theta = 15°$ was adopted according to the experimental results of Warner and Reese.[8] While there is certainly a corner at the connecting part of the slot and the wall, the existence of flow separation

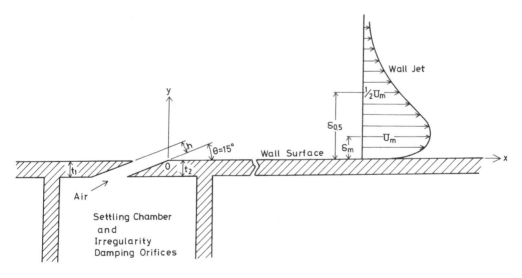

Fig. 1. Small inclined flush slot type injection nozzle.

at this corner could not be detected in the previous investigations even when an external flow was absent. If the fluid is inviscid, flow along the outward deflected wall always exists.

The present investigation was performed to obtain the flow field characteristics in the two-dimensional wall jet of ideal fluid flowing along the deflected surfaces by the method of a conventional Kirchhoff-Helmholtz's free streamline analysis. Then, the results obtained were compared with the experimental results measured previously by the present author for the viscous plane wall jet issuing from the specially designed small inclined flush slot type Coanda nozzle.

II. THEORETICAL CONSIDERATIONS

As the first step in the present investigation, the generalization of the nozzle configuration and flow field is performed. Figure 2 (a) shows the small inclined flush slot type Coanda nozzle, but the coordinate axes are rotated clockwise about the downstream-side corner of the nozzle (0) by an angle of 15°. From this figure, this flow field is considered to be identical to the wall jet issuing tangentially from the new step slot nozzle having same slot height h and located at the upstream section $x' = -h \cot 15°$ from the corner 0.

This flow may be called "two-dimensional upper surface blowing," in which the lower surface of nozzle corresponds to the upper surface of an aircraft wing and the test wall

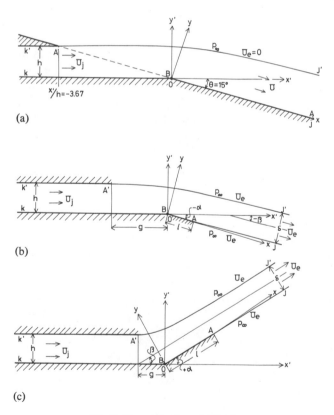

Fig. 2. Generalization of injection nozzle.

surface corresponds to the upper surface of a flat flap deflected downward by 15°. The notations used in the present investigation on the wall jet of ideal fluid are also shown in Fig. 2(a). The generalization of this nozzle configuration and flow field is shown in Fig. 2(b) and (c), in which the deflection angle of wall (α), the length of wall (l), and the location of nozzle (g) are all variable parameters. For the small inclined flush slot Coanda nozzle, $\alpha=(-\theta)=-15°$, $L=l/h=\infty$, and $G=g/h=-3.67$ where h is the slot height of this nozzle.

The flow for positive inclination angle ($\alpha>0$), where the lower wall of the nozzle is deflected inward, has been treated in reports by Sasaki,[9] Birkhoff-Zerantonello,[10] and Gurevich[11] using different methods of comformal mapping. They obtained the respective transformation equations but did not give the practical characteristics of the flow field. On the other hand, the flow for negative inclination angle ($\alpha<0$), where the lower wall of the nozzle is deflected outward, has never been treated because there is no corresponding actual flow field.

Referring to Fig. 2, it is considered in the present investigation that the flow is an ideal fluid jet. Then, injection velocity U_j is uniform far upstream from the slot nozzle with height h. The flow leaving the lips of the upper and lower wall of nozzle forms two free streamlines. The velocity far downstream between these free streamlines is U_e. The jet width far downstream, δ, is defined from the condition of continuity

$$U_j \cdot h = U_e \cdot \delta$$
$$\delta = (U_j/U_e) \cdot h = mh$$
$$m = (U_j/U_e) = \delta/h. \qquad (1)$$

In the previous analyses of the jet of an ideal fluid, the value m was referred to as "the coefficient of contraction." But this term is not suitable for the jet flow discharged from the nozzle $\alpha<0$, because the jet for this case spreads in the downstream sections, the same as for a real viscous jet. Therefore "the ratio of jet width" is used alternatively in the present investigation.

III. CONFORMAL MAPPING AND NUMERICAL CALCULATION

Figure 3 shows the conformal mapping planes to solve the wall jet of an ideal fluid flowing along the deflected plane surface by use of the two-dimensional potential flow theory. In this figure, all lengths are made dimensionless by the slot height h, and all velocities by the far downstream jet velocity U_e. The last transformation equation for this flow problem is written by Sasaki as

$$\zeta^{\pi/2\alpha} = \frac{\sqrt{(b-a)(t-a')} + \sqrt{(b-a')(t-a)}}{\sqrt{(a-a')(t-b)}} \qquad (2)$$

where a, a', b, etc. are the points in the t-plane corresponding to the points A, A', B, etc. in the z-plane. In the t-plane, we can arbitrarily choose the location of three points on the mapping, and we let $j(=j')=0$, $k=\infty$, and $k'=-\infty$, respectively, and the next relation is obtained:

$$k(\infty) > b > a > j = j'(0) > a' > k'(-\infty). \qquad (3)$$

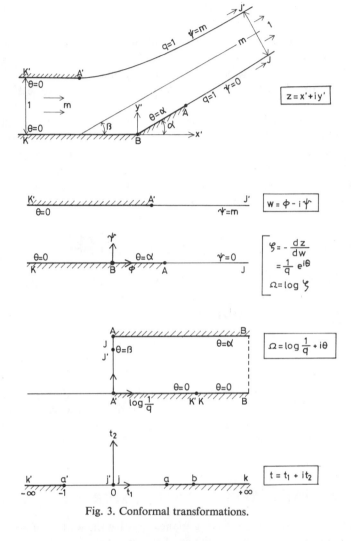

Fig. 3. Conformal transformations.

From the flow configuration, there is next a relation between a, a', and b

$$\left(\frac{1}{m}\right)^{\pi/2\alpha} = \frac{\sqrt{b-a'} + \sqrt{b-a}}{\sqrt{a-a'}} = \frac{\sqrt{b+1} + \sqrt{b-a}}{\sqrt{a+1}} \tag{4}$$

where $a' = -1$ was chosen convieniently in the last equation.

Therefore, the flow in the z-plane is obtained from Eq. (2).

$$z = -\frac{m}{\pi} \int \left\{ \frac{\sqrt{b-a}}{\sqrt{a+1}} \frac{\sqrt{\tau-1}}{\sqrt{b-\tau}} + \frac{\sqrt{b+1}}{\sqrt{a+1}} \frac{\sqrt{\tau-a}}{\sqrt{b-\tau}} \right\}^{2\alpha/\pi} \frac{d\tau}{\tau} e^{i\alpha}. \tag{5}$$

Upon numerical integration, we fix at first the deflection angle of wall α, the length of deflected wall L, and the location of the upper lip of jet nozzle G, and then we assume the initial values for the ratio of jet width m and the parameter a. Putting these values into Eq.

(5) and integrating numerically, we get the wall length $\bar{L}$ and the upper lip location $\bar{G}$. Introducing a new parameter defined by the following equation

$$F = (\bar{L} - L)^2 + (\bar{G} - G)^2 \qquad (6)$$

we can obtain accurate values of m and a by the simplex method to make $F \to 0$.

IV. RESULTS OF COMPUTATION

IV-1. Free Streamline

Figures 4 and 5 show the typical variations of free streamline configurations for $L=1.00$, $G=0.00$, and various inclination angles α. Two streamlines leaving from the upper and lower lips of the slot nozzle show the jet boundaries, and the distance between these streamlines gives the jet width.

For the jet of an ideal fluid issuing from a nozzle with inclination angle $\alpha = -15°$ which is identical to that of the small inclined flush slot type Coanda nozzle, the variations of free streamlines for various location of the upper nozzle lip G are shown in Figs. 6 and 7 for both $L=1.00$ and 2.00, respectively. It is seen from these figures for $\alpha = -15°$ that, for longer deflected wall $L=2.00$ and larger negative lip location $G=-1.00$, the lower free streamline is nearly straight and its inclination angle is identical to that of the deflected wall. That is to say, this flow can approximate the jet flowing along the infinitely long deflected wall ($L=\infty$).

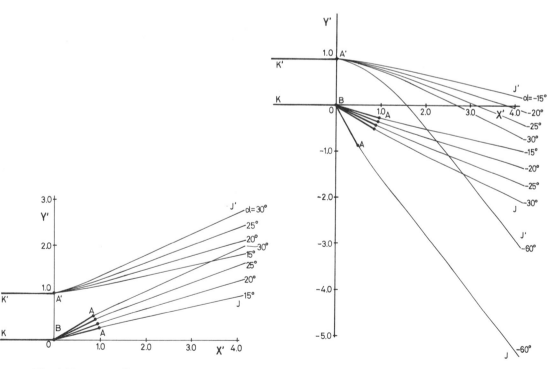

Fig. 4. Free streamlines ($L=1.00$, $G=0.00$, $\alpha>0$).   Fig. 5. Free streamlines ($L=1.00$, $G=0.00$, $\alpha<0$).

IV-2. Ratio of Jet Width

The effect of the variation of L and G on the ratio of jet width m for $\alpha = -15°$ is shown in Fig. 8. The ratio of jet width m has nearly constant values for different L values when $G<0$. For the case of $\alpha = -15°$, $L=2.00$, and $G=-1.00$, which approximate the flow issuing from the small inclined flush slot type Coanda nozzle, $m=1.030$ and the jet of ideal fluid spreads only 3% at the infinitely downstream section.

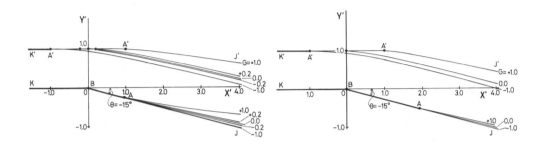

Fig. 6. Free streamlines ($\alpha = -15°$, $L=1.00$).   Fig. 7. Free streamlines ($\alpha = -15°$, $L=2.00$).

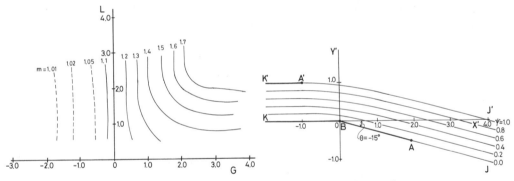

Fig. 8. Jet width or velocity ratio ($\alpha = -15°$).   Fig. 9. Typical streamlines ($\alpha = -15°$, $L=2.00$, $G=-1.00$).

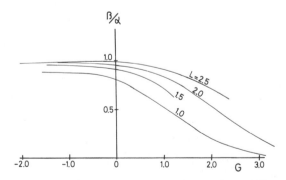

Fig. 10. Jet deflection angle far downstream ($\alpha = -15°$).

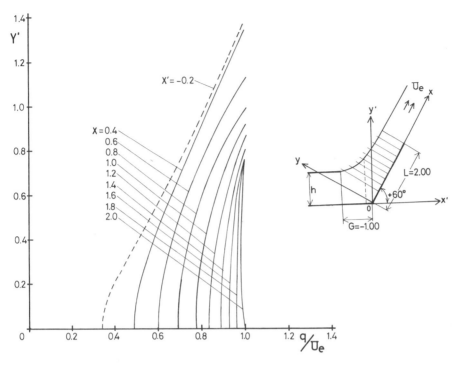

Fig. 11. Velocity distributions ($\alpha = +60°$, $L=2.00$, $G=-1.00$).

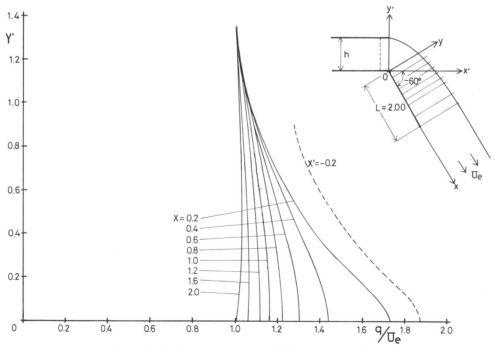

Fig. 12. Velocity distributions ($\alpha = -60°$, $L=2.00$, $G=-0.00$).

IV-3. Streamlines in Jet Flow

Figure 9 shows the typical streamlines in the jet of an ideal fluid obtained for the case $\alpha = -15°$, $L=2.00$, and $G=-1.00$, which corresponds to the flow in the near-region of the wall jet issuing from the small inclined flush slot. This figure shows that all the streamlines in the jet of an ideal fluid for this case are almost parallel to each other in a short distance downstream from the nozzle exit.

IV-4. Direction of Jet Downstream

The results for the jet direction of an ideal fluid in the infinitely downstream section β are given for $\alpha = -15°$ in Fig. 10. From this figure, the jet flow direction β obtained for the case $L \geq 2.00$ and $G \leq -1.00$ is considered to be identical to the deflection angle $\alpha = -15°$.

IV-5. Velocity Distribution in Jet Flow

Figures 11 and 12 show the typical velocity distributions at several sections in the near-region around the corner in the wall jet of ideal fluid issuing from two nozzles: $\alpha = +60°$, $L=2.00$, and $G=-1.00$; and $\alpha = -60°$, $L=2.00$, and $G=+0.00$, respectively. Figure 13 shows the velocity distributions for the case $\alpha = -15°$, $L=2.00$, and $G=-1.00$, the nozzle which corresponds to the small inclined flush slot type Coanda nozzle.

The resultant velocity $q = \sqrt{u^2 + v^2}$ at the outer edge of the ideal fluid jet in still surroundings has always the same value as the velocity at the infinite downstream section U_e, and the maximum velocity in the section appears at the wall surface. In the real viscous wall jet, on the other hand, both velocities at the outer edge of jet and on the wall surface are zero, and the maximum velocity appears at the position between two zero-velocity points.

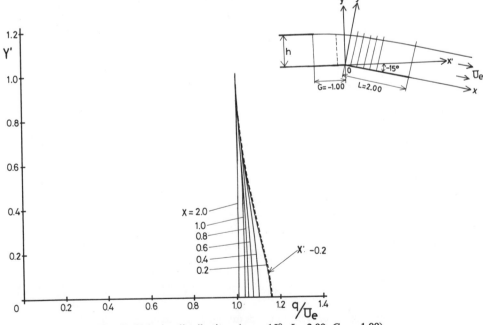

Fig. 13. Velocity distributions ($\alpha = -15°$, $L=2.00$, $G=-1.00$).

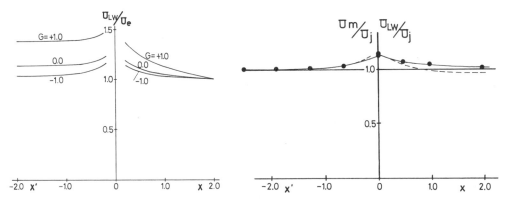

Fig. 14. Velocity distributions along lower wall surface ($\alpha = -15°$, $L = 2.00$, G variable).

Fig. 15. Maximum velocity decay.

IV-6. Velocity Distribution along Wall Surface

Figure 14 shows the distributions of velocity along the lower wall surface in the ideal fluid wall jet, U_{LW}/U_e, for the nozzle $\alpha = -15°$, $L = 2.00$, and different upper lip locations G. In the study on a real viscous wall jet, the decay of maximum velocity is presented in terms of U_m/U_j, where U_j is the injection velocity at the nozzle exit. Figure 15 shows the comparisons between the maximum velocity decay U_m/U_j obtained by the present author for the wall jet of air issuing from the small inclined flush slot into still surroundings with the distributions of U_m/U_j which are reduced from U_{LW}/U_e for the ideal fluid wall jet issuing from the the nozzle $\alpha = -15°$, $L = 2.00$, and $G = -1.00$. The maximum velocity decay around the corner of the nozzle in a real viscous wall jet is much slower than that in the ideal wall jet. It is well known that the maximum velocity decay of a viscous plane wall jet in the far downstream similar flow region is expressed by the following equation

$$\frac{U_m}{V_j} \propto \left(\frac{x - x_0}{h}\right)^{-1/2} \tag{7}$$

where x_0 is the virtual origin of the wall jet. Poor agreement is seen from this figure between the experimental data plotted by the closed circles and the calculated results obtained by the method described here.

IV-7. Wall Pressure Distribution

Figure 16 shows the pressure distributions on the lower wall surface in the ideal fluid wall jet C_p for the nozzle $\alpha = -15°$, $L = 2.00$, and different upper lip locations G. The pressure coefficient in the ideal fluid jet is defined as follows:

$$C_p = \frac{P - P_e}{\frac{1}{2}\rho U_e^2}. \tag{8}$$

where the pressure at the far downstream section P_e is equal to the atmospheric pressure when jet issure into still surroundings. The pressure coefficient used in the study on the real viscous wall jet, on the other hand, is defined in the following relation:

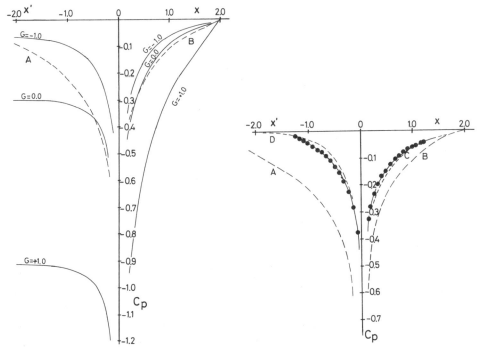

Fig. 16. Static pressure along lower wall ($\alpha = -15°$, $L = 2.00$).

Fig. 17. Static pressure along lower wall.

$$C_p = \frac{P - P_a}{\frac{1}{2}\rho U_j^2}. \tag{9}$$

The experimental data plottd by the closed circles in Fig. 17 were obtained by the present author for the wall jet of air issuing from the small inclined flush slot into still surroundings.

The dashed curves A and B in Fig. 17 show the calculated results of the pressure coefficient for the two-dimensional potential flow around the corner at a deflection angle of $180° + 15°$ to the main stream; curve A is the result obtained when $C_p = 0$ at $X' = x'/h = -3.67$, which corresponded to the nozzle location of the small inclined flush nozzle, and curve B is the result obtained when $C_p = 0$ at $X = x/h = 2.00$, which corresponded to the lip location of the lower deflected wall, respectively. The agreement of these two curves with the experimental data is very poor.

The dotted curve C in Fig. 17 indicates the pressure coefficient downstream from the corner in the ideal fluid wall jet issuing from a nozzle with $\alpha = -15°$, $L = 2.00$, and $G = -1.00$. The dotted curve D indicates the pressure coefficient reduced from the C_p defined by Eq. (8) to that by Eq. (9) in the upstream region from the corner in the ideal fluid wall jet. Good agreement is seen between these two curves for the ideal fluid wall jet and the experimental results obtained in the near-region of a turbulent plane wall jet issuing from the small inclined flush slot type Coanda nozzle.

Acknowledgment

The author is indebted to Dr. Kichiro Takao, Professor of National Defense Academy, for his help in the numerical computations.

REFERENCES

1) Hatano, M., Okamoto, T., and Suwa, S., The experimental investigation on the velocity distribution in the turbulent wall jet issued from a small inclined slot. *Memoirs of Defense Academy*, Vol. **9**, No. 1 (1969), pp. 431–435; Vol. **11**, No. 1 (1971), pp. 71–77.
2) Hatano, M., Experimental investigation of incompressible turbulent jet along lateral convex wall. *Memoirs of Defense Academy*, Vol. **9**, No. 4 (1970), pp. 771–785.
3) Hatano, M., Experimental investigation of incompressible turbulent jet along lateral concave wall. *Memoirs of Defense Academy*, Vol. **9**, No. 4 (1970), pp. 787–796.
4) Hatano, M. and Okamoto, T., The experimental investigation on the three-dimensional wall jet issued from the small inclined slot. *Memoirs of Defense Academy*, Vol. **13**, No. 2 (1973), pp. 305–314.
5) Hatano, M. and Okamoto, T., Flow in the near-region of wall jet issued from a small inclined slot. *Theoretical and Applied Mechanics*, Vol. **29** (University of Tokyo Press, 1980), pp. 429–439.
6) Hatano, M. and Okamoto, T., Experimental investigation on the turbulence characteristics in the near-region of the wall jet issued from a small inclined slot. *Theoretical and Applied Mechanics*, Vol. **32** (University of Tokyo Press, 1984), pp. 57–71.
7) Hatano, M. and Okamoto, T., Experimental investigation on the flow issuing from Coanda nozzles. *Theoretical and Applied Mechanics*, Vol. **35** (University of Tokyo Press, 1987), pp. 259–269.
8) Warner, C.F. and Reese, B.A., Investigation of the factors affecting the attachment of a liquid film to a solid surface. *Jet Propulsion*, Vol. **27**, No. 8 (1957), pp. 877–881.
9) Sasaki, T., *Tōkaku Shazō to Sono Ōyō* (Conformal mapping and its applications) (Gendai Kōgakusha, 1973) (in Japanese).
10) Birkhoff, G. and Zerantonello, E.H. *Jet Wakes, and Cavities* (Academic Press, 1957).
11) Gurevich, M.I., *Theory of Jets in Ideal Fluids* (Academic Press, 1965).

A Bounded Jet Flow Considering the Initial Turbulence

—— Report 3. Effects of the Initial Turbulence on the Shape Parameter ——

Masahiro NAKASHIMA*, Tsutomu NOZAKI[2]*, Keiji HATTA[3]*, and Teruyuki MORI[4]*

* Kagoshima National College of Technology, Kagoshima, [2]* Faculty of Engineering, Kagoshima University, Kogoshima, [3]* College of Engineering, Chubu University, Kasugai, and [4]* Nagasaki Shipyard and Engine Works, Mitsubishi Heavy Industries Limited, Nagasaki

The effects of the initial turbulence on the bounded jet flow are investigated experimentally by an apparatus having the nozzle aspect ratio of 2. The streamwise variation of the shape parameter contained in the velocity distribution function on the jet center-plane is given by a formula including the initial turbulence intensity. Furthermore, the calculation for the mean flow field, which has been proposed already, is modified by considering the effects of the vortices. The calculated results agree well with the experimental ones over a wide range of initial turbulence intensity. The variation of the turbulence intensity on the jet center axis in the streamwise direction is also discussed. At a certain downstream location, this variation is different from that of the two-dimensional jet flow.

I. INTRODUCTION

Calculations of the turbulent flow have considerably been developed using various numerical methods, for example, two-equation models and large eddy simulations. The bounded jet flow, however, is so complex that wall turbulence in the boundary layer flow in the vicinity of the plates and free turbulence in the free jet interact with each other, as has already been reported.[1] Therefore, it seems to be hard to calculate the bounded jet flow using the methods mentioned above. However, the approximate calculation using the momentum integral equation[2] is still valid. For several nozzle aspect ratios and initial turbulence intensities, the approximate calculations of the mean flow field are carried out using the velocity distribution function[3] in the diffusion direction of the jet and that on the jet center-plane of the bounded jet.[4] It has been found[2,4] that the shape parameter contained in the velocity distribution function on the jet center-plane depends considerably upon the aspect ratio and the initial turbulence. It is necessary to prescribe the formula for the shape parameter containing those factors before the approximate calculation is done. In this paper, the relation between the shape parameter and the initial turbulence intensity is obtained for the nozzle aspect ratio of 2.

One of the authors showed the existence of large-scale streamwise vortices in the bounded jet flow using the Laser-Doppler-Velocimeter.[5] This paper attempts to explain the relation between change in the streamwise vortex in the downstream direction and the shape parameter of the velocity distribution qualitatively. Furthermore, the approximate calculation for the mean flow field is modified by considering the effects of those vortices. As a result, the approximate calculations agree well with the experiments over a wide range of the initial turbulence intensity.

II. EXPERIMENTAL RESULTS AND DISCUSSION

The schematic model of the flow and main symbols are shown in Fig. 1 for the case when the distance $2H$ between the bounding plates is comparatively small. The velocity U_3 on the jet center axis and the velocity u_3 and the jet width b_3 on the jet mid-plane are different from those of the two-dimensional jet reported previously.[4] The nozzle aspect ratio is defined as H/b_0 by using the distance $2H$ between the bounding plates and the nozzle width $2b_0$, $H/b_0=2$ in this experiment. The initial turbulence intensity T_0 is given by the streamwise velocity U_0 at the nozzle exit and the root mean square of its fluctuation. The outline of the experimental apparatus, methods of adjustment for the aspect ratio and regulation of the initial turbulence intensity are similar to those described in the previous paper.[1]

II-1. Similarity of the Velocity Distributions on the Jet Center-Plane

The velocity distributions on the jet center-plane at each downstream location are shown in Fig. 2 using the turbulence intensity as a parameter. In order to discuss the similarity of the velocity distributions on the jet center-plane, the distance z^* from the bounding plate is normalized by the thickness δ_2 determined by the interaction between the free turbulence and the wall turbulence, and the velocity U_z on the jet center-plane is also normalized by the velocity U_3 on the jet center axis. The velocity distributions depend upon T_0 in the region of $x/b_0 < 30$. They agree with each other, however, in the region of $x/b_0 > 40$; that is, the bounded jet flow is independent of T_0 in this region, insofar as consideration is restricted to the flow on the jet center-plane.

II-2. Shape Parameter

According to the previous paper,[4] the velocity distribution function on the jet center-plane is given as follows:

$$g(\zeta) = \zeta_1^{1/7}\left(\frac{8}{7} - \frac{\zeta_1}{7}\right) + k\,\zeta_2(1 - \zeta_2)^3, \tag{1}$$

where $\zeta_1 = z^*/\delta_1$, $\zeta_2 = z^*/\delta_2$, and δ_1 is the boundary layer thickness when the jet flows along

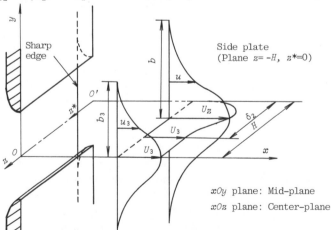

Fig. 1. Schematic model of the flow.

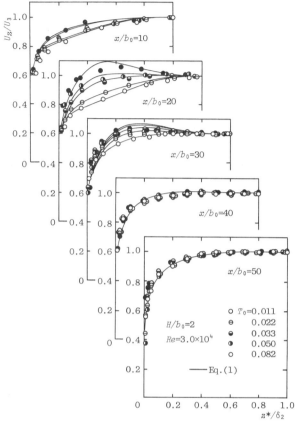

Fig. 2. Velocity distributions on the jet center-plane.

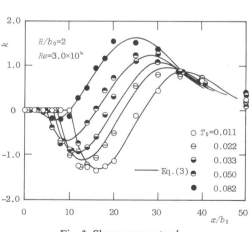

Fig. 3. Shape parameter k.

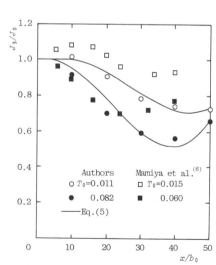

Fig. 4. Momentum on the jet mid-plane.

a plane surface. In this paper the shape parameter k in Eq. (1) is calculated by the same method proposed in the previous paper. The results are shown in Fig. 3, with T_0 as a parameter. As the direction of changes in the values of k with the increase of T_0 is similar to that discussed in the previous report, these discussions are omitted. It is, however, only in this experiment that the values of k are the same regardless of T_0 in the region of $x/b_0 > 40$. This is because the distance between the bounding plates is relatively small so that the jet flow is bounded strongly in this region. This corresponds to the fact that the velocity distributions are similar in this region, as shown in Fig. 2.

It has been clarified by the previous experiments that the shape parameter k contained in Eq. (1) depends upon the aspect ratio H/b_0 and the initial turbulence intensity T_0. Therefore, k is expressed as

$$k = f(H/b_0, T_0, Re, x/b_0). \tag{2}$$

In this experiment the initial turbulence intensity T_0 is only one variable, and its effect on k is investigated. Let k be expressed as

$$k = \sum_{n=1}^{4} \left\{ (C_n + D_n T_0) \left(\frac{x - x_c}{b_0} \right)^n \right\}, \tag{3}$$

where x_c is the length of the zone of flow establishment where the velocity U_0 at the nozzle exit is kept constant. The length x_c varies with the initial turbulence intensity and is expressed as

$$\frac{x_c}{b_0} = 12.2 - 232 T_0 + 1510 T_0^2, \tag{4}$$

according to the experiment. The values of C_n and D_n are shown in Table 1.

Table 1. Values of Cn and Dn.

C_n, D_n \ n	1	2	3	4
Cn	-0.538	5.79×10^{-2}	-1.91×10^{-3}	2.00×10^{-5}
Dn	4.52	-0.300	6.41×10^{-3}	-4.18×10^{-5}

II-3. Momentum on the Jet Mid-Plane

The streamwise variations of the momentum J_3 on the jet mid-plane are shown in Fig. 4. The momentum J_3 decreases in the zone of established flow independently of T_0. The momentum is not conserved in the streamwise direction as far as the jet mid-plane is concerned, and this tendency is different from that of the two-dimensional jet. At locations further downstream J_3 increases a little. This change of tendency has a close relation with the effects of displacement, the so-called marching effect, the relative decrease of the distance between the bounding plates compared with the increased jet width, etc. Therefore, these phenomena about J_3 are peculiarities of the bounded jet flow.

In the experimental results of Mamiya et al.[6] the momentum J_3 tends to become larger than J_0 at the nozzle exit for small turbulence intensities. This is caused by the variation of flow pattern due to the difference of the nozzle shape.[1] Now the solid line shows the momentum decay in the zone of established flow by using the empirical equation

$$\frac{J_3}{J_0} = 1 + E_1\left(\frac{x-x_c}{b_0}\right)^2 + E_2\left(\frac{x-x_c}{b_0}\right)^3, \tag{5}$$

obtained by means of the method of least squares. The values of E_1 and E_2 are -7.01×10^{-4} and 1.32×10^{-5} for $T_0 = 0.011$, and -1.12×10^{-3} and 2.07×10^{-5} for $T_0 = 0.082$, respectively.

II-4. Streamwise Vortex

One of the authors measured the secondary flow on the plane normal to the streamwise direction in the case of $H/b_0 = 1$, using the Laser-Doppler-Velocimeter.[5] These measurements showed that the direction of flow was from the inside to the outside of the jet in the vicinity of the bounding plates and, on the other hand, that there exists an entrainment flow in the middle part between the bounding plates. It was also recognized that two pairs of streamwise vortices exist on the upper/lower part and the right/left part of the flow section because of the two flows mentioned above. These vortices have their axes in the streamwise direction and rotate in clockwise and counterclockwise directions. They probably exist in the case of $H/b_0 = 2$ as well as in the case of $H/b_0 = 1$. The effects of these vortices upon the flow field and the variations of the shape parameter in the downstream direction will be discussed below.

The flow in the case of $T_0 = 0.011$ is discussed as a typical example of the small initial turbulence intensity. At relatively upstream locations, these vortices exist independently of one another and the width of the jet is small. Therefore, the flow on the jet center-plane depends strongly upon the vortices. The fluid on the jet center plane is rolled up by these vortices, and the turbulence intensity of the main flow is small. Due to these effects, the displacement is more amplified than in the ordinary boundary layer flow. The velocity distributions on the jet center-plane have peculiar forms, as shown in Fig. 2, and the shape parameters show considerably lower values, as shown in Fig. 3.

Because the vortices grow gradually in the downstream direction, the whole flow field comes to depend on these vortices and the flow is drawn toward the bounding plates. On the other hand, the turbulence intensity of the flow becomes fairly large even if it is small at the nozzle exit, and the displacement effect by the plates is weakened gradually. Consequently, the flow directs to the bounding plates, marching effects occur, and the shape parameter increases.

At locations further downstream, the jet width increases and the distance between the bounding plates compared with the jet width becomes relatively small. The vortices also interact with one another. Because of these factors, the effect of the vortices on the flow field is weakened by degrees. The velocity distributions on the jet center-plane become similar to those in two parallel flat plates and the shape parameter decreases gradually.

The flow in the case of $T_0 = 0.082$ is discussed as a typical example of the large initial turbulence intensity. The vortex sizes are small and the displacement effect from the bounding plates is weak because the turbulence intensity has already become large as a result of the relatively upstream location. Therefore, the vortices do not substantially affect the flow on the jet center-plane even though there exist vortices in the vicinity of the bounding plates. In the case of $T_0 = 0.082$, therefore, the velocities on the jet center-plane have such distributions as shown in Fig. 2 and the shape parameters show considerable large values compared with those of $T_0 = 0.011$. At locations further downstream, the increase of the jet width is large because of the large turbulence intensity, and then the vortices grow abruptly. As a result, the marching effect mentioned above occurs notably at more upstream locations than in the case of $T_0=0.011$. The velocity distributions have an excess, and the values of k become large. Downstream, the distance between the bounding plates relative to the jet width begins to decrease at locations considerably further upstream than for $T_0=0.011$, and the effect of the vortices on the flow diminishes rapidly. Consequently, the velocity distributions on the jet center-plane become similar to those of the two-parallel flat plates, and the shape parameters decrease gradually.

II-5. Modification of the Approximate Calculations

In the previous approximate calculation for the mean flow field,[2] it was assumed that as the first approximation the momentum outside the interaction thickness of the bounded jet was kept constant like the two-dimensional jet flow. The flow in the case of $H/b_0=1$ was calculated as an example, and the experimental results agreed well with the calculated ones for small turbulence intensity, but they did not agree well for large turbulence intensity. This is due to the fact that streamwise decay of the momentum influenced by the above-mentioned vortices was not considered. This effect is taken into account in this paper. On the basis of the definition in the zone of flow establishment, however, it is assumed that the momentum on the jet mid-plane is kept constant, and the calculation is carried out using the method as reported already.[2]

In that paper the the momentum integral equation is given as

$$\frac{d}{dx}\left\{U_3^2\int_0^{\delta_2}\int_0^{b}f^2(\eta)g^2(\zeta)dydz^*\right\} + \frac{d}{dx}\left\{U_3^2\int_{\delta_2}^{H}\int_0^{b_3}f^2(\eta)dydz^*\right\}$$
$$= -\frac{\tau_{w0}}{\rho}\int_0^{(b)z^*=0}f^2(\eta)dy, \qquad (6)$$

applying the boundary layer approximation to the equation of motion. In this equation, $f(\eta)$ is the velocity distribution function[3] of the jet, $g(\zeta)$ that[4] on the jet center-plane, and τ_{w0} the wall shear stress. In the zone of established flow, Eq. (6) is given as

$$\frac{d}{dx}\left\{J_3\int_0^{\delta_2}h(\zeta_2)g^2(\zeta)dz^*\right\} + \frac{d}{dx}\left\{J_3(H-\delta_2)\right\}$$
$$= -\alpha J_3\left(\frac{\text{Re}}{2}\right)^{-1/4}\left(\frac{U_3}{U_0}\right)^{-1/4}\left(\frac{\delta_1}{b_0}\right)^{-1/4}h(0), \qquad (7)$$

where $h(\zeta_2)$ is the function expressing the variation of the jet width b in z direction and b is expressed as

$$b = b_3 h(\zeta_2), \qquad (8)$$

by using the jet width b_3 on the jet mid-plane. The function $h(\zeta_2)$ is also given in the former paper[2] as

$$h(\zeta_2) = (a-1)(1-\zeta_2)^2 + 1, \qquad (9)$$

considering the boundary conditions. Furthermore, Eq. (5) is used for the momentum J_3 on the jet mid-plane, and Eq. (3) is also used for the shape parameter k included in $g(\zeta)$ in Eq. (7). Finally, this calculation is reduced to the problem to solve the ordinary differential equation for the variable a contained in Eq. (9). This variable a expresses the ratio of the jet width b_3 on the jet mid-plane to the apparent width $(b)_{z^*=0}$ on the bounding plate, and the increase of a shows that the flow is drawn toward the bounding plates by degrees.

The variations of a in the streamwise direction obtained by these calculations are shown

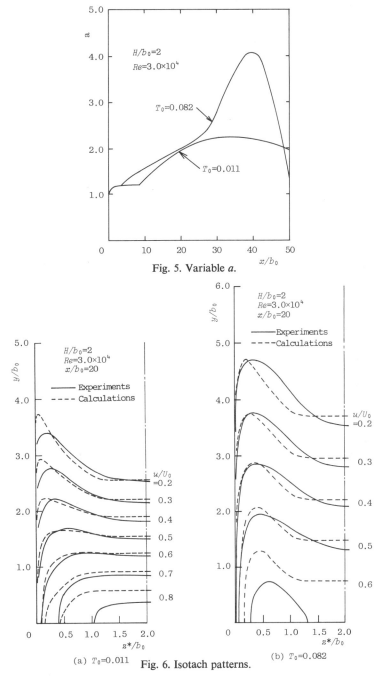

Fig. 5. Variable a.

(a) $T_0=0.011$ Fig. 6. Isotach patterns.   (b) $T_0=0.082$

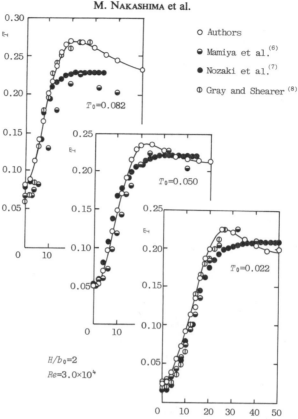

Fig. 7. Turbulence intensity on the jet center axis.

in Fig. 5. The value of a increases gradually in the streamwise direction for any values of T_0. This tendency becomes remarkable for $T_0=0.082$, corresponding to the fact that the shape parameter k shows a considerably large value and the marching effect is strong in this case. Further downstream, the value of a decreases abruptly because of the developing and decaying processes of the vortices mentioned above.

The aspects of the flow obtained using this variable are shown in Fig. 6 by means of the isotach patterns. Over wide range of the flow field the calculations (broken line) agree well with the experiments (solid line) for all turbulence intensities.

II-6. Turbulence Intensity

The turbulence intensity T on the jet center axis is given using the velocity U_3 on that axis and its fluctuation. The variations of the turbulence intensity in the streamwise direction are shown in Fig. 7. At relatively upstream locations, T increases gradually, and this tendency is similar to that of the two-dimensional jet. Further downstream, T decreases for any T_0, whereas in the two-dimensional jet flow T is constant and the flow tends to approach the self-preserving state in this region. It is one of the peculiarities of the bounded jet that the turbulence intensities show this decreasing tendency. In the region in the vicinity of the nozzle exit, the variations in turbulence intensity on the jet center axis of the bounded jet are similar to those of the two-dimensional jet flow because the boundary layer developing on the bounding plates does not reach the jet mid-plane yet. However, at the downstream region where the boundary layer reaches the mid-plane, the velocity fluctuations are also confined. In this region the jet width becomes considerably large and the distance between the bounding plates relative to the jet width becomes small. Therefore, the turbulence intensity decreases because the character of the flow in this region is similar to that of the internal flow such as the flow between the two parallel flat plates or the pipe flow, rather than to an external flow such as the jet flow as long as the flow is in the vicinity of the jet center axis.

III. CONCLUSIONS

The effects of initial turbulence intensity on the bounded jet flow were investigated. The variation of the shape parameter contained in the velocity distribution on the jet center plane is given by the formula including the initial turbulence intensity. The relationship between the streamwise vortices and the variation of the shape parameter in the streamwise direction is discussed. Furthermore, the calculation of the mean flow field, which has been proposed already, is modified by considering the effects of those vortices. As a result, the calculations agree well with the experiments over a wide range of the flow. The variations of the turbulence intensity on the jet center axis are also discussed. The turbulence intensities show the maximum value at a certain downstream location and decrease in the region where the boundary layer reaches the jet mid-plane. This tendency of the turbulence intensity is different from that of the two-dimensional jet, and this is one of the peculiarities of the bounded jet.

REFERENCES

1) Nakashima, M. et al., A bounded jet flow considering the initial turbulence (Experiments for the nozzle having an aspect ratio of 3). *Theoretical and Applied Mechanics*, **32** (University of Tokyo Press, 1984), pp. 47–55.
2) Nakashima, M. et al., Study of a bounded jet flow considering the initial turbulence (3rd report, A method of approximate calculations for the mean velocity field). *Trans. JSME*, **52**, No. 478 (1986-6), pp. 2367–2373.
3) Hatta, K. and Nozaki, T., Two-dimensional and axisymmetric jet flows with finite initial cross sections, *Bull. JSME*, **18**, No. 118 (1975-4), pp. 349–357.
4) Nakashima, M. et al., A bounded jet flow considering initial turbulence (Report 2. In the case of relatively large nozzle aspect ratio). *Theoretical and Applied Mechanics*, **34** (University of Tokyo Press, 1986), pp. 57–69.
5) Nozaki, T. et al., Experimental study of a bounded jet flow (Mechanism of the secondary flow). Proceedings of the 4th International Symposium on Flow Visualization (1986), pp. 495–499.
6) Mamiya, T. et al., The effect of initial turbulence and aspect ratio on internal jet behavior. Proceeding of the 14th Fluidics Symposium (1979), pp. 1–6.
7) Nozaki, T., et al., Reattachment flow issuing from a finite width nozzle (Report 2. Effects of initial turbulence intensity). *Bull. JSME*, **24**, No. 188 (1981-2), pp. 363–369.
8) Gray, R.W. and Shearer, J.L., Effects of upstream disturbances on the spreading of large fluid-amplifier-type jet. *Trans. ASME, Ser. G*, **93**, No. 1 (1971), pp. 53–60.

A Study of Short Fiber Transportation Using a Spiral Flow

——Analysis of Force Acting on a Short Fiber——

Yuji MATSUMAE,* Xiao Ming CHENG,* Masahiro TAKEI,* and Kiyoshi HORII[2]*

Department of Science and Engineering, Waseda Univ., Tokyo, [2] Department of Literature, Shirayuri Women's College, Tokyo

It is well known that fibrous materials are difficult to convey pneumatically, and tend to be deformed into fiber balls conveyance by air. But this deformation does not happen during conveyance by a spiral flow.

The motion of short fibers transported forming a spiral were observed parallel to the pipe axis. The orientation of the fibers is found to be controlled by the spiral flow.

According to the theoretical analysis, the force due to the spiral flow acts on fibers in an orderly manner. This ordered orientation and nonfiber-ball tendency are considered to be closely related with the spiral flow.

Nomenclature

a:	diameter of fiber section	[m]
F:	force acting on the fiber	[m·s^{-2}]
F':	force of axial direction	[m·s^{-2}]
L:	whole length of fiber section	[m]
l:	length of fiber section	[m]
r, θ, z:	cylindrical polar coordinates	[−]
r', θ', z':	new cylindrical polar coordinates	[−]
s:	section area of fiber section	[m^2]
u_r, u_θ, u_z:	velocity components in r', θ', and z' directions, respectively	[m·s^{-1}]
v_r, v_θ, v_z:	velocity components in r, θ, and z directions, respectively	[m·s^{-1}]
δ:	thickness of boundary layer	[m]
ν:	kinematic viscosity	[m^2·s^{-1}]
ρ:	density	[kg·m^{-3}]

I. INTRODUCTION

Few investigators have reported on fibrous materials transported pneumatically. A review of previous work to predict pressure loss in pneumatic conveyance was discussed by I.V. Gould.[1]) But this did not describe the relation between fiber ball phenomena and pneumatic conveyance. It is known that fibrous materials are very difficult to convey by air, because they are deformed into fiber balls. They are not deformed into fiber balls when conveyed by a spiral flow, however.

We found in previous studies[2-4]) that if turbulence moves radially inward through a contraction pipe, it deforms a spiral flow and produces maximum axial velocity at the axis. The motion of short fibers was observed in a spiral flow. These fibers were transported on a constant path parallel to the pipe axis, forming a spiral. In order to explain the fiber

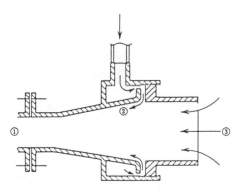

Fig. 1. Coanda spiral flow apparatus.

orientations[5-7] and nonfiber ball phenomena, the theoretical analysis of force acting on a short fiber is discussed below.

II. EXPERIMENTS

II-1. Experimental Apparatus to Produce Spiral Flow and Turbulence

A Coanda slit was made in a ring form, and an apparatus in which a conical part was connected to this ring (Fig. 1) was constructed. Spiral flow was generated by means of this apparatus. Turbulent flow was generated by means of the Coanda ring flow apparatus without a conical part.

II-2. Experimental Conditions

Air flows through the slit along the Coanda curved surface into part 2, then in the direction toward part 1 in Fig. 1. Due to this flow, a negative pressure zone arises at this bottom part. The pressure at the air inflow part was 3.5 kgf·cm^{-2}, the flow rate was 200–300 Nl·min^{-1} and the mean velociy was 35 m·s^{-1} at part 1. The apparatus was horizontally supported, and by inserting the fibers from part 3, the motion of fibers was observed. The glass fibers were 13 mm long.

III. RESULTS

The observed results of fiber motion in a spiral flow and in turbulence are shown in Fig. 2 and Fig. 3, respectively.

III-1. Transportation of Fibers Using a Spiral Flow

Fibers were transported constant parallel to the pipe axis, forming a spiral. In the outlet section, each fiber flew out like an arrow, or remained parallel to the pipe axis. As a result, fibers were parallel to the pipe axis in a uniform array on the plane (Fig. 2).

III-2. Transportation of Fibers Using Turbulence

The air flow made fibers turbulent. Some of the fibers were transported such that their angle to pipe axis changed every moment. Other fibers, as observing from the side of the

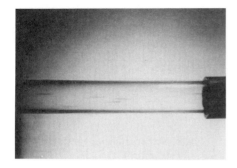

Fig. 2. Fiber motion in a spiral flow.

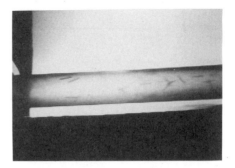

Fig. 3. Fiber motion in turbence.

pipe, showed a spinning motion around a point located on the fiber. This point moved more or less away from the center. In the outlet section, the fibers flew out, and remained in the pipe plane as explained above. As a result, fibers were in a turbulent state on the plane (Fig. 3).

IV. THEORETICAL ANALYSIS

——Force Acting on a Fiber in a Spiral Flow——

IV-1. Assumptions

(1) The properties of elastic solids and fiber mass are ignored in this analysis.

(2) The fiber always moved making an orderly signature curve as observed from the side of the pipe, and rotation occurred around the axis of the pipe as observed from the section of pipe (Fig. 4).

(3) The flow around an infinitesimal section of fiber (length l, radius a) is treated as a parallel flow to the section (Fig. 5).

Fig. 4. Pattern of fiber motion.

Fig. 5. Flow around an infinitesimal section of a fiber.

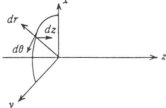

Fig. 6. Coordinate system.

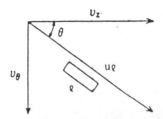 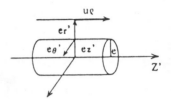

Fig. 7. Fluid velocity on the boundary layer of the fiber section.

Fig. 8. New coordinates system (r', θ', z').

(4) The description, "the flow around the fiber", means the velocity of the surface in the boundary layer (thickness δ).

(5) The friction acting on a fiber is obtained from calculation of the stress tensor and the velocity in the boundary layer of the fiber section which depends on the viscosity of fluid.

IV-2. Coordinate System

The coordinate system (v_r, v_θ, v_z) shown in Fig. 6 is used. v_r is set at zero based on the spiral flow created in the pipe. The velocity of the surface in the boundary layer, that is, point δ m away from the fiber surface, is obtained from the composition of the θ and z components:

$$u_l = \sqrt{v_z^2 + v_\theta^2}. \tag{1}$$

The flow direction of the fluid is parallel to the fiber section l. The boundary layer theorem is applied for the surface of a fiber (sectional area πa^2, length l) (Fig. 7).

A new coordinate system (r', θ', z') depends on fiber section l as shown in Fig. 8. Each component of the new coordinate system (r', θ', z') is set at e'_r, e'_θ, and e'_z in distinction from dr, $d\theta$, and dz of the coordinate system (r, θ, z). The z'-distance suggests the flow direction. In this case, the θ component of the fluid is ignored.

IV-3. Navier-Stokes and Continuity Equations in Incompressible Fluid Field
  Assumptions
u_r and u_z do not depend on the θ component.
u_θ is set at zero.
Components
  Each component is obtained as follows:

$$u_r \frac{\partial u_r}{\partial r} + u_z \frac{\partial u_r}{\partial z} = -\frac{1}{\rho}\frac{\partial p}{\partial r} + \nu\left(\frac{\partial^2 u_r}{\partial r^2} + \frac{1}{r}\frac{\partial u_r}{\partial r} + \frac{\partial^2 u_r}{\partial z^2} - \frac{u_r}{r^2}\right). \tag{2}$$

$$\frac{1}{\rho r}\frac{\partial p}{\partial \theta} = 0. \tag{3}$$

$$u_r \frac{\partial u_z}{\partial r} + u_z \frac{\partial u_z}{\partial z} = -\frac{1}{\rho}\frac{\partial p}{\partial z} + \nu\left(\frac{\partial^2 u_z}{\partial r^2} + \frac{1}{r}\frac{\partial u_z}{\partial r} + \frac{\partial^2 u_z}{\partial z^2}\right). \qquad (4)$$

$$\frac{\partial u_r}{\partial r} + \frac{\partial u_z}{\partial z} + \frac{u_r}{r} = 0. \qquad (5)$$

IV-4. Boundary Layer Thickness
r and z are changed to dimensionless form as follows:

$$\bar{z} = \frac{z}{l}, \quad \bar{r} = \frac{r}{\delta}. \qquad (6)$$

Continuity Eq. (5) is rewritten as

$$\frac{\delta}{l} \approx \sqrt{\frac{\nu}{l\,u_l}} = \frac{1}{\sqrt{R}}, \quad R = \frac{l\,u_l}{\nu}. \qquad (7)$$

IV-5. Boundary Layer Pressure
Equation (2) is changed to dimensionless form using Eq. (6)

$$u_l \frac{\partial u_l}{\partial z} = -\frac{1}{\rho}\frac{\partial p_l}{\partial z}. \qquad (8)$$

The pressure of boundary layer is determined by calculating the velocity u_l of the boundary layer.

IV-6. Boundary Layer Velocity Components
The following fundamental equations explain the flow boundary layer.

$$\frac{1}{r}\frac{\partial}{\partial r} r u_r + \frac{\partial u_z}{\partial z} = 0. \qquad (9)$$

$$u_z \frac{\partial u_z}{\partial z} + u_r \frac{\partial u_z}{\partial r} = u_l \frac{\partial u_l}{\partial z} + \nu \frac{1}{r}\frac{\partial}{\partial r}\left(r \frac{\partial u_z}{\partial r}\right) \qquad (10)$$

where the boundary condition are as follows:
on the surface of the fiber section ($r=a$) $u_r = u_z = 0$;
on the surface of the boundary layer ($r=\delta$) $u_z = u_l$;
and $u_l = $ const. and $u_l \cdot \partial u_l / \partial z = 0$.

$$\bar{u}_z = \frac{u_z}{u_l} = f\left(r\sqrt{\frac{u_l}{\nu z}}\right), \quad \bar{u}_r = \frac{u_r}{u_l}\sqrt{\frac{l\,u_l}{\nu}} = f_1\left(r\sqrt{\frac{u_l}{\nu z}}\right). \qquad (11)$$

The solutions are predicted to be formularized as Eq. (11), since when the flow and fiber are parallel to each other through the whole length L of the fiber, the flow around the infinitesimal fiber has no feature as the flow of boundary layer does through the whole length of the plane. In other words, the solutions of the boundary layer equations are expressed as the functions of $\bar{z}$ and $\bar{r}$ without containing the function l. The solutions depend on $\bar{r}/\sqrt{\bar{z}} = r\sqrt{u_l/\nu z}$ only.

IV-7. Frictional Force in Boundary Layer

From Eqs. (9) and (10)

$$2\xi \phi''' + \left\{\xi\left(1 - \frac{1}{\xi+1}\right)\phi' + \xi\phi - \xi\int \frac{1}{(\xi+1)^2}\phi' d\xi + 2\right\}\phi'' = 0. \quad (12)$$

From Eq. (12). frictional force is obtained on the condition that ϕ and ϕ' are set at zero in the case of $\xi = 0$, and ϕ' at 1 in the case of $\xi = \infty$.

$$\sigma_{rz} = \rho\nu\left(\frac{\partial u_z}{\partial r}\right)_{r=0} = \rho\nu u_l\sqrt{\frac{u_l}{\nu z}}\phi''(0), \quad u_l = \sqrt{v_z^2 + v_\theta^2}. \quad (13)$$

IV-8. Force Acting on Fibers

The force acting through the whole length L of a fiber is obtained by integrating Eq. (13) from zero to L on the condition that the fiber and flow are parallel to each other.

$$F = \int_0^L \sigma_{rz}\,dz = \phi''(0)\,\rho\,u_l\,\sqrt{u_l\,\nu}\int_0^L \frac{1}{\sqrt{z}}\,dz$$
$$= \phi''(0)\,\rho\,u_l\sqrt{u_l\,\nu\,L}.$$

IV-9. Axial Direction Force

The relation between F and coordinate system (r, θ, z) is shown in Fig. 9. The relation between the fiber motion in a spiral at z-distance is shown in Fig. 10. The properties of an elastic solid and mass of a fiber are ignored. The axial pitch of the spiral is considered to result from the gravity acting on the fiber, the tension, and the elastic force of the fiber. But in this analysis, they were ignored. The force of the axial direction is solved on the condition that the axial pitch of the spiral is already arranged, v_θ and v_z are constant along the fiber, and v_r is set at zero.

The angle θ made by the fiber and z-distance is written as $\theta = \tan^{-1}(v_\theta/v_z)$.

The force of axial direction F' is obtained as follows

$$F' = \frac{v_z}{\sqrt{v_z^2 + v_\theta^2}}\,\phi''(0)\,\rho\,u_l\,\sqrt{u_l\,\nu\,L}.$$

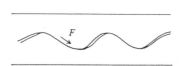

Fig. 9. The relation between F and coordination system (r, θ, z).

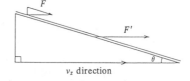

Fig. 10. The force of axial direction F'.

V. CONCLUSIONS

(1) By using a spiral flow, fibers were transported constantly parallel to the pipe axis, forming a spiral. The fibers were not deformed into fiber balls.

(2) By using turbulence, fibers were transported such that they formed an angle to the pipe axis changing at every moment and showing a spinning motion around a point located on the fiber. The fibers were deformed into fiber balls.

(3) The force acting through the whole length L of the fiber is obtained as

$$F = \phi''(0)\rho u_l \sqrt{u_l \nu L}$$

(4) The orientation of fibers was controlled by a spiral flow. The explanation for this is force F acting on the fiber is stronger than the axial force F' in a spiral flow. But in turbulence, force F is considered to be almost the same as the force F'. The fiber in a spiral flow is used for stronger force along the fiber, and the fiber in turbulence is used for the reverse force to the axial flow.

Acknowledgments

The authors wish to thank Professor Yasuhiko Aihara (Tokyo University) and Professor Bunsaku Hashimoto (Waseda University) for their help with this experiment. Sumitomo Coal and Mining Co., Ltd. and Fukuvi Chemical Ind. Co., Ltd. also gave us a great deal of help and cooperation during the experiment.

This project was completed with the aid of the Research Development Corporation of Japan.

REFERENCES

1) Gould, I.V. and Aust, M.I.E., Vacuum-pneumatic conveying of raw wool. *Mech. Eng. Trans.* (1983), pp. 77–83.
2) Horii, K. and Murata, T., A conceptual approach to the spiral movement. *J. Jpn. Petro. Inst.*, Vol. 5 (1984), pp. 452–455.
3) Horii, K., Murata, T., Takarada, M., and Marui, T., A study of spiral flow (Part 1). *Trans. Jpn. Soc. Aero. Space Sci.*, Vol. 28, No. 81 (1985), pp. 123–131.
4) Horii, K., Matsui, S., Yamaguchi, K., and Matsumae, Y., A study of coanda spiral flow. *Theoretical and Applied Mechanics*, Vol. 36 (1988), pp. 185–193.
5) Schulgasser, K., Fibre orientation in machine-made paper. *J. Materials Sci.*, Vol. 20 (1985), pp. 859–866.
6) Kamal, M.R., Song, L., and Singh, P., Fiber and matrix orientations in injectable molded fiber reinforced composites. *ANTEC' 86* (1986), pp. 133–139.
7) Kamal, M.R., Song, L., and Singh, P., Measurement of fiber and matrix orientations in fiber reinforced composites. *Polymer Composites*, Vol. 7, No. 5 (1986), pp. 323–329.

IV
NUMERICAL DYNAMICS

Breakdown of the Karman Vortex Street Due to Natural Convection: Numerical Simulation

Katsuhisa NOTO and Ryuichi MATSUMOTO

Department of Mechanical Engineering, Kobe University, Kobe

The present paper makes clear the thermo-fluid behavior of the wake behind a cylinder for which separation points are approximately fixed. An elliptical cylinder with major axis oriented at right angles to the main stream is chosen. Time-dependent Navier-Stokes and energy equations are solved numerically. The numerical results show that a breakdown of the Karman vortex street occurs due to the natural convection. The streamline, the vorticity, and the isotherm are clarified in a detailed breakdown of the Karman vortex street. The Strouhal number increases due to the natural convection.

Nomenclature

a:	thermal diffusivity	y:	perpendicular coordinate to the x-direction from the center of cylinder
c:	one half of distance between focuses of the elliptical cylinder	β:	thermal expansion coefficient
E:	transformation coefficient, Eq. (17)	Δt:	time step
$\vec{e}$:	unit vector in x-direction	$\Delta\eta, \Delta\xi$:	mesh constants in η- and ξ-directions
g:	gravitational acceleration	δV:	amplitude of the initial disturbance
Gr:	Grashof number, Eq. (15)	ζ:	vorticity
p:	static pressure	η, ξ:	elliptical coordinates, Eq. (10)
Pe:	Péclet number	ν:	kinematic viscosity
Pr:	Prandtl number	ρ:	density of fluid
Re:	Reynolds number, Eq. (14)	Ψ:	stream function
Ri:	Richardson number, Eq. (16)	ω:	circular frequency of the initial disturbance
T:	temperature		
t:	time		
t_m:	time during the initial disturbance is given	*Subscripts*	
		i, j:	grid numbers in η- and ξ-directions
$\vec{u}$:	velocity vector	w:	denotes the value on the surface of cylinder
u, v:	velocity components in x- and y-directions	∞:	denotes the value at infinity
v_η, v_ξ:	velocity components in η- and ξ-directions	*Superscript*	
x:	coordinate with anti-gravitational direction from the center of cylinder	$'$:	denotes quantity with physical dimension

I. INTRODUCTION

The present paper deals with the breakdown of the Karman vortex street due to natural convection formed behind an elliptical cylinder with major axis oriented at right angles to the main stream.

The Karman vortex street has been studied both analytically and experimentally by many researchers.[1] However, their studies have been limited to the Karman vortex street in either the equi-temperature field or the forced convective field. On the other hand, the thermo-fluid behavior in a wake with natural convection behind a circular cylinder has been treated theoretically and numerically by several authors.[2-6] They treated the wake with natural convection as a symmetrical steady state problem. That is, their studies referred to neither the breakdown nor the development of the Karman vortex street due to natural convection.

In general, when the effect of natural convection increases, changes in flow pattern cannot be ignored. Furthermore, unknown flow phenomena due to natural convection may occur in the Karman vortex street. From this point of view, the present authors have studied unsteady thermo-fluid behaviors of the Karman vortex street in a wake with natural convection, the development of which they first found theoretically in 1983.[7] They have since investigated the same problem both numerically and experimentally.[8-13]

However, it has not been made clear theoretically whether or not a breakdown occurs behind a cylinder with separation points that are approximately fixed. The present paper deals with a wake with natural convection behind an elliptical cylinder by numerically solving time-dependent equations. The major axis of this elliptical cylinder is oriented at right angles to the main stream. Separation points are expected to be approximately fixed at given points on this cylinder.

In this paper, a numerical technique is developed for obtaining an asymmetrical thermo-fluid pattern. Preliminary calculations are carried out in detail for accuracy of the present numerical technique. The calculation is restricted for $0 \leq Gr/Re^2 \leq 1.0$. As a result, it was clarified that the natural convection breaks down the Karman vortex behind the cylinder with separation points approximately fixed.

II. PHYSICAL MODEL

An elliptical cylinder is horizontally placed in an infinite space. The lengths of the major and the minor axes of this cylinder are $2c \cdot \cosh\xi_w$ and $2c \cdot \sinh\xi_w$ respectively. The major axis is oriented at right angles to the main stream direction. Let us consider a main stream in the x'-direction with uniform velocity u'_∞ in a region far from this cylinder. If the cylinder is isothermally heated, natural convection occurs from the cylinder. This natural convection and the main stream are combined to form a mixed convective flow. Such a flow shows a very complicated pattern and plays an important role in the wake. Separation phenomena are expected to occur near the side edges of this cylinder. To analyze numerically the wake phenomenon with natural convection, physically reasonable assumptions are made. The flow is kept in a two-dimensional laminar state. Boussinesq's approximation is employed. The viscous dissipation can be neglected.

III. MATHEMATICAL FORMULATION

III-1. Governing Equations

For an unsteady laminar flow with a considerably low velocity, the governing equations can be written as follows:

$$\nabla \cdot \boldsymbol{u}' = 0, \tag{1}$$

$$D\boldsymbol{u}'/Dt' = -(1/\rho')\nabla p' + \nu'\nabla^2 \boldsymbol{u}' + g'\check{e}\beta'(T' - T'_\infty), \tag{2}$$

$$DT'/Dt' = a'\nabla^2 T'. \tag{3}$$

The following dimensionless variables are introduced:

$$\left.\begin{aligned}
(x, y) &= (x', y')/c, \\
(u, v) &= (u', v')/u'_\infty, \\
p &= p'/\rho' u'^2_\infty, \\
t &= t' u'_\infty/c, \\
T &= (T' - T'_\infty)/(T'_w - T'_\infty).
\end{aligned}\right\} \tag{4}$$

Let us introduce the dimensionless stream fuhction Ψ defined as

$$u = \partial\Psi/\partial y, \; v = -\partial\Psi/\partial x, \tag{5}$$

and the dimensionless vorticity ζ defined as

$$\zeta = \nabla \times \boldsymbol{u}. \tag{6}$$

When using Eqs. (4) to (6), the stream function equation, the vorticity transport equation, and the heat balance equation become as follows:

$$\zeta = -\nabla\Psi^2, \tag{7}$$

$$D\zeta/Dt = (2/Re)\nabla^2\zeta - (Ri/2)\partial T/\partial y, \tag{8}$$

$$DT/Dt = (2/Pe)\nabla^2 T. \tag{9}$$

Because the shape of the cross-section of the cylinder is elliptical, the Cartesian coordinate system is transformed into an elliptical coordinate system such as

$$(x, y) = (\cos\eta \cdot \sinh\xi, \sin\eta \cdot \cosh\xi). \tag{10}$$

This is one of the boundary-fit curvilinear coordinate transformations proposed by Thompson et al.[14]

In the new coordinate system, Eqs. (7)~(9) are rewritten as follows:

$$E\zeta = -\nabla^2\Psi, \tag{11}$$

$$E \cdot \partial\zeta/\partial t + \partial(\Psi, \zeta)/\partial(\eta, \xi) = (2/Re)\nabla^2\zeta - (Ri/2)$$
$$\times (\cos\eta \cdot \sinh\xi \cdot \partial T/\partial\xi - \sin\eta \cdot \cosh\xi \cdot \partial T/\partial\eta), \tag{12}$$

$$E \cdot \partial T/\partial t + \partial(\Psi, T)/\partial(\eta, \xi) = (2/Pe)\nabla^2 T, \tag{13}$$

where

$$Re = u'_\infty \cdot 2c/\nu', \tag{14}$$

$$Gr = (2c)^3 g' \beta'(T'_w - T'_\infty)/\nu'^2, \tag{15}$$

$$Ri = Gr/Re^2, \tag{16}$$

$$E = \sin^2\eta + \sinh^2\xi. \tag{17}$$

The problem treated in the present paper is an initial-boundary value problem. The initial conditions are the following potential flow state:

$$\Psi = \exp(\xi_w) \cdot \sinh(\xi - \xi_w) \cdot \sin\eta, \tag{18}$$

$$\zeta = 0, \tag{19}$$

$$T = 0. \tag{20}$$

The boundary conditions of the Dirichlet type become as follows:

$$\xi = \xi_w: \begin{cases} \Psi_w = 0, & (21) \\ \zeta_w = -2\Psi_{2,j}/\{E \cdot (\Delta\xi)^2\}, & (22) \\ T_w = 1, & (23) \end{cases}$$

$$\xi = \xi_\infty: \begin{cases} \Psi_\infty = \exp(\xi_w) \cdot \sinh(\xi_\infty - \xi_w) \cdot \sin\eta, & (24) \\ \zeta_\infty = 0, & (25) \\ T_\infty = 0. & (26) \end{cases}$$

The number of boundary conditions in Eqs. (21) to (26) is mathematically sufficient to solve Eqs. (11) to (13). However, it should also be physically necessary that the Neumann type boundary conditions is satisfied as follows:

$$\xi = \xi_w: \partial\Psi/\partial\xi = 0, \tag{27}$$

$$\xi = \xi_\infty: \begin{cases} \partial\Psi/\partial\xi = 0, & (28) \\ \partial\zeta/\partial\xi = 0, & (29) \\ \partial T/\partial\xi = 0. & (30) \end{cases}$$

The boundary conditions, Eqs. (27) to (30), cannot be mathematically imposed in the calculation procedure. Therefore, it should be checked whether Ψ, ζ, and T satisfy Eqs. (27) to (30) or not after the calculation.

III-2. Time-Dependent Initial Disturbance

The buoyancy becomes an initial disturbance if the buoyancy direction intersects that of the main stream direction. However, in the present thermo-fluid field, the buoyancy direction does not intersect the main stream direction. Therefore, it should be necessary to find and introduce an initial disturbance in the present thermo-fluid field. An initial disturbance should be time-dependent. A forced disturbance with a finite amplitude is given for $0 \leq t \leq t_m$ as follows:

$$(v_\eta)_{\xi=\xi w} = \delta v \cdot \sin(2\pi\omega t), \tag{31}$$

$$(v_\xi)_{\xi=\xi w} = 0. \tag{32}$$

Whether a flow in the wake develops into a vortex street or not is determined by whether the initial disturbance is amplified or diminished with time. Thermo-fluid patterns after the initial disturbance have been amplified or diminished is obtained by the present numerical technique. While the initial disturbance is given, the boundary conditions, Eqs. (22) and (27), are changed as follows:

$$\zeta_w = -2\{\Psi_{2,j} + (v_\eta)_{\xi=\xi_w}E_w \cdot (\Delta\xi)\}/\{E_w \cdot (\Delta\xi)^2\}, \tag{33}$$

$$\{(\partial\Psi/\partial\xi)/E\}_{\xi=\xi_w} = -(v_\eta)_{\xi=\xi_w}. \tag{34}$$

IV. NUMERICAL METHOD

Numerical calculations are carried out with the aid of a combination of the forward-time-and-centered-space (FTCS), successive-over-relaxation (SOR), and alternating-direction-implicit (ADI) schemes. The initial disturbance was given. Unsteady numerical solutions were obtained at every time step.

Preliminary calculations were carried out for the validity of the outer boundary location and the grid size. The outer boundary was located at the position of $(40 \sim 150)c$ from the origin of the cylinder. The validity of the outer boundary location was checked by thermo-fluid patterns and by Eqs. (28) to (30). A much finer grid is used in the present calculation for grid resolution, because a numerical calculation result is not correct if a computational grid is coarser than a real vortex scale. As a result, all of the present numerical calculations satisfied the outer boundary conditions and the grid resolution. Considering the Courant and the Neumann conditions, a time step Δt is determined.

Other preliminary calculations were carried out for checking the initial disturbance. The influences of the initial disturbance were investigated for various values of δv, ω, and t_m. As a result, it was found that, in the same phase, the period and amplitude of the thermo-fluid fluctuations were the same for different initial disturbances. That is, the thermo-fluid behaviors after a long time were not influenced by δv, ω, and t_m. Therefore, the initial disturbances were fixed by the following condition:

$$(\delta v, \omega, t_m) = (2.0, 2.0, 0.5). \tag{35}$$

V. RESULTS AND DISCUSSIONS

V-1. Calculation Conditions

The value of ξ at the wall, ξ_w, is equal to 0.45. Therefore, the thickness ratio of the elliptical cylinder is equal to 0.422. The Prandtl number Pr is equal to 0.72 for air. The Reynolds number in the present calculations is equal to 100. The Grashof number is a plus value, because the temperature of the cylinder is higher than that of the main stream. The Grashof number Gr is restricted to the range $0 \leq Gr \leq 10^4$. Therefore, the Richardson number range is $0 \leq Ri \leq 1.0$.

The representative length of the Reynolds Re, the Grashof Gr, and the Richardson Ri numbers is the length $2c$ between focuses of the elliptical cylinder. The Reynolds, the Grashof, and the Richardson numbers with the major axis length as the representative length are 1.10, 1.34, and 1.10 times the above Re, Gr, and Ri, respectively.

V-2. Flow Pattern in Breakdown of the Karman Vortex Street

Streamlines in the neutral buoyant wake behind the elliptical cylinder with the major axis oriented at right angles to the main stream are shown in Fig. 1(1). The flow pattern in this case is in an unsteady state. Therefore, the flow pattern at $t=40$ is typically shown. Each flow pattern in Figs. 1(1) to 1(4) is at time $t=40$. Streamlines in Fig. 1(1) are shown by every step $\Delta\Psi=0.1$ for $-2.0\leq\Psi\leq 2.0$. Here, we define the streamline whose value is equal to zero as the zero-streamline. The zero-streamline in Fig. 1(1) sways remarkably. As a result, the flow pattern is asymmetrical to the line $y=0$. A vortex attached to the cylinder exists. As time passes, a vortex sheds from the cylinder and moves downwards. The direction of rotation of the vortex is the same as that of the Karman vortex street.[15] Therefore, the flow pattern for $Gr=0$ is like the Karman vortex street. A half period of stream function fluctuation is equal to 11.0. The separation points exist near the side edges of the cylinder.

Next, streamlines in the wakes with natural convection for the same Reynolds number as that of Fig. 1(1) are observed.

Sreamlines for $Gr=10^3$ are shown in Fig. 1(2). The flow pattern is in an unsteady state, too. Streamlines are expressed for $-2.5\leq\Psi\leq 2.5$. The swaying motion of the zero-streamline in Fig. 1(2) is similar to that in Fig. 1(1). However, the flow pattern for $Gr=10^3$ is essentially different from that for $Gr=0$. That is, as time passes, a vortex sheds from the cylinder, moves downward, and is broken down in the wake. A vortex does not exist in the downstream far from the cylinder. The separation points are near the side edges.

For a larger heating rate than $Gr=10^3$, streamlines for $Gr=5624$ are shown in Fig. 1(3). Streamlines are expressed for $-1.5\leq\Psi\leq 1.5$. The flow pattern is in an unsteady state, too. Streamlines at $t=32$ and 38 are shown in Figs. 4(a-1) and 4(a-2). As time passes, the flow pattern in Fig. 4(a-1) changes into those of Figs. 4(a-2) and 1(3). The flow pattern in Fig. 4(a-2) is in an anti-phase state in Fig. 4(a-1). Therefore, a half period of stream function fluctuation is equal to 6.0. The swaying motion of the zero-streamline in Fig. 1(3) is smaller than that in Fig. 1(2). The separation points exist near the side edges, too.

The flow-pattern for $Gr=10^4$ is shown in Fig. 1(4). This value of the Grashof number has more dominant buoyancy than that for Fig. 1(3). The swaying motion of the zero-streamline in Fig. 1(4) is smaller than that in Fig. 1(3). The flow pattern for $Gr=10^4$ is essentially different from that for $Gr=5624$. As time passes, the attached vortex on the cylinder oscillates from right to left and does not shed from the cylinder. That is, the vortex street is broken down by a large dominant buoyant force. The flow pattern is approximately symmetrical to the line $y=0$. The separation points in Fig. 1(4) exist near the same side edges as in Figs. 1(1) to 1(3).

Consequently, even if the buoyant force becomes dominant, the separation points do not move. However, the Karman vortex street is broken down by the natural convection, as shown in Figs. 1(1) to 1(4). The natural convection decreases the period of streamlines. That is, the Strouhal number increases due to the natural convection.

Flow patterns in Fig. 1(1) to 1(4) are in qualitatively good agreement with flow visualization results of breakdown.[9]

V-3. Vorticity and Isotherm in Breakdown of the Karman Vortex Street

Vorticities in breakdown of the Karman vortex street due to the natural convection from the elliptical cylinder with the major axis oriented at right angles to the main stream are shown in Figs. 2(1) to 2(4). The vorticity is shown by each step $\Delta\zeta=0.1$ for $-1.0\leq\zeta\leq -0.3$ and $0.3\leq\zeta\leq 1.0$. Each distribution in Figs. 2(1) to 2(4) is in an unsteady state and

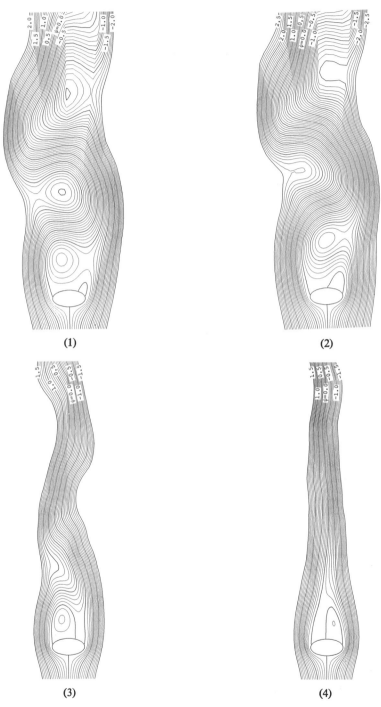

Fig. 1. Streamline in breakdown of Karman vortex from the elliptical cylinder whose major axis is oriented at right angle to main stream ($Re=100$, $t=40$).
(1) $Gr=0$, (2) $Gr=1000$, (3) $Gr=5624$, (4) $Gr=10000$.

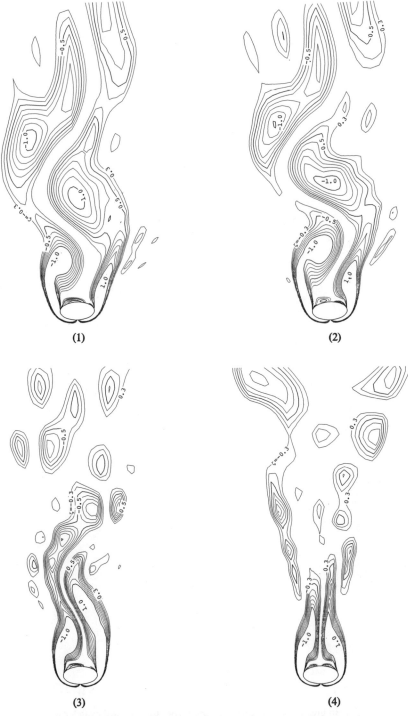

Fig. 2. Vorticity in breakdown of Karman vortex ($Re=100$, $t=40$).
(1) $Gr=0$, (2) $Gr=1000$, (3) $Gr=5624$, (4) $Gr=10000$.

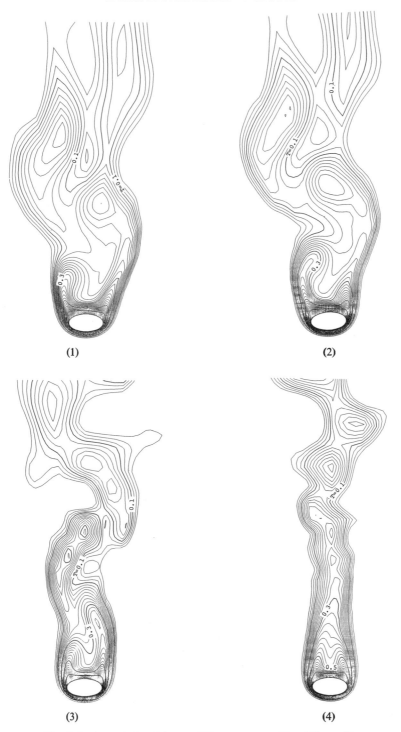

Fig. 3. Isotherm in breakdown of Karman vortex ($Re=100$, $t=40$).
(1) $Gr=0$, (2) $Gr=1000$, (3) $Gr=5624$, (4) $Gr=10000$.

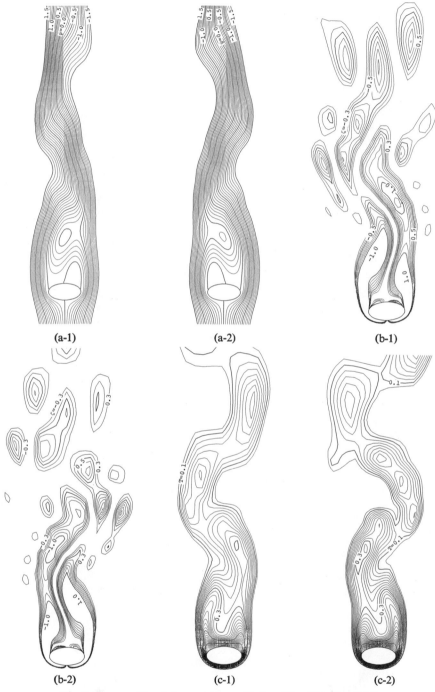

Fig. 4. Change of breakdown pattern with time ($Re=100$, $Gr=5624$).
  (a) Stream line   (1) $t=32$, (2) $t=38$,
  (b) Vorticity   (1) $t=32$, (2) $t=38$,
  (c) Isotherm   (1) $t=32$, (2) $t=38$.

is at $t=40$. The period of vorticity fluctuation is the same value as that of streamline fluctuation.

Vorticities for $Gr=0$ are shown in Fig. 2(1). The absolute value of vorticity is larger near the cylinder. The separating shear layers exist near the side edges of the cylinder. The vorticity is transported downstream. The vorticity for $Gr=10^3$ is shown in Fig. 2(2). The distribution is similar to that in Fig. 2(1). For a larger heating rate than $Gr=10^3$, the vorticity distribution for $Gr=5624$ is shown in Fig. 2(3), and is in an unsteady state, too. Vorticity distributions at $t=32$ and 38 are shown in Figs. 4(b-1) and 4(b-2). As time passes, the vorticity distribution in Fig. 4(b-1) changes into those in Figs. 4(b-2) and 2(3). The distribution in Fig. 4(b-2) is in an anti-phase state of that in Fig. 4(b-1). For a larger heating rate than $Gr=5624$, the vorticity for $Gr=10^4$ is shown in Fig. 2(4). The distribution near the cylinder is appoximately symmetrical in this case. This effect is caused by dominant natural convection.

Consequently, if the natural convection is added to the Karman vortex street, the vorticity distribution becomes symmetrical, and the Karman vortex street breaks down.

The flow patterns in Fig. 1 are influenced by temperature distribution. Therefore, isotherms in breakdown of the Karman vortex street due to the natural convection are shown in Figs. 3(1) to 3(4). Representations of isotherm are expressed by every isotherm step $\Delta T=0.025$ for $0.025 \leq T$. Each distribution is in an unsteady state, and is at $t=40$. The period of isotherm fluctuation is the same value as those of streamline and vorticity fluctuations.

Isotherms for $Gr=0$ are shown in Fig. 3(1). For larger heating rates than $Gr=0$, isotherms for $Gr=10^3$ and 5624 are shown in Figs. 3(2) and 3(3) respectively. The isotherms are in unsteady states. Isotherms at $t=32$ and 38 for $Gr=5624$ are shown in Figs. 4(c-1) and 4(c-2). As time passes, the isotherm in Fig. 4(c-1) changes into those in Figs. 4(c-2) and 3(3). The distribution in Fig. 4(c-2) is in an anti-phase state of that in Fig. 4(c-1). For a larger heating rate than $Gr=5624$, the isotherm for $Gr=10^4$ is shown in Fig. 3(4). The distribution near the cylinder becomes approximately symmetrical. This effect is caused by dominant natural convection.

Consequently, if the natural convection is added to the Karman vortex street, the isotherm distribution becomes symmetrical, and the Karman vortex street breaks down.

Characteristics in the breakdown of the vortex street due to natural convection were clarified in the velocity vector, the velocity components u and v, the stream function fluctuation, the vorticity fluctuation, the temperature fluctuation, the wall shear stress, and the Nusselt number.

VI. CONCLUSIONS

The problem of breakdown of the Karman vortex street due to natural convection from the elliptical cylinder with the major axis oriented at right angles to the main stream is made clear in detail. The time-dependent Navier-Stokes and energy equations are solved numerically by a combination of the FTCS, the SOR, and the ADI finite difference methods. The Cartesian coordinate system is transformed into the elliptical coordinate system. To obtain an asymmetrical flow pattern, a time-dependent initial disturbance was given. Preliminary calculations were carried out for the computational accuracy and the outer boundary location. As a result, numerical solutions are obtained at every time step. From streamline, vorticity, and isotherm, the breakdown of the Karman vortex street due to natural convection is clarified as follows.

(1) Even if natural convection occurs, the separation points are approximately fixed on the wall of the ellipitical cylinder.
(2) Natural convection breaks down the Karman vortex street behind the elliptical cylinder.
(3) The streamline, the vorticity, and the isotherm are clarified in the process of breakdown of the Karman vortex street.
(4) Flow patterns obtained in the present investigation are in qualitatively good agreement with flow visualization results of breakdown.[9]

The results presented here are believed to be the first finding of the breakdown of the Karman vortex street due to the natural convection from the cylinder for which separation points are approximately fixed. These highly useful results provide much information which is relevant to the nature of the Karman vortex street.

REFERENCES

1) E.g., Aref, H. and Siggia, E.D., Evolution and breakdown of a vortex street in two dimensions. *J. Fluid Mech.*, **109** (1981), pp. 435–463.
2) Sparrow, E.M. and Lee, L., Analysis of mixed convection from a horizontal cylinder. *Int. J. Heat Mass Transfer*, **19** (1976), pp. 229–232.
3) Merkin, J.H., Mixed convection from a horizontal circular cylinder. *Int. J. Heat Mass Transfer*, **20** (1976), pp. 73–77.
4) Hatton, A.P., James, D.D., and Swire, H.W., Combined forced and natural convection with low-speed air flow over horizontal cylinders. *J. Fluid Mech.*, **42** (1970), pp. 17–31.
5) Badr, H.M., Laminar combined convection from a horizontal cylinder-parallel and contra flow regimes. *Int. J. Heat Mass Transfer*, **27**, No. 1 (1984), pp. 15–27.
6) Wong, K.L. and Cheng, C.K., The finite element solutions of laminar flow and combined convection of air from three horizontal cylinders in staggered arrangement. *Proc. 2nd ASME/JSME Thermal Engineering Conference*, **5** (1987), pp. 9–15.
7) Noto, K. and Matsumoto, R., Breakdown and development of the Karman vortex street due to the buoyant force. *Proc. 1st ASME/JSME Thermal Eng. Conference*, **3** (1983), pp. 227–234.
8) Noto, K. and Matsumoto, R., Breakdown and development of the Karman vortex due to natural convection. *Trans. Japan Soc. of Mech. Engineers*, Ser. B, **50**, No. 460 (1984), pp. 3015–3023.
9) Noto, K., Ishida, H., and Matsumoto, R., Breakdown phenomenon of the Karman vortex street due to the natural convection. *AIAA Paper* No. 84-1547 (1984), pp. 1–7.
10) Noto, K. and Matsumoto, R., A breakdown of the Karman vortex street due to the natural convection. *Flow Visualization III* (1985), pp. 348–352.
11) Noto, K. and Matsumoto, R., Numerical simulation on development of the Karman vortex due to the negatively buoyant Force. *Numerical Methods in Laminar and Turbulent Flow*, **5** (1987), pp. 796–809.
12) Noto, K., Tsugui, H. and Matsumoto, R., Development of the Karman vortex due to buoyant force in opposing flow. *Flow Visualization IV* (1987), pp. 649–654.
13) Noto, K. and Matsumoto, R., Breakdown of the Karman vortex street due to natural convection. *Numerical Methods in Thermal Problems*, **5** (1987), pp. 484–499.
14) Thompson, J.F., Thames, F.C., and Mastin, C.W., Automatic numerical generation of body-fitted curvilinear coordinate system for field containing any number of arbitrary two-dimensional bodies. *Journal of Computational Physics*, **15** (1974), pp. 299–319.
15) Schlichting, H., *Boundary-Layer Theory*, 7th ed. (McGraw-Hill Book Co., 1979), p. 18.

An Implicit, High-Order, Accurate Upwind Scheme for Unsteady Euler Equations

Kenichi MATSUNO

National Aerospace Laboratory, Chofu, Tokyo

An implicit scheme for unsteady Euler equations is presented in this paper. The scheme employs an unknown quantity of the order of the square of the time step. This choice of the unknown quantity results in a "δ^2(delta square)-form" of the algorithm. The present scheme has second-order time accuracy without any iteration. The iteration procedure, which is the essential algorithm of the scheme, improves both numerical accuracy and robustness, especially for a stiff system. This paper presents the schemes for one- and two-dimensional compressible Euler flows.

I. INTRODUCTION

Today most of the implicit schemes for the Euler/Navier-Stokes equations are based on the so-called "δ-form".[1] The δ-form is a very useful and powerful form for computation of steady flow problems when we seek a steady state solution by a time-dependent approach, because the right-hand side of the scheme has finite difference (or volume) approximations for the governing steady state equations. This fact means that the unknown quantity appearing on the left-hand side of the scheme becomes numerically zero after a sufficient number of time marching steps; then the right-hand side equals zero, and the solution satisfies the steady-state equations. Moreover, the δ-form can avoid dependence of the converged solution on the time step size.

When we deal with essentially unsteady flow computations, however, it seems that the advantage of the δ-form is not operative for this kind of problem. A scheme toward unsteady flow computations is demanded in terms of both time-accuracy and robustness for various kind of computational conditions. For the condition in which the "stiffness" is serious, the available schemes—for example, the Beam-Warming scheme, the TVD scheme, and so on—are forced to proceed in very small time steps in order to do substantially time-accurate computations.

The purpose of this paper is to present a "δ^2-form," which is especially fit for unsteady flow calculations, and to construct an implicit iterative scheme based on the δ^2-form. The detail of the derivation of the δ^2-form is presented in ref. (2). In this paper, emphasis is placed on the extension to a multi-dimension.

II. NUMERICAL SCHEME

We will explain the new scheme for one-dimensional Euler equations. The Euler equations are written in the conservation form as

$$q_t + F_x = 0, \tag{1}$$

where

$$q = \begin{pmatrix} \rho \\ \rho u \\ e \end{pmatrix}, \quad F = \begin{pmatrix} \rho u \\ \rho u + p \\ u(e + p) \end{pmatrix}, \tag{2}$$

where, p, u, and e are density, velocity, and total energy, respectively. The pressure p is evaluated by

$$p = (\gamma - 1)(e - \rho u^2/2), \tag{3}$$

where γ is specific heat ratio.

The conventional three-time-level scheme of the δ-form can be written for Eq. (1) as

$$\left[\frac{1+\tau}{\Delta t} + \theta D_x A^n\right] \delta q^n = -[D_x F^n] + \frac{\tau}{\Delta t} \delta q^{n-1}, \tag{4}$$

where

$$A = \frac{\partial F}{\partial q} \tag{5}$$

and

$$q^{n+1} = q^n + \delta q^n, \quad q^n = q(t^n),$$
$$t^{n+1} = t^n + \Delta t. \tag{6}$$

Here D_x is a finite-difference operator for the space derivative. The quantity δq^n is the unknown in the δ-form and is the order of Δt. We should observe that the leading term of the right-hand side of Eq. (4) is the finite difference approximation to the steady Euler equations.

At least second-order time accuracy will be necessary for an accurate unsteady flow calculation. Thus we will consider a second-order accurate scheme and choose the combination of either ($\tau=0$, $\theta=1/2$: the Crank-Nicolson) or ($\tau=1/2$, $\theta=1$:3-point backward Euler) in Eq. (4).

Now, we derive the "δ^2-form." We define a new unknown quantity $\delta^2 q$ as

$$q^{n+1\langle\nu+1\rangle} = q^{n+1\langle\nu\rangle} + \delta^2 q, \tag{7}$$

where ν is the index designating the number of iterations and the initial guesses are

$$q^{n+1\langle 0\rangle} = q^n + (q^n - q^{n-1}) \quad \text{(1st-order guess)}, \tag{8a}$$

or

$$q^{n+1\langle 0\rangle} = q^n + (q^n - q^{n-1}) + (q^n - 2q^{n-1} + q^{n-2})$$
$$\text{(2nd-order guess)}. \tag{8b}$$

The order of the introduced unknown quantity is $(\Delta t)^2$. Rewriting Eq. (4) by use of Eq. (7), and arranging with the order estimation under the assumption of the second-order time accuracy, we obtain the following implicit scheme[2]:

$$\left[\frac{1+\tau}{\Delta t} + \sigma\theta D_x A^{n+1\langle\nu\rangle}\right]\delta^2 q$$

$$= -\left[\frac{1}{\Delta t}\{(1+\tau)q^{n+1\langle\nu\rangle} - (1+2\tau)q^n - \tau q^{n-1}\}\right.$$

$$\left. + \{\theta D_x F^{n+1\langle\nu\rangle} + (1-\theta)D_x F^n\}\right]$$

$$(\nu = 0, 1, 2, \ldots), \tag{9}$$

where

$$F^{n+1\langle\nu\rangle} = F(q^{n+1\langle\nu\rangle}),$$

and σ is a user-specified parameter of $O(1)$. The following two evaluations for $A^{n+1\langle 0\rangle}$ are possible:

$$A^{n+1\langle 0\rangle} = A(q^n), \tag{10a}$$

or

$$A^{n+1\langle 0\rangle} = A(q^{n+1\langle 0\rangle}). \tag{10b}$$

The right-hand side of Eq. (9) is the finite difference approximation to $[q_t + F_x]$ at $t = t^{n+\theta} = t^n + \theta\Delta t$. Hence we can symbolically rewrite Eq. (9) as

$$\left[\frac{1+\tau}{\Delta t} + \sigma\theta D_x A^{n+1\langle\nu\rangle}\right]\delta^2 q = -[q_t + F_x]^{n+\theta}. \tag{11}$$

Since Eq. (9) employs the unknown quantity of the order $(\Delta t)^2$, we will call this form the "δ^2-form" by analogy to the "δ-form" for Eq. (4). We can see that the δ^2-form is suitable for a calculation of the unsteady equations because its right-hand side is the numerical approximation to the governing unsteady Euler equations. The present scheme is a kind of an iterative scheme, and employs the unknown quantity of the order of $(\Delta t)^2$. Therefore the present scheme is named "δ^2-correction scheme." Though the present scheme has an iterative algorithm, it theoretically has second-order time accuracy without any iteration. That is,

$$q^{n+1} = q^{n+1\langle 1\rangle} = q^{n+1\langle 0\rangle} + \delta^2 q$$

$$(= q^n + (q^n - q^{n-1}) + \delta^2 q). \tag{12}$$

The quantity q^{n+1} in Eq. (12) has second-order time accuracy. In the case where a flow field is moderate, the algorithm without iteration will be efficient. On the other hand, in the case where there is severe "stiffness" or a large time step is used, the iterative algorithm will make the scheme robust as well as numerically more accurate. The iterative algorithm at each time step is a kind of Newton method. Therefore the convergence rate is very high (normally one or two iterations are sufficient).

Upwind finite difference approximations for space derivatives are suitable for a system of nonlinear hyperbolic conservation laws, since they can make the scheme robust and damp unfavorable oscillation near discontinuity. In this paper, the differentiable flux vector splitting by van Leer[3] is utilized. The flux and its Jacobian are partitioned into two parts: a forward flux and a backward flux, i.e.,

$$F = F^+ + F^- \tag{13a}$$

and

$$A = A^+ + A^- \tag{13a}$$

We apply upwind operators according to the directions of waves. For the implicit side (the left-hand side) of Eq. (9), first-order upwind operators are used in order to construct a tri-diagonal system:

$$D_x A^{n+1(\nu)} \delta^2 q = (\nabla_x A^{+\,n+1(\nu)} + \Delta_x A^{-\,n+1(\nu)}) \delta^2 q, \tag{14}$$

where ∇_x is a backward difference operator and Δ_x a forward difference operator. On the other hand, the high-order-accurate upwind operator with van Leer's differentiable limiter[3] is utilized for the explicit side (the right-hand side) of Eq. (9), i.e.,

$$\begin{aligned} D_x F &= D_x^- F^+ + D_x^+ F^- \\ &= \frac{1}{\Delta x}[(F_{i+1/2}^+ - F_{i-1/2}^+) + (F_{i+1/2}^- - F_{i-1/2}^-)]. \end{aligned} \tag{15}$$

Let f be an element of the flux vector F:

$$f_{i+1/2}^+ = f_i^+ + \frac{s}{4}[(1-ks)\nabla^+ + (1+ks)\Delta^+]_i, \tag{16a}$$

$$f_{i+1/2}^- = f_{i+1}^- - \frac{s}{4}[(1-ks)\Delta^- + (1+ks)\nabla^-]_{i+1}, \tag{16b}$$

where

$$\Delta_i^+ = f_{i+1}^+ - f_i^+, \quad \nabla_i^+ = f_i^+ - f_{i-1}^+, \quad s = \frac{2\Delta\nabla + \varepsilon}{\Delta^2 + \nabla^2 + \varepsilon}, \tag{17}$$

and ε is a small number ($\varepsilon = 10^{-6}$).

III. EXTENSION TO MULTI-DIMENSION

Extension to the multi-dimension is straightforward. In this paper, the approximate factorization technique is applied in order to reduce computer storage and to make the computation well-vectorized.

The two-dimensional Euler equations may be written in conservation-law form in Cartesian coordinates as

$$q_t + F_x + G_y = 0, \tag{18}$$

where

$$q = \begin{pmatrix} \rho \\ \rho u \\ \rho v \\ e \end{pmatrix}, \quad F = \begin{pmatrix} \rho u \\ \rho u^2 + p \\ \rho u v \\ u(e+p) \end{pmatrix}, \quad G = \begin{pmatrix} \rho v \\ \rho u v \\ \rho v^2 + p \\ v(e+p) \end{pmatrix} \tag{19}$$

and

$$p = (\gamma - 1)\left[e - \frac{1}{2}\rho(u^2 + v^2)\right]. \tag{20}$$

The δ^2-correction scheme of an approximate factored form for the two-dimensional equations (18) is written as

$$\left[\frac{1+\tau}{\Delta t} + \sigma\theta D_x A^{n+1\langle\nu\rangle}\right]\left[1 + \frac{\Delta t}{1+\tau}\sigma\theta D_y B^{n+1\langle\nu\rangle}\right]\delta^2 q$$
$$= -\left[\frac{1}{\Delta t}\{(1+\tau)q^{n+1\langle\nu\rangle} - (1+2\tau)q^n - \tau q^{n-1}\}\right.$$
$$+ \{\theta(D_x F^{n+1\langle\nu\rangle} + D_y G^{n+1\langle\nu\rangle})$$
$$\left. + (1-\theta)(D_x F^n + D_y G^n)\}\right], \tag{21}$$

or, symbolically,

$$\left[\frac{1+\tau}{\Delta t} + \sigma\theta D_x A^{n+1\langle\nu\rangle}\right]\left[1 + \frac{\Delta t}{1+\tau}\sigma\theta D_y B^{n+1\langle\nu\rangle}\right]\delta^2 q$$
$$= -[q_t + F_x + G_y]^{n+\theta} \tag{22}$$

Here, A and B are flux Jacobians such that

$$A = \frac{\partial F}{\partial q}, B = \frac{\partial G}{\partial q}. \tag{23}$$

In an application, Eq. (18) is transformed into a curvilinear coordinate system, and a finite volume approach is utilized. In this paper, applications to the two-dimensional flow problems are made in order to show the validity of the present "δ^2-form" for the multi-dimensions. Therefore the central difference operator with Jameson's nonlinear artificial dissipation was utilized for simplicity of programming.

IV. NUMERICAL CALCULATIONS

In the following numerical calculations, the time difference method of the Crank-Nicolson type ($\theta = 1/2$, $\tau = 0$) is utilized in the case where the maximum Courant number is less than 1; otherwise, the 3-point backward Euler ($\theta = 1$, $\tau = 1/2$) is utilized. For the initial guess at each time step, the first-order guess, Eq. (8a), was used.

At first, the accuracy and validity of the scheme are investigated and verified for one-dimensional Euler flows. A first example is the shock tube problem, as employed by Sod.[4] The initial pressure and density ratios are 10 and 8, respectively. The number of the grid points is 101 with the diaphragm located at the 1/3 tube length position. The ratio $\Delta t/\Delta x$ is 0.5, where the local maximum Courant number is about 0.93. The calculated result at the 50th time step was compared with the exact solution. Figure 1 is the case in which the scheme employed one iteration at each time step. The positions of the shock and the contact surface are the same as those in the exact solutions. The shapes of the shock and the contact surface are very realistic.

The present scheme is developed for the unsteady flow calculations. However, when there exists a steady-state solution, the present scheme will produce a steady flow as the limit of the time-dependent solution. The next example is a steady nozzle flow calculation. We consider a divergent nozzle[5] where $A(x) = 1.398 + 0.347 \tanh(0.8 \times -4)$.

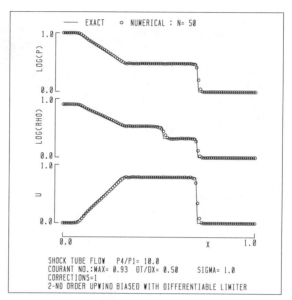

Fig. 1. Comparison between the calculation (with 1-iteration) and the exact solution for the shock tube problem.

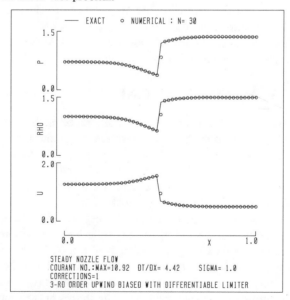

Fig. 2. Comparison between the calculation (with 1-iteration) and the exact solution for the steady nozzle flow.

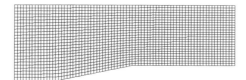

Fig. 3. Computational grid for the supersonic compression duct, 20% slope.

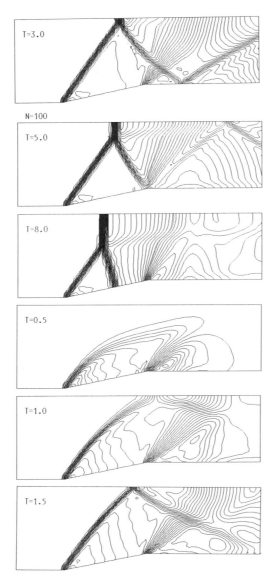

Fig. 4. Iso-Mach lines at the normalized time: 0.5, 1, 1.5, 3, 5, and 8 for the compression duct problem at Mach 1.6 (line interval = 0.03).

We do not examine the accuracy of the transient solution. The calculation was performed with the boundary conditions of supersonic inflow ($M = 1.26$) and subsonic outflow (p-out/p-in $= 1.9$), and with the initial condition of linear interpolation between the exact steady-state boundary values. Fifty-one grid points and the inlet Courant number 10 were used. The comparison between the converged solution and the exact solution is shown in Fig. 2, where the calculated result is at the 30th time step. The present scheme is valid even for the steady flow calculation.

Some of the numerical results for the two-dimensional flows are given in Figs. 3–6.

The first example is an unsteady flow calculation for a supersonic compression duct. Figure 3 shows the computational mesh, composed of 76×26 points. The calculation started from a uniform flow having inlet flow properties (Mach number $= 1.6$). This geometry and the initial conditions are identical with the test case used in ref. [6], except for the number of grid points and the Courant number. In the present calculation, a relatively coarse grid and a rather large Courant number (about 3.5) were used. In this calculation, each time step is iterated until the right-hand side residual of Eq. (21) becomes less than a certain criterion. The number of iterations was 3 on average: at the early stage it was relatively large; later it was less than 2. Figure 4 shows Mach number contours of the transient solutions.

The next example is the application of the present scheme to the transonic steady flow calculation. The flow field in a channel with a circular arc "bump" on the lower wall was calculated. This geometry is identical to the case of ref. (7). We used the same grid points and the same initial/boundary conditions as those of ref. (7). The present scheme used the constant time step Δt throughout the calculation. Figure 5 shows the convergence history of this transonic calculation. In this figure, L2-residual means a residual of the steady part of the flow equations. We should note that this calculation was performed in a time-accurate manner; hence the convergence rate has a physical meaning. Figure 6 shows the C_p-contour of the converged solution, which is identical with that of ref. (7).

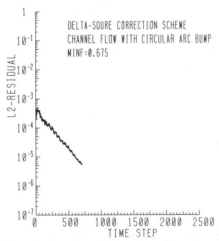

Fig. 5. Convergence history for the transonic flow in the channel with the 10% circular arc bump.

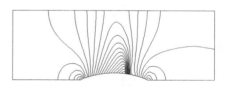

Fig. 6. Iso-C_p lines (line-interval $= 0.1$).

The results shown above indicate the applicability of the present scheme to not only unsteady flow problems but also to steady flow problems.

V. CONCLUSION

The δ^2-form for unsteady equations is introduced in this paper. This δ^2-form is derived as the result of the construction of the finite difference scheme such that the numerical approximation of the whole of the governing unsteady equations appears in the right-hand side of the scheme. Though the scheme of the δ^2-form is essentially iterative, it has second-order time accuracy without iterations. The implicit, upwind scheme of the δ^2-form is presented and applied to one- and two-dimensional flow problems. Numerical results show that the present scheme of the δ^2-form is also promising for the two-dimensional flows.

REFERENCES

1) Pulliam, T.H. and Steger, J.L., Recent improvement in efficiency, accuracy, and convergence for implicit approximate factorization algorithm. *AIAA Paper* 85-0360, (1985).
2) Matsuno, K., δ^2-correction scheme for unsteady Euler equations. *NAL TR*-948, (1987). (in Japanese).
3) Anderson, W.K., Thomas, J.L., and van Leer, B., A comparison of finite volume flux vector splittings for the Euler equations, *AIAA Paper* 85-0122 (1985).
4) Sod, G.A., A Survey of several finite difference methods for systems of nonlinear hyperbolic conservation Laws. *J. Computational Phys.*, **27** (1978), pp. 1-31.
5) Yee, H.C., Beam, R.M., and Warming, R.F., Stable boundary approximations for a class of implicit schemes for the one-dimensional inviscid equations of gas dynamics. *AIAA Paper* 81-1009 (1981).
6) Borrel, M. and Montagne, J.L., Numerical study of a non-centered scheme with application to aerodynamics. *AIAA Paper* 85-1497 (1985).
7) Chima, R.V. and Johnson, G.M., Efficient solution of the Euler and Navier-Stokes equations with a vectorized multiple-grid algorithm. *AIAA J.*, **23**, No. 1 (1985), pp. 23-32.

Numerical Simulation of Flow Ejected from a Nozzle by Electric Force

Yusuke TAKEDA,* Daisuke TAKAHASHI,[2]* Katsuya ISHII,[3]* and Hideo TAKAMI[2]*

* Imaging Technology Research Center, Ricoh Company Limited, Tokyo, [2]* Department of Applied Physics, University of Tokyo, Tokyo, [3]* Institute of Computational Fluid Dynamics, Tokyo

A numerical method that makes use of equations of incompressible flow with a free surface and an electric field equation is discussed and applied to the study of a flow ejected from a nozzle by electrostatic force.

In this study, the fluid is treated as electrically perfect conductor. The improved marker and cell method[6] and charge simulation method[7] are used to solve the motion of fluid and electric field, respectively. The surface configuration is determined by the balance between the surface tension of the fluid and electrostatic force.

Calculations are performed for several variations of fluid characteristics, strength of the electric field, and nozzle dimension. Computational results are reasonable for each variable.

I. INTRODUCTION

It is a very important and interesting problem to form small liquid droplets in combustion or ink jet technology. Typical methods apply a pressure pulse or electrostatic force to fluid in a nozzle. Numerical simulations of such flow using axisymmetrical Navier-Stokes equation have been reported.[1-3] For ink jets, some experimental or analytical studies of fluid deformation by electrostatic force have been performed.[4,5] However, few numerical studies of this phenomenon have been carried out because it is difficult to determine the surface configuration by taking account of both surface tension of the fluid and the electrostatic force. Therefore, quantitative evaluation of the contribution of variables in this phenomenon (surface tension of the fluid, kinematic viscosity, nozzle dimension, etc.) to deformation has not yet been performed.

In this paper, we calculate the axisymmetrical Navier-Stokes equation and the axisymmetrical Laplace equation which govern the motion of the fluid and the electric field, respectively. Here, in order to simplify the calculation, we assume that the fluid is an electrically perfect conductor. We use the improved marker and cell method[6] for the fluid motion and use the charge simulation method[7] for the electric field which can treat the surface configuration more easily than the finite difference or finite element method. At each time step, the Navier-Stokes equation is integrated and the Poisson equation for pressure is solved iteratively using the boundary condition at the free surface which determines the surface tension of the fluid and the electrostatic force.

Computations are performed for several combinations of surface tension of the fluid, kinematic viscosity of the fluid, length of the nozzle, radius of the nozzle, and strength of the electric field.

II. BASIC EQUATIONS

The governing equations that describe the motion of the fluid are the equation of continuity and the Navier-Stokes equation. In an axisymmetric coordinate system (z, r), they are expressed in the following forms:

$$\frac{\partial u}{\partial z} + \frac{\partial v}{\partial r} + \frac{v}{r} = 0, \tag{1}$$

$$\frac{\partial u}{\partial t} + (\boldsymbol{u} \cdot \nabla) u = -\frac{\partial p}{\partial z} + \nu \Delta u, \tag{2}$$

$$\frac{\partial v}{\partial t} + (\boldsymbol{u} \cdot \nabla) v = -\frac{\partial p}{\partial r} + \nu \left(\Delta v - \frac{v}{r^2} \right), \tag{3}$$

where (u, v) are (z, r) components of the velocity, p the pressure, and ν the kinematic viscosity. By taking the divergence of the Navier-Stokes equation, we get the Poisson equation for the pressure:

$$\Delta p = -\frac{\partial D}{\partial t} + \nu \nabla D - \Delta \cdot ((\boldsymbol{u} \cdot \nabla) \boldsymbol{u}), \tag{4}$$

$$D = \nabla \cdot \boldsymbol{u}. \tag{5}$$

The governing equation of the electric field is the Laplace equation:

$$\Delta \phi = 0 \tag{6}$$

where ϕ is the electric potential.

III. BOUNDARY CONDITIONS

The boundary conditions are as follows:
1) On the wall of the nozzle, we impose no-slip condition for velocity:

$$u = 0, \quad v = 0. \tag{7}$$

The pressure on the wall is derived from the Navier-Stokes equations.

2) At the entrance of the nozzle, the velocity is determined by extrapolation and the pressure is assumed to be constant with its value p_0 determined by

$$p_0 = \frac{2\gamma}{R_0} \tag{8}$$

where γ is the surface tension of the fluid, and R_0 the radius of the nozzle.

3) On the free surface, we impose the condition of continuity of stress. If we neglect the effect of the external fluid (air), this condition is expressed as

$$\sigma^*_{i,j} \cdot n_j = \sigma^+_{i,j} \cdot n_j \tag{9}$$

where $\sigma^*_{i,j}$ and $\sigma^+_{i,j}$ are fluid and electric stress tensor, respectively, and n_j the normal vector to the surface. Each stress tensor is expressed as follows:

$$\sigma^*_{i,j} = -p_s \delta_{i,j} - \gamma\left(\frac{1}{R_1} + \frac{1}{R_2}\right)\delta_{i,j} + \nu\left(\frac{\partial u_i}{\partial x_j} + \frac{\partial u_j}{\partial x_i}\right), \tag{10}$$

$$\sigma^+_{i,j} = D_i \cdot E_j - \frac{1}{2} D_k E_k \, \sigma_{i,j} \tag{11}$$

where $\delta_{i,j}$ is Kronecker's delta, D_i the electric flux, and R_1 and R_2 the principal radii of curvature of the surface. Here, we neglect the viscous term in Eq. (10) to simplify the calculation. Finally, the surface pressure p_s is given by the following equation:

$$p_s = -\gamma\left(\frac{1}{R_1} + \frac{1}{R_2}\right) + \frac{1}{2}\varepsilon E^2. \tag{12}$$

The velocity on the free surface is determined by extrapolation from the velocity inside the fluid.

IV. NUMERICAL METHOD

The improved marker and cell method was described in detail by Takahashi et al.[6] so that only a brief discussion of the basic features will be presented here. In the finite difference calculation, the Navier-Stokes equation (Eqs. (2) and (3)) and the Poisson equation for pressure (Eq. (4)) are approximated as follows:

$$u^{n+1} = u^n + \Delta t\left[-(\boldsymbol{u}^n \cdot \nabla)u^n - \left(\frac{\partial p}{\partial z}\right)^n + \nu \Delta u^{n+1}\right], \tag{13}$$

$$v^{n+1} = v^n + \Delta t\left[-(\boldsymbol{u}^n \cdot \nabla)v^n - \left(\frac{\partial p}{\partial r}\right)^n + \nu\left(\Delta v^{n+1} - \frac{v^{n+1}}{r^2}\right)\right], \tag{14}$$

$$\Delta p^n = -\frac{D^n}{\Delta t} - \nabla \cdot ((\boldsymbol{u}^n \cdot \nabla)\boldsymbol{u}^n) \tag{15}$$

where Δt is the interval of time difference and the index n denotes the number of the time step. These equations are calculated in the form of the central difference. A unit of rectangular mesh (which we call "cell") is classified into empty (E), fluid (F), surface (S), or wall (W) cell. The physical quantities of fluid (velocity and pressure) are represented in the center of each cell. Markers are put only in S cells and moved with the velocity of S cells. The typical classification of the cells and the arrangement of the markers are shown in Fig. 1. The stress due to surface tension at the surface cell is calculated from the principal radii of surface curvature that are determined from the positions of markers. Markers are rearranged at each time step and the number of markers in each cell is always controlled to be constant. Using Eq. (15), pressure at F cells is calculated iteratively with the boundary condition of the pressure of S cells. Using Eqs. (13) and (14), velocity at F cells is integrated by an iterative method and the velocity of S cells is determined by extrapolation.

As regards calculations of electric field, we use the charge simulation method.[7] Since the fluid is an electrically perfect conductor, we only consider the nonfluid region. In this calculation, the electric potential on the surface of the fluid and the nozzle is 0 and the potential at the electrode is V. We choose typical points (equipotential points) on the free surface of the fluid, the nozzle surface, and the electrode. Moreover, we put ring-shaped virtual charges inside the fluid, the nozzle, and the electrode. Figure 2 shows the typical arrangement of equipotential and virtual charge points. The charge at each virtual charge point Q_j is calculated by solving the following equation:

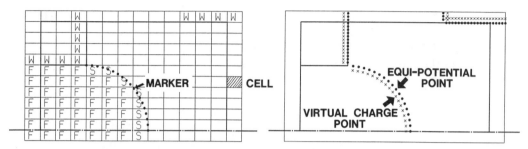

Fig. 1. Typical classification and arrangement of markers. F: Fluid cell; S: surface cell; W: wall cell; □: empty cell.

Fig. 2. Typical arrangement of equipotential and virtual charge points.

$$\sum_{j=1}^{m} P_{i,j} \cdot Q_j = \begin{cases} 0 \text{ (on the fluid or nozzle)} \\ V \text{ (on the electrode)} \end{cases}$$

$$(i = 1, 2, \ldots, m) \qquad (16)$$

where m is the number of equipotential points and the coefficient $P_{i,j}$ is the electric potential at the i-th position induced by a unit charge at the j-th position. After solving Eq. (16), the electric potential at the equipotential point ϕ_i is calculated by

$$\phi_i = \sum_{j=1}^{m} P_{i,j} \cdot Q_j. \qquad (17)$$

Electric field E_i is given by

$$E_i = \sum_{j=1}^{m} F_{i,j} \cdot Q_j \qquad (18)$$

where the coefficient $F_{i,j}$ is the electric field at the i-th position by a unit charge at the j-th position.

Finally, the computation procedure is as follows:

1) At the n-th time step, markers are placed as they approximate the configuration of the fluid region.

2) The values of stress due to the surface tension of S cells are calculated from the principal radii of curvature.

3) Using Eqs. (16)–(18), the values of electrostatic force of S cells are calculated.

4) The values of pressure of S cells are determined by Eq. (12).

5) Using Eq. (15), the values of pressure of F cells at the n-th time step are calculated by an iterative method.

6) Using Eqs. (13) and (14), the values of velocity of F cells at the $(n+1)$st time step are calculated by an iterative method.

7) Markers are moved with these velocities by time interval Δt.

8) The configuration of the fluid region at the $(n+1)$st time step is determined by the positions of markers and every cell is newly classified.

9) Return to step 2.

V. NUMERICAL RESULTS AND DISCUSSION

Figure 3 is a schematic illustration with the parameters used in the calculation. Initially, the surface configuration of the fluid is a half-sphere without electric field. In this situation, the fluid does not move since the surface stress of the fluid is equal to the pressure at the entrance of the nozzle. At $t=0$, an electric field is suddenly applied and the fluid begins to move. We calculate 6 cases with the scheme mentioned in Section 4. The condition in each case is shown in Table 1. Here, we regard case 1 as the standard case. Kinematic viscosity of the fluid (ν), surface tension of the fluid (γ), electric potential of the electrode (V) length of the nozzle (Ln), and the radius of the nozzle (R_0) used in case 1 are changed in case 2—case 6, respectively.

Figures 4–9 show the numerical results of each case. In these figures, velocity vectors

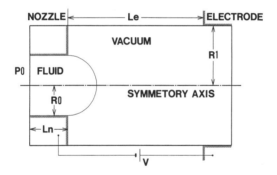

Fig. 3. Schematic illustration and parameters.

Table 1. Characteristics in cases 1–6.

Characteristics	Case 1	Case 2	Case 3	Case 4	Case 5	Case 6
Surface tension (dyn/cm)	25.0	25.0	15.0	25.0	25.0	25.0
Kinematic viscosity (cm^2/sec)	0.02	0.04	0.02	0.02	0.02	0.02
Electrode potential (V)	2000	2000	2000	4000	2000	2000
Nozzle radius ($\times 10^{-3}$mm)	100	100	100	100	100	150
Electrode radius (mm)	0.35	0.35	0.35	0.35	0.35	0.35
Nozzle length (mm)	0.2	0.2	0.2	0.2	1.2	0.2
Distance from nozzle exit to electrode (mm)	1.3	1.3	1.3	1.3	1.3	1.3
Pressure at entrance of nozzle (dyn/cm^2)	5000	5000	3000	5000	5000	5000
Mesh points (z)	200	200	200	200	300	200
Mesh points (r)	70	70	70	70	70	70
Mesh interval (z) (mm)	0.01	0.01	0.01	0.01	0.01	0.01
Mesh interval (r) (mm)	0.005	0.005	0.005	0.005	0.005	0.005
Time interval ($\times 10^{-6}$ sec)	0.75	0.75	0.75	0.75	0.75	0.75

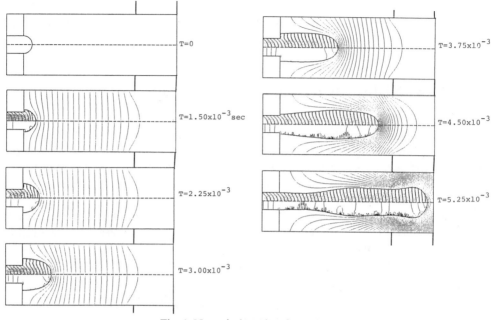

Fig. 4. Numerical results of case 1.

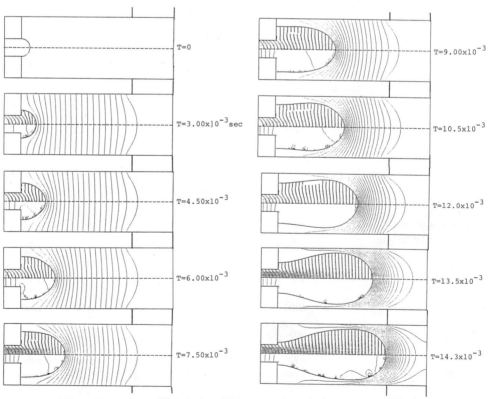

Fig. 5. Numerical results of case 2.

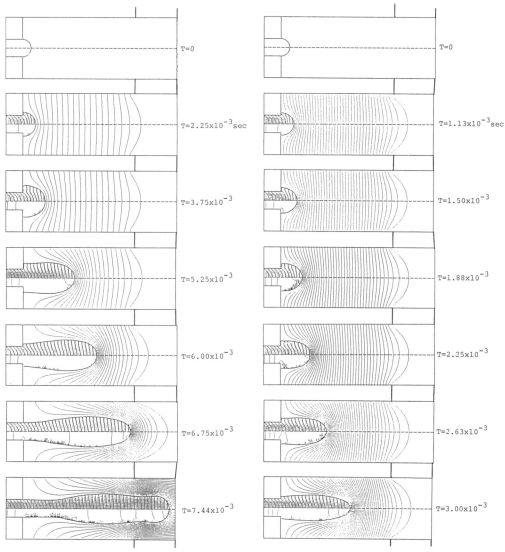

Fig. 6. Numerical results of case 3.   Fig. 7. Numerical results of case 4.

and pressure contours are plotted on, respectively, the upper and lower side of the symmetry axis in the fluid region and the contours of electric potential are plotted on the non-fluid region. Furthermore, Fig. 10 shows the relationship between time and the distance from the exit of the nozzle to the head of the jet. In Fig. 4, when the electric field is applied, the fluid is pulled in the z-direction expanding in the r-direction along the surface of the nozzle. Subsequently the electric field concentrates at the head of the jet and the z-component of velocity near the axis increases. Consequently, the fluid pulled out in the z-direction passes through the hole of the electrode without separating into droplets. In the case of larger kinematic viscosity (case 2), deformation speed of the fluid is slower than that in case 1. The fluid configuration is thicker in comparison with that in case 1. It is suspected

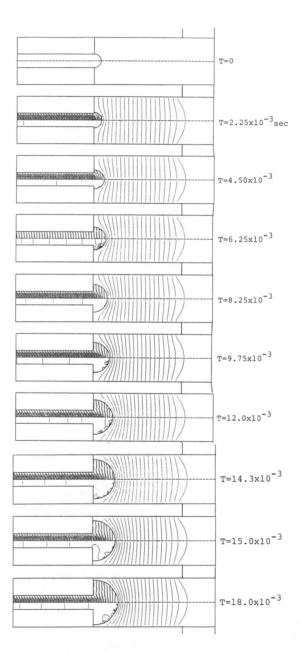

Fig. 8. Numerical results of case 5.

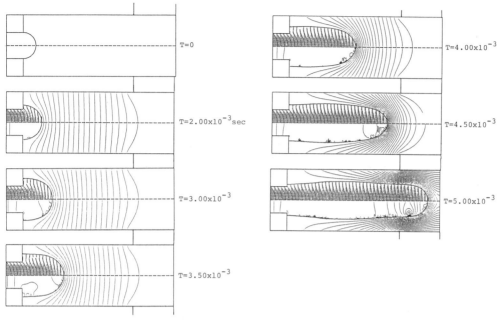

Fig. 9. Numerical results of case 6.

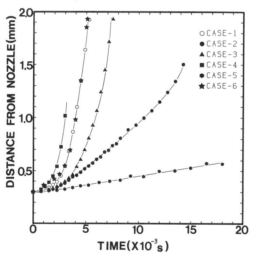

Fig. 10. Relationship between time and distance from nozzle exit to head of jet.

that these differences are caused by the pressure gradient in the nozzle which is proportional to the kinematic viscosity. In the case of smaller surface tension (case 3), the fluid configuration is much the same as that in case 1. However, deformation speed is lower than that in case 1 since the pressure at the entrance of the nozzle is smaller than that in case 1. In case 4, with stronger electric field, moving speed of the fluid is higher than in case 1, as expected. Moreover, the fluid configuration is sharpened by the effect of stronger concentration of the electric field. In the case of a longer nozzle (case 5), the pressure at the exit of the nozzle decreases since the length of the nozzle is longer. Deformation speed is very small and the fluid expands radially for a long time. In this case, the electrostatic force is dominant. In the case of larger radius of the nozzle (case 6), deformation speed is about the same as that in case 1.

It is known experimentally that the fluid configuration is similar to a circular cone. However, this was not observed in the present calculations. It is suspected that this difference is caused by the pressure at the exit of the nozzle and the strength of the electric field. Under the conditions that we used in the calculation, the stress due to surface tension and the pressure at the entrance of the nozzle determined by the surface tension play a dominant part in the deformation of the fluid. Therefore, we expect that the fluid configuration will resemble a cone under conditions of longer nozzle and stronger electric field.

VI. CONCLUSIONS

In this paper, we tried a numerical simulation of a flow ejected from a nozzle by electrostatic force. We used the improved marker and cell method for the motion of the fluid and the charge simulation method for the electric field, respectively. At the free surface, the pressure is determined by the calculation of the surface tension of the fluid and the electrostatic force. Numerical results show reasonable behavior for changes of parameters. Furthermore, the effect of various parameters on fluid deformation can be observed.

Computations were carried out using a HITAC M260D computer. Typical computational time was about 5 hours per case.

REFERENCES

1) Fromm, J.E., Numerical calculation of the fluid dynamics of drop-on demand jets. *IBM. Res. Devel.*, Vol. **28** (1984), pp. 322–333.
2) Katano, Y., Kawamura, T., and Takami, H., Numerical study of drop formation from a capillary jet using a general coordinate system. *Theoretical and Applied Mechanics*, Vol. **34** (University of Tokyo Press, 1986), pp. 3–14.
3) Ebi, Y. and Kawamura, T., Numerical study of droplet formation from liquid jet. *Ricoh Tech. Rep.*, Vol. **5** (1981), pp. 4–11 (in Japanese).
4) Taylor, G., Disintegration of water drops in an electric field. *Proc. Roy. Soc. London. Ser. A*, Vol. **280** (1964), pp. 383–397.
5) Macky, W.A., Some investigations on the deformation and breaking of water drops in strong electric fields. *Proc. Roy. Soc. London, Ser. A*, Vol. **133** (1931), pp. 565–587.
6) Takahashi, D., Takeda, Y., and Takami, H., Numerical simulation of Collision of liquid droplets. *Theoretical and Applied Mechanics*, Vol. **36** (University of Tokyo Press, 1988), pp. 367–380.
7) Kohno, T. and Takuma, T., *Numerical Calculation Method of Electric Field*, 1st ed., Corona Publishing Co. (1980), p. 38 (in Japanese).

Effect of Ground on Wake Roll-Up behind a Lifting Surface

Keqin ZHU* and Hideo TAKAMI[2*]

* University of Science and Technology of China, [2*] Department of Applied Physics, University of Tokyo, Tokyo

The superconvergence of an improved vortex lattice method for a two-dimensional flat plate has been verified theoretically. The improved vortex lattice method is used to predict the ground effect on wake roll-up behind a lifting surface. The wake geometry is obtained through iteration by satisfying the condition that free vortex lines are consistent with the local streamlines. The ground effect is simulated by use of an image vortex system. Numerical results show rapid convergence of the iterative procedure. The effect of height-above-ground is discussed. The wake vortex becomes unstable as the lifting surface descends too close to the ground.

I. INTRODUCTION

When a wing is flying very close to the ground, the ground has the effect of producing significantly large lift and changing the position and strength of the wake vortex. Good prediction of these phenomena is important to aircraft designed for STOL operations.

Hackett and Evans[1] studied the problem using a two-dimensional, time-dependent analogy, assuming that all quantities in the wake region vary slowly in the freestream direction, thus replacing the three-dimensional steady flow by a two-dimensional, time-dependent one. Their procedure failed because vortices could penetrate the ground. To avoid this, a buffer layer next to the ground is used. Goetz et al.[2] predicted the lift and drag characteristics of a wing subjected to the ground effect using a higher-order panel method without wake roll-up. They concluded that proper modeling of the wing wake would be required to improve accuracy. Deese and Agarwal[3] calculated the flow over such a wing using the Euler methods. Three-dimensional Euler computations require large amounts of computer time and memory. Consequently, only one example was calculated in their paper without providing the ground effect on wake roll-up.

In this paper an improved vortex lattice method is developed to predict the effect of the ground on wake roll-up behind a lifting surface. The vortex lattice method (VLM), the simplest of the panel methods, is already a classic approach and has wide application, but up to now it has not fully been understood. One of the problems is the choice of the locations of vortices and control points, which is mainly based on empirical considerations but has little theoretical basis. On the other hand, an excellent study by Lan[4] used the quasi-vortex-lattice method (QVLM), in which wing edge and Cauchy singularities are accounted for. The two-dimensional numerical results for a flat plate reproduce the exact solution to the vorticity density at a finite number of points. In the following sections, superconvergence of the QVLM solution for two-dimensional plate is first verified rigorously. Then an improved VLM, which shows the same superconvergence as the QVLM, is presented. Finally, this improved VLM is used to predict the effect of the ground on wake roll-up behind a lifting surface. The numerical results are compared with experimental results.

II. FORMULATION

II-1. Mathematical Model

Let us consider a steady, incompressible, inviscid, and irrotational flow in the region R outside a lifting surface, its infinitesimally thin wake, and the ground. Let v be the velocity field expressed as the gradient of a velocity potential φ satisfying the Laplace equation

$$\Delta \varphi = 0. \tag{1}$$

On the boundary ∂R, the following conditions are to be satisfied: 1) the ground is impermeable; 2) the lifting surface is impermeable; 3) the pressure is continuous and there is no fluid flow across the wake vortex sheet; 4) the Kutta condition is satisfied at the trailing edge; 5) the disturbance velocity vanishes at an infinite distance from the lifting surface and in its wake. Although the finite difference and finite element methods may be used to solve this, such treatment would still suffer from the need to construct a three-dimensional grid throughout region R. The panel methods make use of a special property of the Laplace equation. According to Green's third identity, it is not necessary to construct such a grid. Instead, it is possible to formulate this problem in such a way that the unknown function is limited to the lifting surface, and then the solution of the Laplace equation can be obtained using an integral equation.

In all panel methods, the lifting surface is divided into a number of panels, and the continuous vorticity distribution representing the wing is replaced by discrete vortices. Then the integral equation is approximately replaced by a system of algebraic equations. Hunt[5] presented a comprehensive survey of the mathematics underlying panel methods. The discretization of the integral equation is applied at three different levels: satisfaction of boundary conditions, approximation of geometric surfaces, and approximation of surface distributions. All of these approximations lead to numerical error. In order to improve the accuracy of the numerical calculation using panel methods, we need to increase the number of panels and use higher-order panel methods. These require more computer time and memory. It is important to study how to reduce the errors by the choice of panels and control points without increasing the number of panels. Our attention will focus on the vortex lattice method in this paper.

II-2. Superconvergence Characteristics

For simplicity, the flow past a two-dimensional flat plate is discussed in this section. In thin airfoil theory, the integral equation reduces to the following equation:

$$\int_0^1 \frac{\gamma(x')}{x - x'} \, dx' = 2\pi, \tag{2}$$

where γ is the vorticity density, and x represents the distance from the leading edge to a point on the plate. The exact solution of the equation can be shown to be

$$\gamma(x) = 2\sqrt{\frac{1-x}{x}}. \tag{3}$$

The lift and moment coefficients are 2π and $-\pi/2$, respectively. In view of knowing the exact form of the solution, it is appropriate to study the accuracy of the numerical solutions for this case.

In the conventional VLM, the length of every element is the same. A point vortex is placed at each elemental quarter-chord. The condition of tangency of the flow is satisfied at each elemental three-quarter chord. Integral equation (2), when discretized, becomes

$$\sum_{i=1}^{N} a_{ij}\gamma_i = 2\pi \quad (j = 1, 2, \ldots, N), \tag{4}$$

where N is the total number of the elements on the plate, and

$$a_{ij} = \frac{\Delta x}{x_j - x_i} = \frac{1}{0.5 + j - i}. \tag{5}$$

Here $\gamma_i \Delta x$ represents the discrete vortex strength at the quarter-chord point of the i-th element and γ_i is the vorticity density at the same location.

Table 1 shows the calculated results for five elements. The second column is the vorticity density calculated by the conventional VLM. The third column represents fractional errors. Obvious errors of the numerical solution can be found, especially near the leading edge, where the predicted pressures are too low. If the number of elements is increased, the numerical solution will improve. However, as mentioned above, we are interested in improving numerical solutions without increasing the number of the elements.

According to the QVLM developed by Lan,[4] we transform the x-coordinate to the θ-coordinate through the relation

$$x = \frac{1}{2}(1 - \cos\theta). \tag{6}$$

Then the chord is divided into small elements according to the semicircle scheme. In this case, the positions of control points are determined by

$$x_j = \frac{1}{2}\left(1 - \cos\frac{j\pi}{N}\right) \quad (j = 1, 2, \ldots, N), \tag{7}$$

and the positions where vorticity density is calculated are given by

$$x_i = \frac{1}{2}\left\{1 - \cos\frac{(i - 0.5)\pi}{N}\right\} \quad (i = 1, 2, \ldots, N). \tag{8}$$

Table 1. Comparison of thin airfoil vortex distribution (VLM, 5 elements).

X	γ by VLM	Exact γ	Fra. error
0.05	7.73126	8.71780	−11.3%
0.25	3.43612	3.46410	− 0.8%
0.45	2.20893	2.21108	− 0.1%
0.65	1.47262	1.46760	0.3%
0.85	0.85903	0.84017	2.2%

Table 2. Comparison of thin airfoil vortex distribution (QVLM, 5 elements).

X	γ by QVLM	Exact γ	Fra. error
0.02	12.62750	12.62750	0
0.21	3.92522	3.92522	0
0.50	2.00000	2.00000	0
0.79	1.01905	1.01905	0
0.98	0.31677	0.31677	0

Then relation (5) can be written as

$$a_{ij} = \frac{\pi}{N} \frac{\sin((i-0.5)\pi/N)}{\cos((i-0.5)\pi/N) - \cos(j\pi/N)}. \tag{9}$$

The calculated results show that the vorticity density predicted by QVLM are exact (Table 2). In general, this fact can be verified for any N and is called superconvergence (the terminology is borrowed from papers on finite element methods).

In order to verify the superconvergence of QVLM, it is necessary to prove that the solutions of the system of algebraic equations (4) and (9) are equal to the values of exact solutions at the positions of point vortices for any N. Since the values of the exact solutions at the positions of these discrete vortices are given by

$$\gamma_i' = 2\frac{1 + \cos((i-0.5)\pi/N)}{\sin((i-0.5)\pi/N)}, \tag{10}$$

it is sufficient to prove that the following equations are identities:

$$\frac{1}{N}\sum_{i=1}^{N} \frac{1 + \cos((i-0.5)\pi/N)}{\cos((i-0.5)\pi/N) - \cos(j\pi/N)} = 1 \quad (j = 1, 2, \ldots, N). \tag{11}$$

The left-hand side of Eq. (11) can be rewritten as

$$\frac{1}{N}\sum_{i=1}^{N} \frac{1 + \cos((i-0.5)\pi/N)}{\cos((i-0.5)\pi/N) - \cos(j\pi/N)}$$

$$= \frac{1}{N}\sum_{i=1}^{N}\left\{1 + \frac{1 + \cos(j\pi/N)}{\cos((i-0.5)\pi/N) - \cos(j\pi/N)}\right\}$$

$$= 1 + \frac{1}{N}(1 + \cos(j\pi/N))\sum_{i=1}^{N}\frac{1}{\cos((i-0.5)\pi/N) - \cos(j\pi/N)}.$$

According to the theory of Chebyshev polynomials,[6] we have the identity

$$\sum_{i=1}^{N}\frac{1}{\cos((i-0.5)\pi/N) - \cos(j\pi/N)} = 0 \quad (j = 1, 2, \ldots, N-1).$$

On the other hand, we obviously have for $j = N$

$$\begin{cases} 1 + \cos(j\pi/N) = 0 \\ \sum_{i=1}^{N}\frac{1}{\cos((i-0.5)\pi/N) - \cos(j\pi/N)} \neq \infty \end{cases} \quad (j = N).$$

Therefore Eqs. (11) are identities and it is shown that the numerical solution of QVLM, determined by Eqs. (4) and (9), is exact for any N.

Thus we conclude that, although the boundary condition is not necessarily satisfied on the surface of the plate, except at the limited control points, it is still possible that the numerical solution gives exact values, and the leakage between control points does not cause loss of accuracy. This is quite different from Hunt's conclusion.[5] In his paper on the mathematics of panel methods, he pointed out that the leakage is probably the primary cause of loss of accuracy in most existing panel methods. The superconvergence of QVLM suggests that the positions of the point vortices and the control points are more important than leakage for the improvement of the numerical solution, at least for the flow past a two-dimensional flat plate.

II-3. Numerical Method

The QVLM was developed to calculate a three-dimensional lifting surface without wake roll-up by Lan.[4] In the present paper it is extended to calculate the effect of the ground on wake roll-up behind a lifting surface. In the QVLM, the continuous vorticity distribution over a wing is replaced by a quasicontinuous one, being continuous chordwise but stepwise constant in the spanwise direction. Calculation of induced velocities is more complicated than in the VLM. To retain both the simplicity of the VLM and the superconvergence of the QVLM, we propose an improved VLM. In this method, the semicircle scheme is used to determine the positions of vortex segments and control points, but the vortex model is the same as in the VLM, that is, the wing circulation is represented by a set of line vortex segments. For a two-dimensional plate, the positions of the point vortices are determined by Eq. (8). The unknown function is Γ_i (the strength of the i-th point vortex), instead of vorticity density. Thus Eqs. (4) and (5) are replaced by

$$\sum_{i=1}^{N} A_{ij} \Gamma_i = 2\pi \qquad (j = 1, 2, \ldots, N), \tag{12}$$

and

$$A_{ij} = \frac{2}{\cos((i - 0.5)\pi/N) - \cos(j\pi/N)}. \tag{13}$$

When γ_i values are determined from discrete vortex strengths, the length of the local element must be taken as $(\pi/2N) \sin((i-0.5)\pi/N)$. Using an approach similar to the previous one, the superconvergence of the improved VLM can be proved. Numerical results obtained by the improved VLM agree completely with the values of the exact solution at the positions of point vortices for the flow past a two-dimensional flat plate. The paneling scheme of the right half of the lifting surface, which is symmetric with respect to the x-axis, is shown in Fig. 1. First, the semispan is divided into N strips by

$$y_j = -\frac{b}{2} \cos\left((j + N)\frac{\pi}{2N}\right) \qquad (j = 0, 1, 2, \ldots, N), \tag{14a}$$

where b is the span of the wing. Then each strip is divided into M panels by

$$x_{ij} = x_{0j} + \frac{c_j}{2}\left\{1 - \cos\left((i - 0.5)\frac{\pi}{M}\right)\right\} \tag{14b}$$

$$(i = 1, 2, \ldots, M)$$

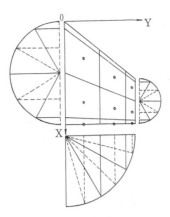

Fig. 1. Paneling scheme of a wing.

where x_{0j} is the x-coordinate of the leading edge on the local chord c_j at $y = y_j$. The positions of the control points are determined by

$$y'_j = -\frac{b}{2}\cos\left((j + N - 0.5)\frac{\pi}{2N}\right) \quad (j = 1, 2, \ldots, N), \tag{15a}$$

$$x'_{ij} = x'_{0j} + \frac{c'_i}{2}\left(1 - \cos\left(\frac{i\pi}{M}\right)\right) \quad (i = 1, 2, \ldots, M), \tag{15b}$$

where x'_{0j} is the x-coordinate of the leading edge on the local chord c'_j, at $y = y'_j$.

In the numerical calculations, the lifting surface and its wake sheet are represented by constant doublet panels, and Eq. (1) and condition (5) are automatically satisfied. For a wake sheet, constant doublet panels are equivalent to using free vortex lines. Every vortex line is composed of a series of straight segments joined head to tail. A free vortex line ends in a semi-infinite straight vortex line with freestream direction. The first segment of each vortex line emanating from the trailing edge lies in the plane of the wing to satisfy condition (4). The strength of a free vortex line equals that of the bound panel which is joined with this vortex line. To determine the doublet distribution of the lifting surface, condition (2) must be satisfied simultaneously at all the control points. The wake geometry is obtained using condition (3). The ground effect is simulated by using an image vortex system to satisfy condition (1).

Our problem is nonlinear since the position of the wake is not known in advance. To obtain the first solution of vortex strength, the free vortex lines are assumed to be parallel to the freestream. Under the condition that a free vortex segment must be parallel to the local velocity at its upstream end to satisfy (3), wake relaxation is repeated, until the maximum change in the positions of the free vortex segments is less than a specified limit.

Induced velocities are calculated by using the Biot-Savart law. Small Rankine vortex cores, in which the induced velocity at any point is proportional to the distance from the center of the core, are placed on the vortex lines to avoid too large velocity at close proximity of vortex lines. Such a large induced velocity is obviously nonphysical and may cause the solution to diverge. The specification of the core radius is made on the basis of empirical considerations. Another problem with induced velocities is whether or not the influence of the entire free vortex line should be ignored when computing the velocity at a point on the line. The influence was ignored by Maskew[7] to prevent divergence of his numerical calculations. Jepps[8] emphasized that since the velocity induced by a smoothly curved line vortex at a point on itself is infinite, it is meaningless to include the influence of the rest of the vortex line after ignoring the infinite influence of the local segment. The fact is that the infinite self-induced velocity is nonphysical, and indicates that the line vortex model is over-simplified. After introducing the vortex core, this infinite velocity disappears. The influence ignored by Maskew is included in the present, but the numerical calculations still converge rapidly.

III. NUMERICAL RESULTS

A computer program based on the method described in the preceding section was developed and applied. Parameters used were: AR (aspect ratio of the wing), α (angle of attack), h (distance from the midpoint of the root chord of the wind to the ground), r (core radius of line vortices), and d_i (maximum change of the positions of the free vortex seg-

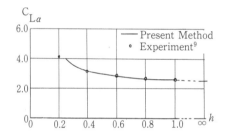

Fig. 2. Ground effect on the lift of a rectangular wing ($AR = 2.0$).

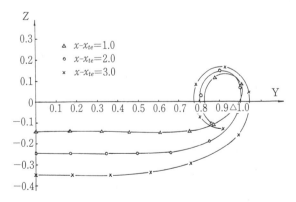

Fig. 3. Wake roll-up behind a rectangular wing next to the ground ($AR = 2$, $\alpha = 10°$, and $h = 1.0$).

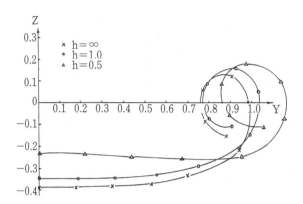

Fig. 4. Ground effect on wake roll-up behind a rectangular wing ($AR = 2$, $\alpha = 10°$, and $x\text{-}x_{te} = 3.0$).

ments between the results of the i-th and the (i-1)-th iterations). Length parameters are nondimensionalized by the wing root chord.

Figure 2 shows the variation of $C_{l\alpha}$ for a flat rectangular wing of $AR=2$ at values of h from 1.0 to 0.25. The value for no ground effect is also shown. It is seen that the ground has little effect at h greater than 1.0 and the wing produces significantly larger lift when flying very close to the ground than when flying in infinite space under otherwise the same conditions. The experimental results from Fig. 6 of Jepps[8] are also shown in Fig. 2 and the agreement is excellent.

Figure 3 shows the rolled up wake sheet for a flat rectangular wing of $AR=2$ at $h=1.0$ and $\alpha=10°$ for three different downstream positions, $x\text{-}x_{te}=1.0$, 2.0, and 3.0. Subscript "te" represents trailing edge. The other relative parameters are $M=3$, $N=10$, and $r=10^{-5}$. The computation converges rapidly with $d_1=0.08809$, $d_2=0.00463$, $d_3=0.00046$, and $d_4=0.00006$. In these calculations, the numerical solution is considered to be convergent as soon as d_i becomes less than 10^{-4}. The stretching and spiral motions of the wake sheet are well simulated, especially near the tip region.

Figure 4 shows the ground effect on the shape and position of the wake sheet behind the same wing at $x\text{-}x_{te}=3.0$ for three different values of h. It is seen that the ground has little effect on wake roll-up at h values greater than 1.0. For h smaller than 1.0, the ground effect becomes obvious. The vortex position moves both outward and upward due to the ground effect, as shown in Fig. 4. The closer the wing is to the ground, the stronger the effect is until h decreases to 0.2. The solution is divergent at $h=0.2$. In other words, the wake vortex will become unstable as the wing descends too close to the ground.

REFERENCES

1) Hackett, J.E. and Evans, M.R., Vortex wakes behind high-lift wings. *J. Aircraft*, Vol. **8**, No. 5 (1971), pp. 334–340.
2) Goetz, A.R., Osborn, R.F., and Smith, M.L., Wing-in-ground effect aerodynamic predictions using panair. *AIAA Paper*, 84–2429 (1984).

3) Deese, J.E. and Agarwal, R.K., Euler calculation for flow over a wing in ground effect. *AIAA Paper*, 86-1765 (1986).
4) Lan, C.E., A quasi-vortex-lattice method in thin wing theory. *J. Aircraft*, Vol. **11**, No. 9 (1974), pp. 518–527.
5) Hunt, B., The panel method for subsonic aerodynamic flows: a survey of mathematical formulations and numerical models and an outline of the new British aerospace scheme. *Computational Fluid Dynamics* (W. Kollman, ed., Hemisphere Publ. Corp., 1980), pp. 99–166.
6) Luke, Y.L., *The Special Functions and Their Approximations* (Academic Press, 1969), Vol. **2**, Chap. 15.
7) Maskew, B., *Ph.D. Thesis*, Loughborough University (1972).
8) Jepps, S.A., Computation of vortex flow by panel methods. *Computational Fluid Dynamics* (W. Kollman, ed., Hemisphere Publ. Corp., 1980), pp.505–542.
9) Saunders, G.H., Aerodynamic characteristics of wings in ground proximity. *Canadian Aeronautics and Space Journal*, Vol. 11, No. 6 (1965), pp. 185–192.

Macroscopic Dynamic Simulations of the Impact of Flexible Beams

Kazuo YAMAMOTO

Central Research Laboratory, Mitsubishi Electric Corporation, Amagasaki, Hyogo

A simplified mathematical model and numerical procedures for the impact phenomena of flexible beam-like structures for space use are proposed for the purpose of application in the initial design phases. The inertia model is made by a discretized system and the stiffness model is formulated by a continuous system using the finite element approach. The concept of coefficient of restitution that is used in classic dynamics is introduced between the concentrated mass points that collide with each other. The equations of motion derived from energy-based Lagrangean formulation are solved numerically using the extended mass matrix that contains the constraint condition due to collision. As a result of numerical simulations of a few physical configurations, some of basic knowledge of the response of the structural system and the transfer characteristics due to collision are made clear.

I. INTRODUCTION

Effective means for the realization of large space structural systems in orbit are the process of automatic extension/deployment or remote manipulator-based assembly. This process is usually followed by impact phenomena such as latching of mechanisms at the final stage of extension/deployment and docking of a plural number of substructures under construction. Consequently, the design of these structural systems requires effective prediction of transient dynamics due to impact.

The impact phenomenon of structures should essentially be understood as a contact problem of continua which have highly nonlinear properties such as large deformation of structural members and elasto-plastic behavior of materials.[1] In order to obtain the solutions of the problems, troublesome large-scale calculations with extremely high precision are necessary. In this paper, to avoid such troublesome calculations and obtain the macroscopic dynamic properties in initial design phases, a simplified mathematical model for the beam and analysis procedures are proposed. In the model, the inertia property is described by the discretized system and the stiffness property is described by the continuous system using the finite element approach. The concept of the coefficient of restitution is introduced between the plural pieces of concentrated mass that collide with each other. By performing some numerical simulations of the impact using this proposed model, basic knowledge was obtained of the response of structural systems and transfer properties of various physical amounts from one structure to another.

II. DESCRIPTION OF THE PROBLEM

Figure 1 shows the mathematical model of the system. The model consists of a plural number of (two as a rule) straight beams. For simplicity of analysis without loss of generality, the motion is restricted within the plane of paper (two-dimensional motion). The

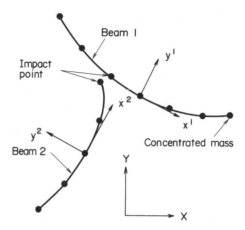

Fig. 1. Mathematical model of the system.

formulation of the inertia model uses a concentrated system, and that of stiffness model a continuous system. That is, the k-th beam ($k = 1,2$) is divided into n_k finite elements and the i-th grid point has the concentrated mass M_i^k and moment of inertia J_i^k ($i=1,\ldots, n_k+1$), and the i-th element has Young's modulus E_i^k, area of cross section A_i^k, and moment of cross section I_i^k. Transfer of the force from one beam to another is assumed to exist only between the concentrated mass points that collide with each other. The coefficient of restitution, which is usually used in the area of rigid-body dynamics, is established between the two mass points.[2] This concept means that microscopic characteristics based on the deformation and material properties of the areas that collide with each other are replaced by an equivalent macroscopic parameter. Large flexural deformation is assumed for the elastic behavior of the beam. At the ends of beams various boundary conditions are introduced such as the elastic constraint, rigid constraint, and no constraint. As the coordinate systems to describe the present problem, we employ the basic coordinate system o-XY, which is fixed to the inertial space, and a local coordinate system o-$x^k y^k$, which is attached to the center of each beam.

III. FORMULATION OF DYNAMICS

III-1. Equations of Motion for Beams

Figure 2 shows the geometry of a representative i-th finite element of the k-th beam. An element coordinate system o-$\xi\eta$ having its origin at the center of the element is adopted. Let m_i^k and j_i^k be the concentrated mass and moment of inertia of the first and the second grid points of the i-th element, l_i^k be the length of the i-th element, and L_i^k be the distance between the element center and the origin of the coordinate system o-$x^k y^k$ of the k-th beam. The locations of the first and the second grid points $\bar{r}_{i1}^k$ and $\bar{r}_{i2}^k$ in the basic coordinate system are expressed by the following equations:

$$\bar{r}_{i1}^k = r_0^k + E^k \cdot \tilde{r}_{i1}^k$$
$$\bar{r}_{i2}^k = r_0^k + E^k \cdot \tilde{r}_{i2}^k \tag{1}$$

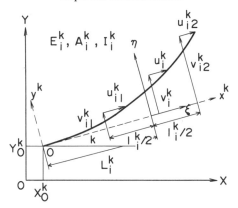

Fig. 2. Geometry and deformation of beam element.

where r_0^k is the location vector of the origin of local coordinate system in the basic coordinate system, $\bar{r}_{i1}^k$ and $\bar{r}_{i2}^k$ are the location vectors of the first and the second grid points in the local coordinate system, and E^k is the direction cosine matrix between the axes of the basic and k-th local coordinate systems. Each is expressed in detail in the following form:

$$\bar{r}_{i1}^k = \{\bar{X}_{i1}^k, \bar{Y}_{i1}^k\}^T, \bar{r}_{i2}^k = \{\bar{X}_{i2}^k, \bar{Y}_{i2}^k\}^T$$

$$r_0 = \{X_0^k, Y_0^k\}^T$$

$$\tilde{r}_{i1}^k = \{L_i^k - \frac{1}{2}l_i^k + \bar{u}_{i1}^k, \bar{v}_{i1}^k\}^T$$

$$\tilde{r}_{i2}^k = \{L_i^k + \frac{1}{2}l_i^k + \bar{u}_{i2}^k, \bar{v}_{i2}^k\}^T$$

$$E^k = \begin{bmatrix} \cos \alpha^k & -\sin \alpha^k \\ \sin \alpha^k & \cos \alpha^k \end{bmatrix} \qquad (2)$$

where $\bar{u}_{i1}^k$, $\bar{u}_{i2}^k$ are the axial displacements and $\bar{v}_{i1}^k$, $\bar{v}_{i2}^k$ are the transverse displacements of the first and the second grid points of the i-th element, and α^k is the rotation angle of the k-th local coordinate system. Let $\bar{\theta}_{i1}^k$ and $\bar{\theta}_{i2}^k$ be the elastic rotational angle of the first and the second grid points of the element respectively, and the kinetic energy of the i-th element T_i^k is described as follows:

$$T_i^k = \frac{1}{2}m_i^k(\dot{\tilde{r}}_{i1}^k \cdot \dot{\tilde{r}}_{i1}^k + \dot{\tilde{r}}_{i2}^k \cdot \dot{\tilde{r}}_{i2}^k) + \frac{1}{2}j_i^k\{(\dot{\alpha}^k + \dot{\bar{\theta}}_{i1}^k)^2 + (\dot{\alpha}^k + \dot{\bar{\theta}}_{i2}^k)^2\}. \qquad (3)$$

On the other hand, we approximate the elastic displacement in the axial and transverse directions within the i-th element u_i^k, v_i^k using the polynomials of element coordinate ξ as follows:

$$u_i^k = B_0 + B_1\left(\frac{\xi}{l_i^k}\right)$$

$$v_i^k = C_0 + C_1\left(\frac{\xi}{l_i^k}\right) + C_2\left(\frac{\xi}{l_i^k}\right)^2 + C_3\left(\frac{\xi}{l_i^k}\right)^3. \qquad (4)$$

The coefficients B_0, B_1, C_0, ..., C_3 are constants to be determined and are expressed by the first and the second grid point values for the axial, transverse, and rotational displacements as follows:

$$B_0 = \frac{1}{2}(\bar{u}_{i1}^k + \bar{u}_{i2}^k), \qquad B_1 = \bar{u}_{i2}^k - \bar{u}_{i1}^k$$

$$C_0 = \frac{1}{2}(\bar{v}_{i1}^k + \bar{v}_{i2}^k) - \frac{1}{8}l_i^k(\bar{\theta}_{i2}^k - \bar{\theta}_{i1}^k)$$

$$C_1 = \frac{2}{3}(\bar{v}_{i2}^k - \bar{v}_{i1}^k) - \frac{1}{4}l_i^k(\bar{\theta}_{i1}^k + \bar{\theta}_{i2}^k)$$

$$C_2 = \frac{1}{2}l_i^k(\bar{\theta}_{i2}^k - \bar{\theta}_{i1}^k)$$

$$C_3 = -(\bar{v}_{i2}^k - \bar{v}_{i1}^k) + l_i^k(\bar{\theta}_{i1}^k + \bar{\theta}_{i2}^k).$$

The axial strain of a two-dimensional beam under large flexural deformations can be approximated using the displacement gradient up to the second order as follows:

$$\varepsilon_x = \varepsilon_{x0} + \frac{\partial u_i^k}{\partial x} - \frac{\partial^2 v_i^k}{\partial x^2} y + \frac{1}{2}\left(\frac{\partial v_i^k}{\partial x}\right)^2 \tag{5}$$

where ε_{x0} is the uniform initial strain in the axial direction and y is the distance from the neutral axis. Consequently, the strain energy of the i-th element U_i^k has the following expression:

$$U_i^k = \int_{(\text{vol})} \frac{1}{2} E_i^k \varepsilon_x^2 \, d(\text{vol})$$

$$= \int_{(\text{vol})} \frac{1}{2} E_i^k \left\{\varepsilon_{x0} + \frac{\partial u_i^k}{\partial x} - \frac{\partial^2 v_i^k}{\partial x^2} y + \frac{1}{2}\left(\frac{\partial v_i^k}{\partial x}\right)^2\right\}^2 d(\text{vol}). \tag{6}$$

Where gravitational potential exists, the potential energy of the i-th element P_i^k can be expressed as

$$P_i^k = -m_i^k g(\bar{r}_{i1}^k + \bar{r}_{i2}^k) \cdot e \tag{7}$$

where e is a unit vector that defines the direction of gravity force. Defining β as the angle between the direction of gravity and minus the Y-axis of the basic coordinate system, e is written in the following relation:

$$e = \{-\sin\beta, \cos\beta\}^T.$$

Summing up each for all elements of all beams, we obtain total kinetic energy T, total strain energy U, and total potential (gravity) energy P.

We define a set of variables that describes the present problem in the following:

$$\boldsymbol{q} = \{q_j\} = \{X_0^k, Y_0^k, \alpha^k, \bar{u}_1^k, \bar{v}_1^k, \bar{\theta}_1^k, \ldots, \bar{u}_{N_k}^k, \bar{v}_{N_k}^k, \bar{\theta}_{N_k}^k\}^T \ (k = 1,2). \tag{8}$$

Summarizing the constraint conditions according to contact between the beams and due to various supports at the beam ends, we write the following equation:

$$F_i(q_j) = 0. \tag{9}$$

Consequently, the equations of motion for the system can be formulated by the following Lagrange equations with the constraint conditions:

$$\frac{d}{dt}\left(\frac{\partial T}{\partial \dot{q}_j}\right) - \frac{\partial T}{\partial q_j} + \frac{\partial (U+P)}{\partial q_j} + \lambda_l \frac{\partial F_l}{\partial q_j} = 0 \qquad (10)$$

where λ_l is Lagrange's multiplier. Adding Eq. (10) and the second time derivative of Eq. (9) simultaneously, we obtain the final form of the equations of motion as follows:

$$[M] \cdot \{\hat{\ddot{q}}\} = \{f\} \qquad (11)$$

where $[M]$ is the extended mass matrix and $\{f\}$ is the extended force vector, each of which contains the constraint conditions in itself. In order to obtain the numerical solutions of nonlinear equations of motion (11), we employ Newmark's iterative time integration scheme.[3]

III-2. Computational Procedures for Collision

The constraint condition followed by the collision of two beams is proposed as follows: first of all, the plane of contact (the line of contact) is defined. In the case of collision of a beam end with the side of another beam, the longitudinal axis of the latter beam is defined as the plane of contact, and in the case of collision of a beam with a rigid wall the plane of the wall is defined as the plane of contact. Thus, we treat only the configuration in which the plane of contact can be defined definitely. We assume that transfer of forces through the plane of contact is possible only in the direction normal to the plane, that is, the coefficient of friction on the plane is assumed to be zero.

When the mass points of beams that collide are separated from each other, the uncoupled equations of motion for each beam are solved independently. When they are in contact with each other, the equations of motion (coupled with each other by the constraint force) and a constraint equation $v_{col}^1 - v_{col}^2 = 0$ are solved simultaneously. In the above equation v_{col}^1 and v_{col}^2 are the displacements of mass points to collide in the direction normal to the plane of contact.

Detailed procedures for the analysis of collision/repulsion processes are shown as follows:
A) Processes into contact
1) Solve the uncoupled equations and obtain the time when the distance of two mass points to collide (D) becomes zero, and calculate the velocity $\dot{\bar{v}}_{col}^1$, $\dot{\bar{v}}_{col}^2$ of the mass points normal to the plane of contact.
2) Calculate the velocity $\dot{\hat{v}}_{col}^1$, $\dot{\hat{v}}_{col}^2$ of mass points after collision using the law of conservation of momentum and the equation of restitution shown as follows:

$$M_{col}^1 \dot{\hat{v}}_{col}^1 + M_{col}^2 \dot{\hat{v}}_{col}^2 = M_{col}^1 \dot{\bar{v}}_{col}^1 + M_{col}^2 \dot{\bar{v}}_{col}^2$$

$$\dot{\hat{v}}_{col}^2 - \dot{\hat{v}}_{col}^1 = \mu_L (\dot{\bar{v}}_{col}^1 - \dot{\bar{v}}_{col}^2)$$

where μ_L is the local coefficient of restitution.
3) Calculate the uncoupled equations by one time step and continue to solve the equation; if $D > 0$, return by one time step and continue to solve the coupled equations if $D < 0$.

B) Processes from contact to restitution

1) Solve the coupled equations and obtain the time when the constraint force of contact (f) changes from a positive to negative value.
2) Solve the uncoupled equations using the velocity of mass points in the time obtained above as the initial value.

IV. NUMERICAL RESULTS AND DISCUSSION

Numerical examples for a few typical special configurations of the beams are examined and discussed.

IV-1. Collision of a Beam and a Rigid Mass

Transverse collision of a rigid mass to the center of the beam with both ends fixed in the transverse direction and spring supported in the axial direction is numerically analyzed. Typical physical constants used in the analysis are shown in Table 1.

Figure 3 shows the time-history displacement of the rigid mass and beam center with the condition that the initial velocity of rigid mass $\dot{V}_0 = 1$ m/s and the local coefficient of restitution $\mu_L = 0.5$. In the nonlinear formulation (large flexural deformation), the maximum displacement of the beam center is surpressed due to generation of more axial forces

Table 1. Typical physical constants (I).

Number of beam elements	4	—
Length of beam	1	m
Young's modulus	2.942×10^{10}	N/m²
Specific mass	9.807×10^3	kg/m³
Sectional area	7.854×10^{-5}	m²
Moment of cross section	4.909×10^{-10}	m⁴
Constant of support spring	2.311×10^6	N/m
Mass of rigid body	7.700×10^{-1}	kg

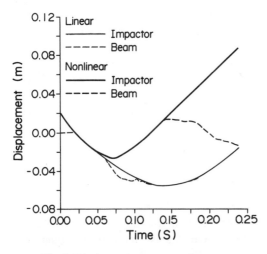

Fig. 3. Displacement response of colliding points.

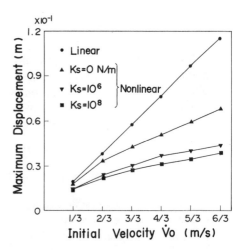

Fig. 4. Maximum displacement of the beam center.

than in the linear formulation. Larger axial forces due to flexural deformation are generated with a larger spring constant at the beam ends.

Figure 4 shows the relation of the maximum displacement and the velocity of the rigid mass with the parameter of the spring constant. In the linear formulation the maximum displacement does not depend on the spring constant and is almost in direct proportion to the initial velocity of the rigid mass. In the formulation with large deformation, on the other hand, the maximum displacement increases with increasing initial velocity $\dot{V}_0$, but the rate of increase is reduced with increasing spring constant.

Figure 5 shows the dependence of the maximum displacement and restituted velocity upon the local coefficient of restitution. Both for the linear and nonlinear formulations, the restituted velocity increases by about 20% near the region of perfect elastic collision ($\mu_L = 1.0$). It is recognized that no marked difference of restituted velocity can be seen between the linear and nonlinear formulations.

IV-2. Collision of a Beam and a Bar

Transient behavior was analyzed in the case where a bar moving in its axial direction with a constant velocity collided normally against the center of a beam side at rest without any constraint in its ends. Here, the term "bar" is used for a member which deforms only in the axial direction, the term "beam" being used for a member that can undergo flexural deformation. In this configuration, no change exists in the attitude angle of either member, and no essential difference is recognized between the linear and nonlinear formulations. Table 2 lists typical physical constants of the system.

Figure 6 shows the time-history locations of colliding mass points in the direction of the collision where the initial velocity of bar $\dot{V}_0 = 1$ m/s and the local coefficient of restitution $\mu_L = 0.5$. The time-history constraint force between two members due to contact is superimposed in Fig. 6. When the collision starts between the beam and bar, the contact points repeat the detailed processes of contact and repulsion and finally reach perfect separation as a whole.

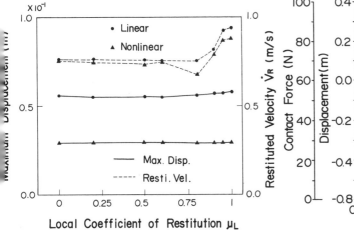

Fig. 5. Maximum displacement and restituted velocity.

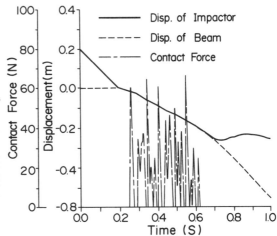

Fig. 6. Displacement response and contact force of colliding points.

Table 2. Typical physical constants (II).

Beam		
Number of elements	4	–
Length	20	m
Young's modulus	8.131×10^{10}	N/m^2
Specific mass	2.746×10^4	kg/m^3
Sectional area	3.642×10^{-3}	m^2
Moment of cross section	7.354×10^{-5}	m^4
Bar		
Number of elements	4	–
Length	5	m
Young's modulus	1.178×10^7	N/m^2
Specific mass	2.746×10^4	kg/m^3
Sectional area	7.284×10^{-3}	m^2

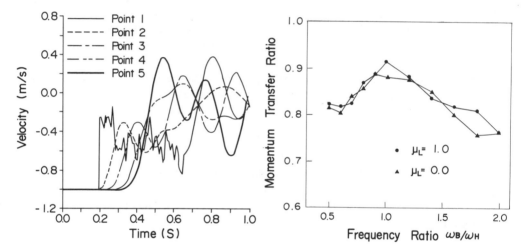

Fig. 7. Velocity response of several points of a bar.

Fig. 8. Momentum transfer characteristics.

Figure 7 shows the time-history velocity response of each point on the bar. The disturbance generated in the end of bar due to collision propagates through the bar, reaches the other end, and returns to the contact point after reflecting at the end. The final separation of the beam and the bar takes place in this instant.

Figure 8 shows the characteristics of momentum transfer from bar to beam with the ratio of lowest eigenfrequency of two members. When the ratio of eigenfrequency is nearly equal to 1, we can see the peak of momentum transfer ratio in the characteristic curves for both large and small coefficients of restitution. This is probably because the response level of the beam increases due to the dynamic impedance of the beam and bar. The difference in the local coefficient of restitution appears not to have marked influence on the characteristics of momentum transfer. The characteristics of energy transfer reveal a tendency identical to those of momentum transfer.

IV-3. Collision of a Free Beam against a Rigid Wall

Transient dynamics were examined in the case where a beam-like free structure with variable stiffness collides against a rigid wall with several values of the local coefficient of restitution and incident angle. The typical physical constants used are identical to those of the beam shown in Table 2 except that Young's modules is 8.131×10^9 N/m².

The dependence of the transient deformation of the beam upon the stiffness is examined. The beam stiffness (Young's modulus E) ranges from 8.131×10^9 to 8.131×10^7 N/m². The initial velocity is $\dot{V}_0 = 3$ m/s, the initial attitude angle is 45° and the angle of incidence is 90°. Figure 9 shows the superimposed time-history configuration of typical (high and low stiffness) cases. For a beam with high stiffness, the beam is in contact with the rigid wall only at its ends. For low stiffness, on the other hand, the beam is in contact with the rigid wall not only at its ends but also at the intermediate part due to the large bending displacement. Figure 10 shows the time-history attitude angle of the beam center. For lower Young's modulus, the attitude angle of the beam center has its larger peak at a later time due to the lower speed at which the disturbance generated at the end of beam propagates to the center.

Collisions with different angles of incidence were analyzed for two different coefficients of restitution. For all cases the initial attitude angle was 45°. In Fig. 11, the time-history of beam configuration is compared between transverse incidence (angle of incidence =

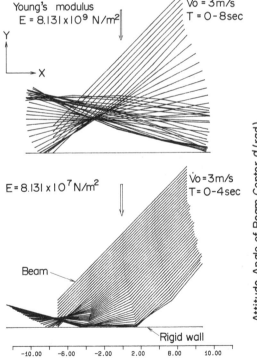

Fig. 9. Difference of time-history configuration due to beam stiffness.

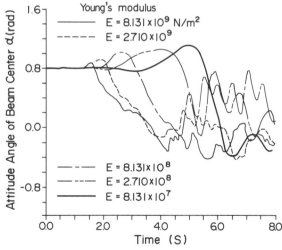

Fig. 10. Time-history attitude angle of beam center.

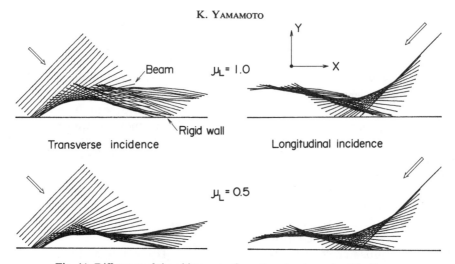

Fig. 11. Difference of time-history configuration due to angle of incidence.

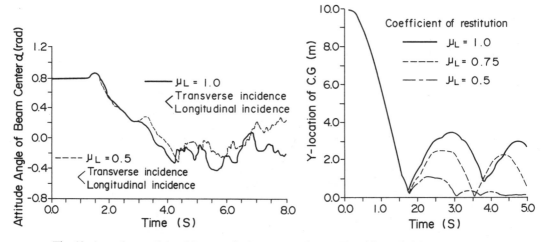

Fig. 12. Accordance of time-history attitude angle of beam center.

Fig. 13. Time-history height of C.G. of a beam.

315°) and the longitudinal incidence (225°) for the local coefficients of restitution $\mu_L = 1.0$ and $\mu_L = 0.5$. The time-history of the beam configurations is different in the beams with different incident angles both for $\mu_L = 1.0$ and $\mu_L = 0.5$. Figure 12 shows that the time-history attitude angles is independent of incident angle. This is probably because the friction of the rigid wall is assumed to be zero, i.e., no transfer of forces occurs in the direction paralell to the wall in the course of collision, and the same magnitude of normal forces are applied in the both cases. The same observation was also obtained in the elastic displacement in each point of the beam.

Finally, collisional behavior of the beam is examined in the existence of gravitational force. When a beam-like structure with variable values of the local coefficient of restitution falls in the 1G gravity field from an initial configuration with a 45° of attitude angle at rest and collides with a horizontal rigid wall, the time-history height of C.G. was obtained as shown in Fig. 13. With a decreasing local coefficient of restitution, the peak value of the

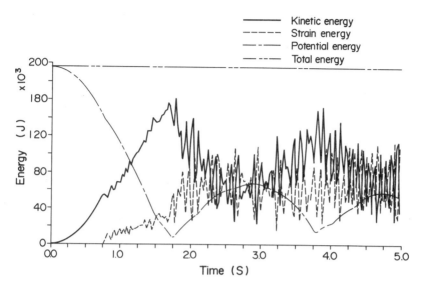

Fig. 14. Various kinds of time-history energy.

mean height of the beam after each collision becomes small, and the mean time of collision after a collision is also reduced. This can be explained by energy dissipation due to collision. If the local coefficient of restitution μ_L is 1.0, the value of total energy of the beam should remain constant. Figure 14 shows the time-history of various kinds of energy. It is confirmed that the calculated total energy of the beam remains constant in the course of sequential collisions. This verifies the present simulations.

V. CONCLUSIONS

A simplified mathematical model and numerical procedures were proposed for the simulation of collisional phenomena of flexible beam-like structures for space construction. Simulation analyses were performed for several typical configurations of the geometry and physical properties of the structure. The following conclusions were made:

1) The macroscopic model using local coefficients of restitution can simulate the actual collision of beam-like structures with fair precision.

2) Basic knowledge was gained of the influence of the proposed local coefficients of restitution and other parameters of structures upon the structure responses and transfer characteristics of various types of collisions.

REFERENCES

1) Zukas, J.A. et al., *Impact Dynamics*. John Wiley & Sons (1982).
2) Yamamoto, K., Numerical analysis of low-velocity transverse impact of flexible beamlike structures. *Proc. JSASS/JSME Structures Conference*, Vol. **28** (1986), pp. 22–25.
3) Newmark, N.M., A method of computation for structural dynamics. *Proc. ASCE, J. Engng. Mech. Div.*, **EM-3** (1959), pp. 67–94.

Study of Response Phenomena to the Change of External Field in Vortex-Like Phase for XY-Type Spin Glass State by Computer Simulation

Yuhei NATSUME, Kenji FUJIMOTO, and Tomoki YOSHIHARA

Department of Physics, Faculty of Science, Chiba University, Chiba

Dynamic properties, in particular, the response phenomena of spin glass posssessing a short range XY-type interaction on the square lattice to the adiabatic change of the external magnetic field are discussed on the base of spin dynamic simulation (SDS) which has been proposed and developed by the present authors. Calculated field-induced alteration of a spin pattern from the state obtained by zero field cooling (ZFC) is explained in detail in combination with the behavior of chirality by the introduction of the new concept for "the dissociation of chirality-pair". Furthermore, the investigation of the change by the switching off the field from the state obtained by field cooling (FC) lead to the classification of the response behavior into reversible and irreversible processes.

I. INTRODUCTION

For the last decade, the method of computer simulation has been extensively adopted to the theoretical investigation of a complex system on physics of condensed matters. As a typical example, the spin system called spin glass (SG)[1,2] including the competition between ferromagnetic and antiferromagnetic interactions in mixed compounds has been studied by many authors[3] on the basis of Monte Carlo simulation (MCS). Such a work by MCS has provided us fruitful results. However, obtained results provide us information mainly as to static aspects of SG, because MCS is introduced on the basis of a stochastic process in a equilibrium state. While, we have, for the last few years, realized that dynamic aspects of SG originating from the character for non-equilibrium state plays an important role: We become faced with the problems how the state of SG is stable and how it responses to the adiabatic change of the external field.

In order to discuss such problems, we apply spin dynamic simulation (SDS) proposed by the present authors[4-7] to the response phenomena of SG state for the system including the competition of XY-type ferromagnetic and antiferromagnetic nearest-neighbor interactions between spins localized at the sites on the square lattice.

On the basis of the results for this SDS in the response phenomena, we make clear the dynamic properties of SG for a 2-dimensional XY-type mixed compound in combination with the characteristic features for a metastable state.

The Hamiltonian treated here is

$$H = - \sum_{\langle ij \rangle} J_{ij}(\vec{M}_i \cdot \vec{M}_j)/(\hbar\gamma)^2 - \vec{H}_0 \sum_i \vec{M}_i, \tag{1}$$

where $\vec{M}_i$ the magnetic moment for i site. (Spin S_i correspond to $\vec{M}_i/(\hbar\gamma)$.) Here, $\vec{H}_0$ is the external field. The sign for the constant of exchange interaction J_{ij} between S_i and S_j is determined to be + or − according as the kind of interaction. Namely, the value of J_{ij} is fixed to be + or −, according as the interaction is ferromagnetic or antiferromagnetic.

In calculation, the antiferromagnetic spin is distributed randomly on a few decades percents of sites. It is supposed that interaction between ferromagnetic and antiferromagnetic spins is antiferromagnetic.

By the procedure discussed in ref. 2), we obtain the following equation for $\vec{M}_i$;

$$d\vec{M}_i/dt = (\vec{M}_i^{eff} - \vec{M}_i)/T_1, \tag{2}$$

where

$$\vec{M}_i^{eff} = \vec{H}_i^{eff} |\vec{M}_i|/|\vec{H}_i^{eff}|. \tag{3}$$

In eq. (3), the local effective field H_i^{eff} at i site is written as

$$\vec{H}_i^{eff} = \sum_{n(i)} J_{i,n(i)} \vec{M}_{n(i)}/(\hbar\gamma)^2 + \vec{H}_0, \tag{4}$$

where $n(i)$ denotes the nearest neighbor for i site. Namely, $\vec{H}_i^{eff}$ expresses the effective field caused by exchange interaction and external field. In combination of eq. (3) with eq. (4), it becomes clear that $\vec{M}_i$ has the tendency to be directed parallel to the $\vec{H}_i^{eff}$ by the process of energy dissipation. In other words, equation (2) lead the system to the local equilibrium state which has been discussed in detail by Palmer.[8] On the basis of the above model and formulation, we make computer simulation. After introducing the concept of chirality and mentioning its properties in section II, we discuss the calculated results for response phenomena in section III in the light of dynamic aspects for SG.

II. CHIRALITY IN XY-SYSTEM

We treat the 2-dimensional XY-type mixed compounds whose Hamiltonian is

$$H = - \sum_{\langle ij \rangle} J_{ij} \cos(\theta_i - \theta_j), \tag{5}$$

where J_{ij} is a nearest-neighbor exchange interaction. The sign of J_{ij} is determined according to the distribution of ferromagnetic and antiferromagnetic ions on a square lattice.

Here, we introduce the concept of a "frustrated plaquette" as follows; if the products of signs for J as to four sides, i.e. $\text{sign}(J_{ij}) \text{sign}(J_{jk}) \text{sign}(J_{kl}) \text{sign}(J_{li})$ for a plaquette $\overline{ijkl}$, is −1, such a plaquette is called a "frustrated plaquette". While if the production is +1, the plaquette is called a "non-frustrated plaquette".

In this paper, two processes of changing the external field are treated as shown in Fig. 1, where the symbol $M(H_0)$ denotes total magnetization (static external field). The one process is illustrated in Fig. 1(a); the system is sufficiently cooled under zero-field ($H_0=0$), until $t=0$. (We call this procedure "zero-field cooling (ZFC)".) Following such a ZFC, H_0 is switched on at $t=0$. In contrast to this, another process is shown in Fig. 1(b); H_0 is switched off after field-cooling (FC). In general, M increases or decreases according to switching on or off the field, respectively.

We investigate these two response phenomena in the above two processes on the basis of SDS, in particular, focusing on kinetics of "chirality". Therefore, we mention the general property of chirality for pure and mixed compounds in this section. Though the concept of chirality in itself has been introduced by Villain[9], we adopt here the following definition

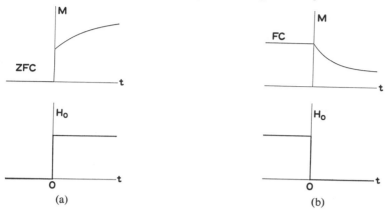

Fig. 1. Two process of changing the external field. (a): After zero-field cooling (ZFC), the external field H_0 is switched on at $t=0$. (b): After field cooling (FC), H_0 is switched off at $t=0$. The behavior of total magnetization M is schematically drawn for both process.

proposed by Kawamura et al.[10]; for a plaquette composed by four spins, the chirality is defined by

$$C = \sum \text{sign}(J_{ij}) \sin(\theta_i - \theta_j), \qquad (6)$$

where the summation is taken along the closed-loop of the corresponding plaquette. In a pure ferromagnetic XY system, the chirality expresses the center of a vortex introduced by Kosterlitz and Thouless.[11] As for XY system including the competition between exchange interaction, the chirality becomes more important especially when the corresponding plaquette is frustrated. In short, the following property should be noted as discussed by Villain[9]; if two plaquettes consisting of six spins are considered as a unit system, the ground state for the two frustrated plaquettes has two-fold degeneracy as shown in Fig. 2(a). In such a ground state, the chirality pair has the opposite sign. However, this system has also two types of a metastable excited state whose energy has a minimum value locally in the configuration space for these six spins. The chirality pair for these states has the same sign as shown on Fig. 2(b). Further, Villain has pointed out that the chirality pair with same sign $++$, $(--)$ describes a vortex (antivortex) on the square lattice and that the effective interaction between two chiralities C_i, C_j located on each plaquette $\vec{r}_i$, $\vec{r}_j$ has the following form;

$$H_{eff} \propto C_i C_j \log(|\vec{r}_i - \vec{r}_j|). \qquad (7)$$

Therefore, the interaction between chiralities $+$ and $-$ is attractive. As discussed in the next section, this interaction plays an essential role in response phenomena.

III. CALCULATED RESULTS AND DISCUSSION

We apply SDS to mixed XY compounds in order to explore response phenomena. As a typical example to SG state, we adopt a system where 25% of lattice site are randomly substituted by antiferromagnetic ions.

(a) First, we discuss the response process in which we are switching on the external field to the state obtained by ZFC(Fig. 1(a)). The initial state obtained under ZFC is shown in

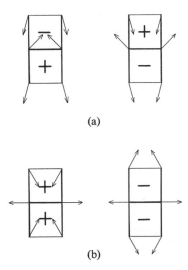

(a)

(b)

Fig. 2. Spin configuration in six spin system consisting of two frustrated plaquettes. (a): Two-fold ground state with chirality pair + −. (b): Two-fold metastable excited state with chirality pair + + or − −.

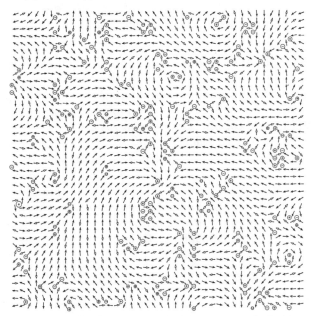

Fig. 3. The given state by ZFC. The black (white) arrow represents the direction of the ferromagnetic (antiferromagnetic) spin. Sign + or − in circle means that the corresponding plaquette has a large absolute value of the chirality. We can find the vortex-structure whose center is formed by the chirality pair + + or − −.

Fig. 3, where the chirality pair with the same sign localizes on the frustrated plaquette. Consequently such a pair tends to form a vortex-structure. Namely, a chirality pair + + (− −) fills the role of the center of a vortex (antivortex), respectively. For this initial state, we switch on the field to the direction of the total magnetization. Then, the slowly change in the whole spin pattern occurs.

In Fig. 4, the final state is shown, where the vortices seen in Fig. 3 almost disappear. If our attention is paid to the chirality, it is clearly seen that the number of the chirality pair with same sign becomes less than that in the initial state. Furthermore, the chirality with opposite sign ($+-$) tends to appear for the frustrated plaquettes. As a consequence, the whole spin pattern is characterized with a domain-structure. This remarkable change in the whole spin pattern is made more clearer by picking up the detail time-evolution of the local spin pattern. In fact, Fig. 5 denotes the subsequent change in the local spin pattern in some region of the lattice; in the first place, a chirality belonging to a pair with same sign separates each other. Namely, it leaves the original frustrated plaquette. This phenomena can be called "the dissociation of the chirality pair". Subsequently, an isolated chirality gets near the chirality with the opposite sign. As a result, we obtain the spin pattern including chirality pairs with opposite sign. In other words, a dissociation of the chirality pair with same sign corresponds to disappearance of a vortex. In the response process, this annihilation of vortices occurs at various places on the lattice. Consequently, the spin pattern in the final state (Fig. 4) becomes quite different from that in the initial state (Fig. 3). Corresponding to this remarkable change, both the total magnetization and the time-correlation show a complicated behavior. From these features, we can expect that this response is irreversible, in combination with the fact that the total energy of the domain-structure is lower than that of the vortex-structure.[5] In summary, we would like to emphasize that

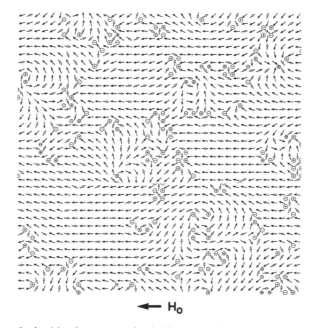

Fig. 4. The final state obtained by the process of switching on the field. (The initial state for this process is shown in Fig. 3.) Here, the field whose direction is indicated at the bottom is parallel to the magnetization of the state in Fig. 3. The chirality pair $++$ or $--$ tends to disappear, while the pair $+-$ appears. This change leads the state to a domain-structure.

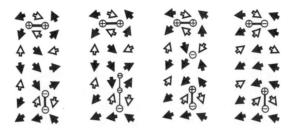

Fig. 5. The typical subsequent change in the local spin pattern from Fig. 3 to Fig. 4. The temporal change is shown from left to the right hand side. The dissociation of chirality pair − − can clearly be shown as discussed in detail in the text.

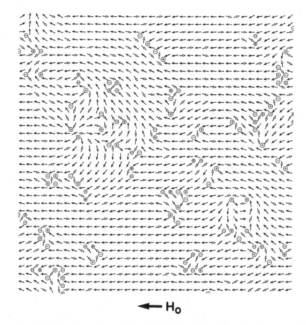

Fig. 6. The given state by FC. The direction of the field is indicated at the bottom. This pattern is characterized with a domain-structure.

the annihilation of the vortex plays an essential role in this response phenomenon making the process irreversible.

(b) Secondly, we treat the another response process in which we switch off the external field on the state obtained under FC (Fig. 1(b)). As a result, we can classify it into following two types of the response according as the spin pattern obtained under FC: One type is reversible and the other is irreversible. The reversible response occurs in the case that the initial spin pattern is characterized by the domain-structure, as shown in Fig. 6. The reversible behavior is caused almost by the change of the spins in non-frustrated plaquettes. In contrast to this, when the vortex-structure strongly remains in the initial state, the response contains both reversible and irreversible processes. For such a case, the time-cor-

Computer Simulation for XY-Type Mixed Magnetic Compound 143

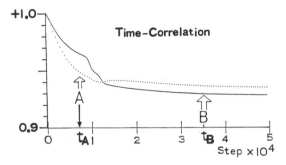

Fig. 7. The time-correlation function for spins in the process of switching off the field. Here, dotted (solid) curve means the correlation averaged by spins in non-frustrated (frustrated) plaquettes. Arrows A and B are use in the restoration of spin pattern discussed in the text and Fig. 9.

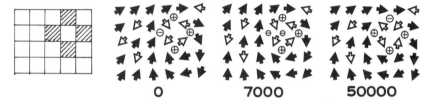

Fig. 8. The subsequent change in the local spin pattern. At the left hand side, frustrated plaquettes are indicated by hatched squares. The change of spin directions in frustrated plaquette, which occurs with significant delay in comparison with that in non-frustrated plaquettes, takes place the determinate alternation of the chirality for frustrated plaquettes.

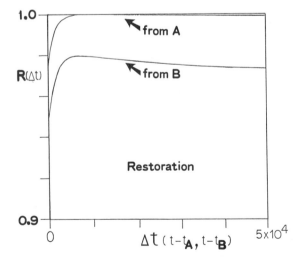

Fig. 9. Restoration of spin pattern. Here, we show how the spin pattern is restored by switching on the field at t_A or t_B indicated in Fig. 7 in response to switching off the field. Restoration is defined as the time correlation between $t=0$ and $t-t_A$ or $t-t_B$.

relation is shown in Fig. 7. Further, the time-evolution of the spin pattern is illustrated in Fig. 8. The following features are seen from Fig. 7 and Fig. 8: the decrease of the time-correlation for spins in frustrated plaquettes (denoted by the solid curve in Fig. 7) reflects the change of the chirality for those plaquettes. Moreover, its decrease occurs slower than that for spins in non-frustrated plaquettes (the dotted curve). At the end of this subsection (b), we make the following new procedure in order to investigate the irreversibility of this response; At the middle of the response process denoted by t_A or t_B in Fig. 7, we switch on the external field. Namely, we investigate the restoration of spin pattern after A or B by this procedure. In order to discuss this process in detail, the following correlation $R(\Delta t)$ is introduced;

$$R(\Delta t) = \sum' S_i(\Delta t) \cdot S_i(t=0)/N, \tag{8}$$

where $\Delta t = t - t_A$ or $t - t_B$. As easily seen from the behavior of $R(\Delta t)$ shown in Fig. 9, the response process consists both of reversible and irreversible ones: the former mainly occurs before A, while the latter occurs after A. As a conclusive remark, we would like to emphasize that the irreversible process of this system appearing with delay in comparison with reversible process originates in the motion of spins for frustrated plaquettes accompanied by the determinate change of the chirality.

Acknowledgments

The authors wish to thank Dr. K. Katsumata of the Institute of Physical and Chemical Research (RIKEN) for valuable discussion and for providing the experimental data of the response phenomena in $Rb_2Mn_{1-x}Cr_xCl_4$.

The numerical calculation was performed on a HITAC M260-K in the Data Processing Centre of Chiba University and on a HITAC M680-H/S-810 in the Computer Centre of the University of Tokyo. This work was supported by the Grant-in-Aid for Scientific Research from the Ministry of Education, Science and Culture. Further, this work was supported by the Grant-in-Aid for Research and Education from Nihon Suido Co. Ltd., Japan.

REFERENCES

1) Suzuki, M., Theory of spin glasses. *Solid State Physics* (Agne Gijutsu Center) **19**, No. 7 (1984), pp. 387–395, No. 1 (1985), pp. 30–37.
2) Ishikawa, Y., Spin glass in condensed random system. *Solid State Physics* (Agne Gijutsu Center) **20**, No. 4 (1985), pp. 229–246.
3) Morgenstern, I. and Binder, K., Evidence against spin glass order in the two-dimensional random-bond Ising model. *Physical Review Letter*, **47** (1979), pp. 1615–1618.
4) Fujimoto, K., Yoshihara, T., and Natsume, Y., Simulation of spin dynamics in the system including competition between ferro and antiferromagnetic interactions: I. Method. *Theoretical and Applied Mechanics*, **36** (University of Tokyo Press, 1987), pp. 413–420.
5) Yoshihara, T., Fujimoto, K., and Natsume, Y., Simulation of spin dynamics in the system including competition between ferro and antiferromagnetic interactions: II. Application. *Theoretical and Applied Mechanics*, **36** (University of Tokyo Press, 1987), pp. 421–427.
6) Natsume, Y., Fujimoto, K., and Yoshihara, T.: Simulation of spin dynamics in the system of competition between ferro and antiferromagnetic interactions. *The Proceeding of International Symposium on Physics of Magnetic Materials*, Sendai, 1987 (World Scientific, Shingapole, 1987), 11A–1d01, pp. 318–321.

7) Natsume, Y., Yoshihara, T., and Fujimoto, K., Computer simulation of spin dynamics for frustration spin glass state in mixed compounds including the competition between ferro- and antiferromagnetic exchange interactions. *IEEE Transactions on Magenetics* MAG-23, No. 5 (1987), pp. 2239–2241.
8) Palmer, R.G., Broken ergordicity spin glasses. *Heidelberg Colloquium on Spin Glasses, Lecture Notes in Physics*, **192** (Springer-Verlag, 1983), pp. 234–251.
9) Villain, J., Two-level system in a spin-glass model: I. General formalism and two-dimensional model. *Journal of Physics C: Solid State Physics*, **10** (1977), pp. 4793–4803.
10) Kawamura, H. and Tanemura, M., Reentrance phenomena in two-dimensional XY spin glass. *Journal of the Physics Society of Japan*, **55**, No. 6 (1986), pp. 1802–1805.
11) Kosterlitz, K.M. and Thouless D.J., Ordering, metastability and phase transitions in two-dimensional systems. *Journal of Physics C: Solid State Physics*, **6** (1973), pp. 1181–1203.

Addendum

The present authors have recently applied this spin dynamic simulation to the study of the ferromagnetic resonance phenomena in mixed magnetic compounds. Such work will be appear in Natsume, Y., Simulation of spin dynamics for FMR in compounds with competing anisotropies. *Cooperative Dynamics in Complex Physical Systems (Springer Series in Sinergetics)* ed. Takayama, H. (Springer, Heidelberg, 1988), and in Natsume, Y., Yamamoto, A., Fujimoto, K. and Yoshihara, T., Spin dynamic simulation for magnetic resonance phenomena in mixed ferromagnetic compounds. *Reviews of Solid State Science* (World Scientific).

V
MECHANICS OF NEW MATERIALS

Mechanics and Design of Composite Materials and Structures

Isao KIMPARA

Department of Naval Architecture, University of Tokyo, Tokyo

"Composite Materials" is a general term for the material that has remarkably superior properties and functions made by combining different kinds and phases of materials artificially. Composite materials are considered as a system which is a material having structure, and tailoring can be made by selection of materials and lamination design. The problems that composite materials face are extended over a wide range such as study on deformation and failure mechanisms, development of testing methods and non-destructive evaluation techniques, reliability assessments, optimum materials and structural design, etc. This paper overviews the present status of study and technology of analysis, evaluation and design of mechanical behaviors of composite materials and structures.

I. INTRODUCTION

Composite materials are generally defined by a material which exhibits some objective effectiveness (composition effect) by forming more than two phases with different physical and chemical properties (composite structure) by means of combining (composition process) more than two different constituent materials (raw materials). Composite materials are quite different from conventional monolithic materials in the fact that the material itself is generally a system having structure so that the resulting properties can be designed.

It is often difficult in composite materials to distinguish clearly between the material and the product. The composition process can be utilized in the fabrication process as it is which is often profitable in composite materials. For this reason the term "Composite Material Structures" is used. The major characteristics of composite material structures are "materials inhomogeneity and anisotropy", "integration of materials and structures" and "indivisibility of materials design and structural design".

Although these aspects must be the source of a great potentiality of composite materials, they often lead to a demerit that design, analysis and inspection are much more difficult in composite material structures than in conventional homogeneous material structures. It is, so to speak, a double-edged sword. For this reason, it is indispensable to develop the techniques for mechanical analysis, strength estimation and structural design taking the effects of inhomogeneity, anisotropy and composites properties into account, in order to utilize effectively the significant properties of composite materials.

In this paper, the status of study and technology of analyzing and designing fiber reinforced composite materials and structures are overviewed and discussed.

II. VIEW-POINTS OF CHARACTERIZATION OF COMPOSITE MATERIALS

Composite materials are composed of reinforcement and matrix, which have essentially

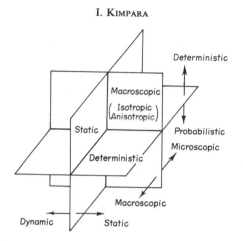

Fig. 1. View-points of characterization of composite materials.

an inhomogeneous structure. Their constitutions are divided into fiber reinforced and particle dispersed types depending upon the configuration of reinforcements. Among the former, there are continuous and discontinuous fiber reinforced types, whose mechanical properties are dependent on fiber orientations and fiber volume fractions. The characterization of fiber reinforced composite materials have been developed mainly on fiber reinforced plastics (FRP), in which the viewpoints are classified as shown in Fig. 1.

II-1. Microscopic/Macroscopic

The method for analyzing the mechanical properties of composite materials taking the effects of different phases of reinforcement and matrix into account is called "Micromechanics". On the other hand, the method regarding composite materials as homogeneous anisotropic bodies is referred to "Macromechanics". One of the major characteristics of composite materials is that the magnitude of macroscopic properties and the degree of anisotropy can be varied depending on fiber orientations and fiber volume fractions. The materials design used in composite materials is mainly based on the macroscopic materials design in which the machanical properties are adjusted to the desired values by changing or combining the kinds of reinforcing fibers, orientation angles or volume fractions.

II-2. Static/Dynamic

The mechanical properties of composite materials need to be characterized as the response to various loads, which are divided into static and dynamic problems depending on the modes of applied load. They are also classified into mechanics of elastic or viscoelastic deformation and failure behaviors, which are related one another since innumerable cumulative damages are generated even during deformation process of composite materials. It is important to clarify the characteristics of composite materials as a multi-load-path-structure examining the total path from initial yield to catastrophic fracture. As the failure modes of composite materials depend on the loading modes, in particular, under static and dynamic loads, a detailed examination is required on the material testing methods.

II-3. Deterministic/Probabilistic

The number of factors to affect a scatter of properties is generally larger in composite materials, which are composed of many different phases of constituents, than in conventional monolithic materials. From the deterministic points of view, only the average of properties is noted so that all the related parameters are treated in a deterministic manner. From the probabilistic points of view, on the other hand, a scatter of properties is taken into account so that the other parameters inevitably vary, which are analyzed in a probabilistic manner. Such a reliability engineering approach has recently been adopted extensively in the evaluation of composite materials.

There are various stages of characterization of composite materials, as shown in Fig. 1, depending on the combination of each view-point (a), (b) and (c) mentioned above: from the most simplified static/macroscopic/deterministic approach to the most sophisticated dynamic/microscopic/probabilistic approach. The last one has, however, scarcely been investigated, which would be one of topics in the field of dynamics of composite materials.

III. CHARACTERIZATION OF COMPOSITE MATERIALS

The most fundamental configuration of composite materials is an unidirectional fiber reinforced composite, in which fibers are embedded parallel in a matrix. An unidirectional composite has a remarkable orthotropy in macroscopic properties, in which fiber-direction (L), transverse-direction (T) and thichness-direction (R) are three principal axes in elastic symmetry. The macroscopic characterization of an unidirectional composite is based on the uniaxial stress-strain response in the principal directions as shown in Fig. 2. It should be noted that the deformation and failure mechanisms of an unidirectional composite are quite different according to the kind (sign) of uniaxial stresses.

These fundamental properties can be derived by means of a micromechanics model as shown in Fig. 3. Most of such micro-mechanics models are based on an idealization of an unidirectional composite into regularly-spaced arrays, typically square or hexagonal

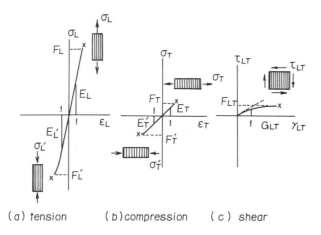

(a) tension   (b) compression   (c) shear

Fig. 2. Macroscopic response of unidirectional lamina.

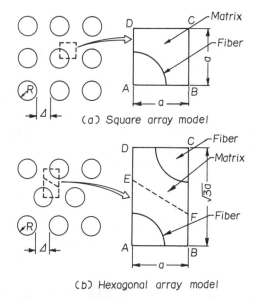

(a) Square array model

(b) Hexagonal array model

Fig. 3. Micromechanics model of unidirectional composite.

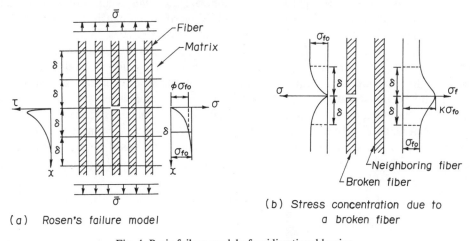

(a) Rosen's failure model

(b) Stress concentration due to a broken fiber

Fig. 4. Basic failure model of unidirectional lamina.

arrays, of elastically-dissimilar circular inclusions in a matrix. Because of the assumed periodic array spacing, a typical repeating unit can be isolated by the broken lines as shown in Fig. 3. The analysis model may be regarded as a rectangular prism, so that all surfaces should remain plane under any loading. To simulate a real uniaxial loading condition, the normal resultant forced in the surfaces of another two directions must be equal to zero and in all surfaces must be equal to zero for thermal stress analysis. These conditions can be satisfied by a generalized plane strain condition.[1]

The failure process of composite materials is a very complicated probabilistic phenomenon due to random failures in fibers, matrix and interfaces leading to a catastrophic fracture. A Monte Carlo simulation is one of the most effective methods to analyze such a

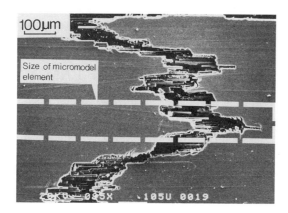

Fig. 5. Observed tensile failure pattern of unidirectional CFRP.

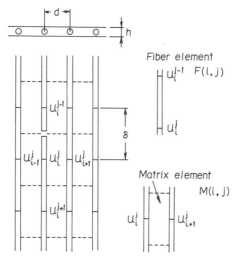

Fig. 6. New failure simulation model of unidirectional lamina.

stochastic process and several studies have so far been carried out. In most of the past studies, however, a simple failure model, as shown in Fig. 4, has been applied, in which only random fiber breaks and stress concentration in the nearest fiber to the broken one are taken into account leading to no more than a flat cleavage plane of a specimen.[2,3]

In an actual tensile failure process of unidirectional composites, not only fiber break, but also interfacial debonding between fibers and matrix and pull-out of fibers are frequently observed. Such a complex failure mechanism often leads to complicated zigzag cleavage plane of a specimen as shown in Fig. 5.[4]

Figure 6 shows a new failure simulation model of an unidirectional composite considering the effect of matrix shear failure as well as fiber breaks based on a shear-lag theory, in which a failure can occur randomly not only in fiber elements but also in matrix elements.[5] This micro model was successfully applied in simulating a complicated failure pattern as shown by the broken line in Fig. 5. A macro model composed of elements of such micro models is also proposed which was effectively used in estimating the statistical nature of strength properties of unidirectional composites of actual size.[6]

IV. MATERIALS DESIGN OF LAMINATES

An unidirectional composite has a large anisotropy so that it is generally not effectively utilized except for a special purpose. The design flexibility as a structural member is much increased if thin unidirectional composites (laminae) with arbitrary orientation angles are bonded together to form a laminate. The degree of anisotropy can be controlled by means of a laminate. For an example, if laminae are oriented at equal angles, $180°/n$ ($n \geq 3$), to the reference axis of a laminate, it is shown that the in-plane properties of a laminate is isotropic, which is called "pseudo-isotropic": in an actual application, a basic constitution of $(0°/\pm 45°/90°)_s$ is often adopted in a laminate design.

As each lamina of a laminate has generally different anisotropic elastic properties and is subjected to deformation constraints one another due to bonding, peculiar deformations such as a coupling and interlaminar stresses arise in a laminate which can not be seen in

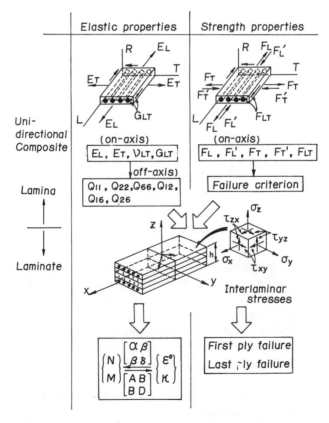

Fig. 7. Macromechanical behavior of lamina and laminate.

a lamina itself. Figure 7 shows a schematic explanation of macromechanical behavior of a laminate composed of different anisotropic laminae.

The stiffness of such a laminated composite material is obtained from the properties of the constituent laminae by a classical lamination theory. The strength of a laminate should also be based on that of its laminae. However, because of various characteristics of composite materials, it is difficult to determine a strength criterion in which all failure modes and their interactions are properly accounted for. In particular, for a laminated composite, failure of the weakest lamina (First Ply Failure: FPF) does not necessarily imply ultimate failure of the entire laminate (last ply failure).

Designing with multi-directional laminates is not straight-forward because ply angles can completely be arbitrary. However, as far as the effective rigidity is concerned, it is shown that lamination parameters can play an important role in materials design problems.[7] The key feature of the method lies in that design specifications or constraints can be graphically represented in a feasible region of lamination parameters. Optimum fiber orientation angles and stacking sequence can be obtained easily by using a lamination parameter diagram.[8] As for the strength design of a laminate, on the other hand, the situation is much more complicated as described above.

V. RELIABILITY ASSESSMENT OF COMPOSITE MATERIAL STRUCTURES

Composite material structures present an unique problem to certificate their quality and reliability. The most essential point is the fact that the material and the component are produced in the same operation, which involves a large number of manufacturing variables and opportunities for introducing defects. As unintentional and undetected deviations from the standard specifications are possible, there can be no guarantee that any two composite components are identical.

From the microscopic point of view, composites should generally contain innumerable internal defects. However, even if they have some defects, many of them are sometimes harmless depending on the defect types, locations and applied stress fields. On the other hand, in a stress-concentrated region, even a very minute defect often causes a crack growth and then arrives at a fatal size. Therefore it is impossible to estimate the strength reliability of composites until the extent that every type of defect takes part in a catastrophic fracture is predicted and examined.

Figure 8 shows a general concept of quality and reliability assessment of composite material structures. "Stress analysis" by means of micro- and macromechanics, "material strength" measured by material testings and the evaluation of effects of "environmental condition" are the three key factors in this system. The importance of non-destructive testing (NDT), inspection (NDI) and evaluation (NDE) should also be included in this system as shown in Fig. 8.

The challenge of using NDE for composites is great. The effective use of signal-processing techniques, the feasibility of incorporating several NDE techniques in one system with the aid of computerized evaluation and the development of quantitative NDE techniques can all help to overcome obstacles presently encountered in the NDE of composites.

For the above mentioned circumstances, the future technology of advanced composites

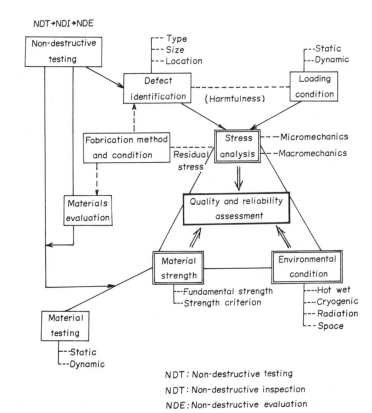

Fig. 8. Concept of quality and reliability assessment of composite materials.

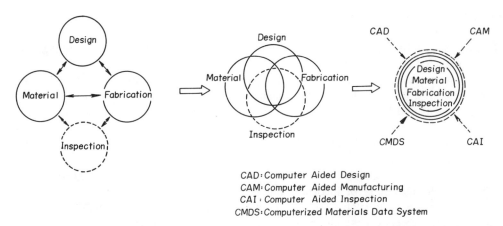

CAD: Computer Aided Design
CAM: Computer Aided Manufacturing
CAI: Computer Aided Inspection
CMDS: Computerized Materials Data System

Fig. 9. Integration of design, material, fabrication and inspection.

will have to be evolved as shown in Fig. 9. In the ultimate ideal goal, "design", "material" and "fabrication" will be incorporated into single stage as well as "inspection".

VI. CONCEPTS OF STRUCTURAL DESIGN

Composite materials have so far been treated in a similar manner as isotropic materials in most cases to substitute conventional metal structures. The directional properties of fibers have seldom been utilized effectively for the weight saving of structures. In exploiting application of composite materials, an elaborate design technique would be required so that materials design, structural design and fabrication technology should closely be entangled.

Design of composite material structures should start from materials design itself as described above. It is necessary to make the material properties well balanced for pursuing the weight saving of structures. For example, it is recognized that the current CFRP (Carbon Fiber Reinforced Plastics) has rather unbalanced properties which should be improved: in particular, higher elongation of transverse direction is needed. Figure 10 depicts the effectiveness of improvement in the transverse elongation.[9] A weight saving of more than 25% would be attained by improvement of transverse elongation from the current 0.5% to 1.0%, while only around 12% weight saving would be attained by that of longitudinal elongation from the current 1.0% to 2.0%.

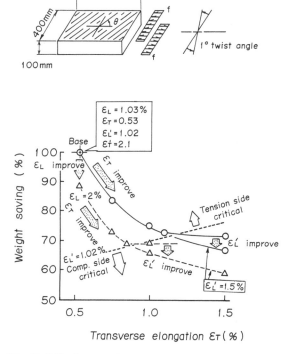

Fig. 10. Effectiveness of transverse elongation improvement.

There is also a potential for exploiting the unique directional properties of composites to achieve aeroelastic tailoring of wings for flight performance optimization beyond the practical limits of metal design. For the structural design of a tailored wing which makes the most use of an optimization computer program is mandatory to achieve a minimum weight structure while taking account of both strength and aeroelastic requirements.[9]

VII. CONCLUSIONS

Composite materials can be a kind of "new materials" if they are regarded as materials having "new functions". However it would be more appropriate that composite materials are looked upon as "materials system technology" in which different conventional materials (including new materials) are combined and united into a useful configuration of materials.

For this reason, the accumulation of detailed "know-how" such as utilization of anisotropy according to application fields would be required more and more for composite materials to be made a full use of in various fields. The systematization of composite materials technology would also be important, because the study of composite materials has a general character in itself that composes the existing science and technology fields parallel.

REFERENCES

1) Hamamoto, A., Kimpara, I., and Takehana, M., Elastic and failure behaviors in the transverse plane of unidirectional fiber reinforced composites. *J. Fac. Eng., Univ. of Tokyo (B)*, **34**, No. 2 (1977), pp. 395–407.
2) Rosen, B.W., Tensile failure of fibrous composites. *AIAA J.*, **2** (1964), pp. 1985–1991.
3) Zweben, C., Tensile failure of fiber composites. *AIAA J.*, **6** (1968), pp. 2325–2331.
4) Sato, N., Kurauchi, T., and Kamigaito, O., Fracture mechanism of unidirectional carbon-fibre reinforced epoxy resin composites. *J. Mater Sci.*, **21** (1986), pp. 1005–1010.
5) Kimpara, I., Ozaki, T., and Tsuji, N., Study on reliability assessment system of composite materials (1st report). *J. Soc. Nav. Archit. Japan*, **158** (1985), pp. 571–579.
6) Kimpara, I., Ozaki, T., and Inoue, K., Study on reliability assessment system of composite materials (4th report). *J. Soc. Nav. Archit. Japan*, **163** (1988), pp. 397–406.
7) Tsai, S.W. and Hahn, H.T., *Introduction to Composite Materials*. (Technomic Pub. Co., 1981), pp. 128.
8) Miki, M., Stiffness-based material design and Optimization for laminated composites. *Materials System*, **3** (1984), pp. 83–98.
9) Ugai, T., and Kikukawa, H., Aeroelastic tailoring study in aircraft design. *Proc. 5th Intern. Conf. Compos. Mater. (ICCM-V)*, (San Diego, 1985), pp. 1221–1232.

Optimum Design of Composite Structures

Yoichi HIRANO

Faculty of Science and Engineering, Chuo University, Kasuga, Bunkyo-ku, Tokyo

It is important to know the deformation and buckling characteristics of composite structures in detail before design. Deformations peculiar to composite structures are known to be due to anisotropy, but the physical reasons for the phenomena have not been fully explained. The present paper attempts to elucidate such phenomena, and buckling characteristics are demonstrated with regard to fiber direction. It is concluded that optimization techniques are necessary for the design of composite structures. Some examples of optimization of composite structures are presented.

I. INTRODUCTION

There are many kinds of composite materials, but only unidirectionally fiber-reinforced plastics are considered in the present paper. Reinforcing fibers are generally glass fibers, carbon fibers, and aramid fibers; common plastics are polyester and epoxy. Unidirectionally fiber-reinforced plastics have strong orthotropy in stiffness and strength, and their mechanical characteristics are very different from those of isotropic materials; therefore, an understanding of orthotropic characteristics is necessary in the design of composite structural elements and structures.

In the present paper several peculiar deformations of composite materials and structures are described and explained physically. Buckling characteristics of composite plates and cylindrical shells are discussed in terms of fiber direction. High dependence upon fiber direction of the deformations and buckling characteristics is found. It is difficult to anticipate the characteristics of composite structures before design and analysis, making optimization techniques helpful in the design of composite structures. Several examples of optimum design of composite structures are presented and discussed below.

II. MECHANICAL CHARACTERISTICS OF COMPOSITE MATERIALS AND STRUCTURAL ELEMENTS

An example[1] of mechanical characteristics of carbon fiber-reinforced epoxy are shown in Table 1. In the table the suffixes "1" and "2" indicate fiber direction and transverse direction, respectively. $F_{1,2}$ and F_{12} are tensile and shear strength, respectively. Table 1 shows that the stiffness and strength in the fiber direction are much higher than in the transverse direction and that the material has strong orthotropy. The specific mass of the material is about 1.5; therefore, the specific stiffness and the specific strength of the material in the fiber direction are higher than those of aluminum. This means that composite structures can be made lighter than aluminum structures, if they are well designed. In the present section several peculiar deformations due to orthotropy are presented and discussed.

Table 1. Material properties of carbon/epoxy.[1]

Stiffness	E_1	1.28×10^4 kgf/mm²
	E_2	9.58×10^2 kgf/mm²
	G_{12}	5.86×10^2 kgf/mm²
	ν_{12}	0.378
Strength	F_1	134 kgf/mm²
	F_2	4.64 kgf/mm²
	F_{12}	8.7 kgf/mm²

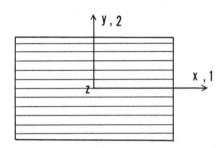

Fig. 1. Unidirectionally fiber-reinforced plastics.

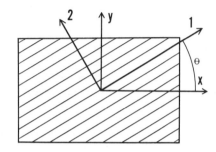

Fig. 2. Unidirectionally fiber-reinforced plastics.

II-1. Stress-Strain Relations and Coupling

A thin composite plate (Fig. 1) has a fiber direction coinciding with the x-axis. The stress-strain relations of the plate can be written as follows[2]

$$\begin{bmatrix} \sigma_1 \\ \sigma_2 \\ \tau_{12} \end{bmatrix} = \begin{bmatrix} Q_{11} & Q_{12} & 0 \\ Q_{12} & Q_{22} & 0 \\ 0 & 0 & Q_{66} \end{bmatrix} \begin{bmatrix} \varepsilon_1 \\ \varepsilon_2 \\ \gamma_{12} \end{bmatrix} \tag{1}$$

where

$$Q_{11} = E_1/(1 - \nu_{12}\nu_{21}),$$
$$Q_{12} = \nu_{12}E_2/(1 - \nu_{12}\nu_{21}),$$
$$Q_{22} = E_2/(1 - \nu_{12}\nu_{21}),$$
$$Q_{66} = G_{12}. \tag{2}$$

Eqs. 1 and 2 show that the material is orthotropic. When the fiber direction is rotated by θ around the z-axis (Fig. 2), the stress-strain relations are written as in Eq. (3):

$$\begin{bmatrix} \sigma_x \\ \sigma_y \\ \tau_{xy} \end{bmatrix} = \begin{bmatrix} \bar{Q}_{11} & \bar{Q}_{12} & \bar{Q}_{16} \\ \bar{Q}_{12} & \bar{Q}_{22} & \bar{Q}_{26} \\ \bar{Q}_{16} & \bar{Q}_{26} & \bar{Q}_{66} \end{bmatrix} \begin{bmatrix} \varepsilon_x \\ \varepsilon_y \\ \gamma_{xy} \end{bmatrix} \tag{3}$$

where

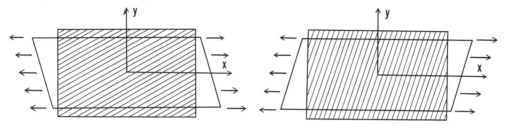

Fig. 3. Minus shear deformation.   Fig. 4. Plus shear deformation.

$$\bar{Q}_{11} = Q_{11} \cos + \theta + 2(Q_{12} + 2Q_{66}) \sin^2 \theta \cos^2 \theta + Q_{22} \sin^4 \theta$$
$$\bar{Q}_{12} = (Q_{11} + Q_{22} - 4Q_{66}) \sin^2 \theta \cos^2 \theta + Q_{12}(\sin^4 \theta + \cos^4 \theta)$$
$$\bar{Q}_{22} = Q_{11} \sin^4 \theta + 2(Q_{12} + 2Q_{66}) \sin^2 \theta \cos^2 \theta + Q_{22} \cos^4 \theta$$
$$\bar{Q}_{16} = (Q_{11} - Q_{12} - 2Q_{66}) \sin \theta \cos^3 \theta + (Q_{12} - Q_{22} + 2Q_{66}) \sin^3 \theta \cos \theta$$
$$\bar{Q}_{26} = (Q_{11} - Q_{12} - 2Q_{66}) \sin^3 \theta \cos \theta + (Q_{12} - Q_{22} + 2Q_{66}) \sin \theta \cos^3 \theta$$
$$\bar{Q}_{66} = (Q_{11} + Q_{22} - 2Q_{12} - 2Q_{66}) \sin^2 \theta \cos^2 \theta + Q_{66}(\sin^4 \theta + \cos^4 \theta). \quad (4)$$

Equation (3) states that shear strain γ_{xy} is observed when the plate is extended in the x-direction and that normal strains ε_x and ε_y are observed under shear force. These facts mean that there is coupling between shear strain and normal stresses, and between normal strains and shear stress.

There can be plus or minus shear deformation under uniaxial tension. Minus shear deformation shown in Fig. 3 is observed in most cases. Plus shear deformation shown in Fig. 4 is observed when composite materials have the following material characteristics[3]

$$G_{12} > \frac{E_1}{2(1 + \nu_{12})} \quad (5)$$

$$G_{12} < \frac{E_2}{2(1 + \nu_{21})}. \quad (6)$$

For most fiber-reinforced plastics actually used Eq. (5) is not satisfied. Some composites may satisfy Eq. (6). If the material properties of composites satisfy Eq. (6) and the fiber direction θ is greater than θ^* given by Eq. (7), plus shear deformation is observed under uniaxial tension.

$$\theta^* = \sin^{-1} \sqrt{\frac{2/E_1 + 2\nu_{12}/E_1 - 1/G_{12}}{2(1/E_1 + 1/E_2 + 2\nu_{12}/E_1 - 1/G_{12})}}. \quad (7)$$

II-2. Lamination Theory and Coupling

A single layer of unidirectionally fiber-reinforced plastics is seldom used in actual cases. Several or many layers are stacked and used as a laminate. Their fiber directions are not generally oriented in the same direction. When the fiber direction of the k-th layer is denoted by θ_k, the stress-strain relations of the layer are obtained by taking θ as θ_k in Eq. (3). Resultant forces (N) and resultant moments (M) for the laminate are given in terms of strains (ε^0) and curvatures (κ) under the Kirchhoff-Love hypothesis.[2)]

$$\begin{bmatrix} N_x \\ N_y \\ N_{xy} \end{bmatrix} = \begin{bmatrix} A_{11} & A_{12} & A_{16} \\ A_{12} & A_{22} & A_{26} \\ A_{16} & A_{26} & A_{66} \end{bmatrix} \begin{bmatrix} \varepsilon_x^0 \\ \varepsilon_y^0 \\ \gamma_{xy}^0 \end{bmatrix} + \begin{bmatrix} B_{11} & B_{12} & B_{16} \\ B_{21} & B_{22} & B_{26} \\ B_{16} & B_{26} & B_{66} \end{bmatrix} \begin{bmatrix} \kappa_x \\ \kappa_y \\ \kappa_{xy} \end{bmatrix} \tag{8}$$

$$\begin{bmatrix} M_x \\ M_y \\ M_{xy} \end{bmatrix} = \begin{bmatrix} B_{11} & B_{12} & B_{16} \\ B_{12} & B_{22} & B_{26} \\ B_{16} & B_{26} & B_{66} \end{bmatrix} \begin{bmatrix} \varepsilon_x^0 \\ \varepsilon_y^0 \\ \gamma_{xy}^0 \end{bmatrix} + \begin{bmatrix} D_{11} & D_{12} & D_{16} \\ D_{12} & D_{22} & D_{26} \\ D_{16} & D_{26} & D_{66} \end{bmatrix} \begin{bmatrix} \kappa_x \\ \kappa_y \\ \kappa_{xy} \end{bmatrix} \tag{9}$$

where A_{ij}, B_{ij}, and D_{ij} are called extensional, coupling, and bending stiffnesses, respectively. Coupling matrix (B) in Eqs. (8) and (9) causes various coupling deformations.

An example of bending deformations of a laminate under axial tension is shown in Fig. 5. The plate has two layers with fiber orientations 0° and 90°. The coupling phenomenon can be physically explained as follows. Let us separate two layers. Both layers have the same extensional strain before separation; therefore, the 0° layer which is stiffer than the 90° layer attracts more extensional force, as shown in Fig. 6. The forces in the 0° and 90°

Fig. 5. Coupling between tension and bending deformation.

Fig. 6. Two separated layers.

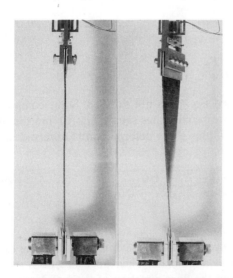

Fig. 7. Coupling between tension and twist (left, unloaded condition; right, loaded condition).[3)]

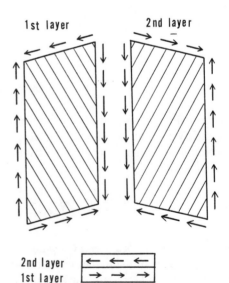

Fig. 8. Two separated layers.[3)]

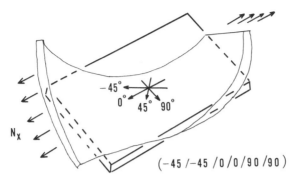

Fig. 9. Coupling between tension, bending and twist.[4]

layers in Fig. 6 compose a bending moment, but in the unseparated condition the bending moment is not applied externally. In order to obtain the original state we have to apply the bending moment in the opposite direction from the one composed of two forces in Fig. 6. Because of this applied bending moment, coupling between bending and extension can be explained.

An example of coupling between twist and extension is next given. Figure 7 shows a two-layered plate made of nylon fiber-reinforced rubber. The fiber orientation of one layer is 30° with respect to the longitudinal axis, and that of the other layer is −30°. The upper end of the plate can be rotated easily along the longitudinal axis. When the laminate is extended, the plate twists because of coupling stiffnesses in Eqs. (8) and (9). The coupling can be understood physically in the following way. If the laminate under tension is separated into two layers, shearing deformations of the two layers under tension have different signs as shown in Fig. 8 because of the different signs of the fiber directions. The two layers are bonded together in reality; therefore, we must apply shearing force to the two layers as shown in Fig. 8 to get the original state. The applied shearing force composes twisting moments at the four side edges. In the bonded state there are no twisting moments at the side edges; therefore, we have to apply twisting moments to cancel the twisting moments introduced above. The twisting moments applied to cancel the moments due to shearing force make the laminate twist.

Another example[4] of coupling is shown in Fig. 9. The laminated plate shown in the figure has six layers whose fiber orientations are −45°, −45°, 0°, 0°, 90°, and 90° from the upper surface. The laminate shows bending and twist under tension. This coupling phenomenon can be physically understood in the same way as explained above.

III. BUCKLING CHARACTERISTICS AND FIBER DIRECTIONS

It has been shown that deformation of composite structures are dependent upon fiber orientation. Buckling characteristics of composite structural elements and structures are also dependent upon fiber direction. In this section laminated plates and circular cylindrical shells are selected as examples to show the dependence of buckling stress on fiber direction.

III-1. *Uniaxial Buckling of Laminated Plates*[5]

Uniaxial buckling characteristics of boron/epoxy laminated plates are shown in Fig.

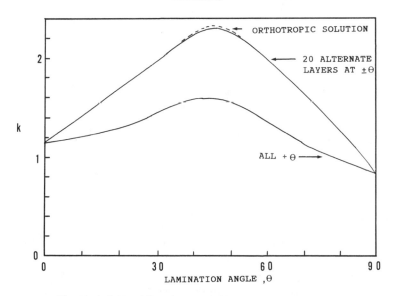

Fig. 10. Axial buckling characteristics of laminated plates.[5]

10. The aspect ratio of the plates a/b is 1.13, and the plates are simply supported at their four edges. Three curves are shown in the figure. The first curve shows the case when all 20 layers have the fiber direction in $+\theta$ with respect to the x-axis. The fiber orientation of the second one are $\pm\theta$ (angle-ply) and are symmetrical with respect to the middle surface. The third curve is the one for the case when the coupling terms D_{16} and D_{26} are taken to be zero. Nondimensional buckling loads ($k = -N_x a^2/E_1 h^3$; h is thickness) are shown versus lamination angles in Fig. 10. The figure indicates that $\theta = 45°$ gives the highest buckling load and that angle-ply laminates give higher buckling loads than all $+\theta$ laminates.

III-2. Buckling of Laminated Circular Cylindrical Shells (A)[6]

Axial buckling loads of angle-ply laminated circular cylindrical shells are shown with respect to fiber direction in Fig. 11. Two kinds of materials, glass/epoxy and boron/epoxy, are selected for calculations. In Fig. 11 S and U denote symmetrical and unsymmetrical buckling loads, respectively. All curves in the figure are symmetrical with respect to $\pm 45°$. The maximum buckling loads are obtained when θ is equal to about $\pm 20°$ and $\pm 70°$ for both materials. It is anticipated that the highest buckling load will be obtained for the laminate configurations that give the same buckling load for both symmetrical and unsymmetrical buckling modes.

III-3. Buckling of Laminated Circular Cylindrical Shells (B)[7]

Experimental axial buckling loads of laminated circular cylindrical shells are obtained and compared with the analytical ones. The length, the radius, and the thickness of the laminated cylinders are 300 mm, 100 mm, and 0.814 mm, respectively. The material used is carbon/epoxy. Three kinds of laminate configurations are chosen for comparison and shown in Table 2. The experiments were carried out by using three specimens for each laminate configuration. Experimental and analytical buckling loads are compared in Table

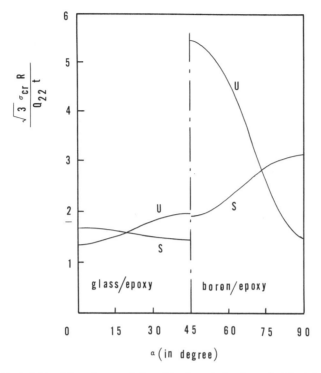

Fig. 11. Axial buckling characteristics of laminated circular cylindrical shells.[6]

Table 2. Laminate configurations.[7]

	Innermost	Outermost
No. 1	+20°/−20°/0°/0°/+40°/−40°	
No. 2	+20°/−20°/+40°/−40°/0°/0°	
No. 3	+40°/−40°/+20°/−20°/0°/0°	

Table 3. Theoretical and experimental buckling loads.[7]

	Theoretical	Experimental
No. 1	8.41 tonf	6.37 tonf
No. 2	6.46 tonf	4.89 tonf
No. 3	4.30 tonf	3.44 tonf

3. The experimental values shown are the mean values of three specimens. It is observed that the buckling load of specimen no. 1 is almost twice that of specimen No. 3 and that the difference in stacking sequence gives a large difference.

The three examples above show that laminate configurations have great influence upon buckling load. Optimum fiber directions for angle-ply laminates are obtained in the first two examples by parametric surveys. For more general and complex composite structures we must utilize optimization techniques to find the best laminate configuration.

IV. OPTIMUM DESIGN OF COMPOSITE STRUCTURES

Optimization techniques can be used to design good composite structures. Reference 8 is a review of optimum structural design. Methods of structural optimization can be separated in two techniques: direct and indirect optimization techniques. Direct optimization techniques are mathematical optimization techniques that minimize or maximize objective functions under various constraints with respect to design parameters. Indirect optimization techniques are the methods that determine design parameters so that certain optimality criteria are satisfied. Fully stressed design and simultaneous failure mode design are examples of optimality criteria methods. Direct methods are logically clear, but can generally treat only 20 or 30 design variables for structural optimization problems.

IV-1. Optimum Design of Laminated Plates under Uniaxial Compression[9]

The laminated plates considered are composed of N layers. It is assumed that 50% of fibers in the i-th layer are oriented in the $+\alpha_i$ direction and that the other fibers are oriented in the $-\alpha_i$ direction. The thickness of each layer is assumed to be the same. When the plates are simply supported at their four edges, a buckling formula is obtained in a closed form. The formula is a function of fiber directions α_i ($i=1, 2, \ldots, N$) and buckling wave numbers m and n. The laminate configuration that gives the highest uniaxial buckling load is found by a mathematical optimization technique (Powell's method). An iterative process is shown in Fig. 12 for the case of a four-layered square plate. The ordinate of the figure is the fiber directions α_i and the nondimensional buckling load ϕ. The initial values of the fiber directions α_i are taken to be 10°, 20°, 30°, and 40°. At the optimum point all fiber

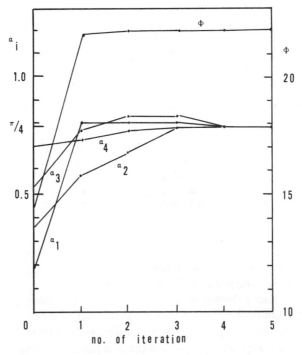

Fig. 12. Variation of α_i and ϕ with number of iteration.[9]

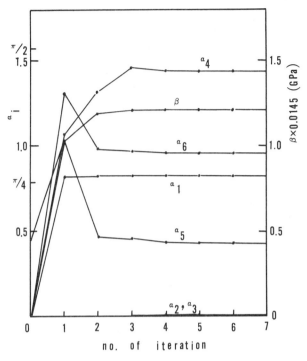

Fig. 13. Variation of α_i and β with number of iteration.[7]

Table 4. Summary of optimum fiber directions for the case of axial compression.[9]

No. of layers	k	a/b	α_1	α_2	α_3	α_4	α_5	α_6	ϕ
3	0	1.0	45°	45°	45°				22.000
4	0	1.0	45°	45°	45°	45°			22.000
6	0	0.5	0°	0°	0°	0°	0°	0°	42.171
6	0	0.8	38°	38°	38°	38°	38°	38°	23.154
6	0	1.0	45°	45°	45°	45°	45°	45°	22.000
6	0	1.25	49.9°	51.0°	48.6°	48.8°	51.0°	49.9°	23.116
6	0	2.0	45°	45°	45°	45°	45°	45°	22.000
6	0.5	0.5	7.9°	3.8°	22.1°	0.5°	7.4°	12.3°	37.024
6	0.5	1.0	45°	45°	45°	45°	45°	45°	14.667
6	0.5	2.0	67.1°	56.4°	56.2°	55.5°	64.0°	61.4°	12.556
6	1.0	1.0	45°	45°	45°	45°	45°	45°	11.000
6	1.0	2.0	71.6°	68.1°	77.5°	61.2°	71.1°	74.1°	8.051

directions converge to 45° which gives much higher buckling stress than the initial values.

Calculated results for various cases are shown in Table 4, where k and a/b are the ratio of the loads in the y-direction with respect to the x-direction and the aspect ratio of the plates, respectively. Similar calculations for shear buckling will be found in Reference 10. Optimization of hat-stiffened composite plates is treated in Reference 11.

IV-2. Optimization of Laminated Cylindrical Shells under Axial Buckling[7]

The circular cylindrical shells considered are made of boron/epoxy. It is assumed that the number of layers is six and that the thickness of the cylinders is 0.254 mm. Fifty percent of fibers in the i-th layer assumed to be oriented in the $+\alpha_i$ direction with respect to the axis of the cylinders. The rest are assumed to be oriented in the $-\alpha_i$ directions. Under this assumption buckling formulae for axisymmetrical and unsymmetrical modes can be obtained in closed forms. The formulae are the functions of fiber directions and buckling wave numbers. Fiber direction that give the highest buckling load is obtained by Powell's method. Figure 13 shows an example of the convergence of numerical calculations. In the figure β is the reduced buckling stress. All initial values of α_i are taken to be equal to $0°$. The suffix 1 of α_1 denotes the innermost layer. After iterations each fiber direction does not converge to the same value as for the case of flat plates. Numerical results for various initial values are shown in Table 5. The table shows that the optimum fiber directions and the buckling loads are different from each other in dependence on the initial values of the fiber directions. This is caused by the fact that the buckling formulae have local minima. In Table 5 β_s and β_u denote symmetrical and unsymmetrical reduced buckling stress. For almost all cases the values of β_s are equal to those of β_u at the optimum points.

IV-3. Aeroelastic Tailoring[12]

Composite structures have peculair deformation characteristics not observed in structures of isotropic materials. If we use these characteristics positively, we can build structures with delicate performances. In other words the characteristics make tailoring design possible. An example will be shown for the design of forwardswept wings of aircraft.

Table 5. Optimum fiber directions for cylindrical shells.[7]

		α_1	α_2	α_3	α_4	α_5	α_6	β_s	β_u
		(in degree)						(in GPa)	
1	S	0.0	0.0	0.0	0.0	0.0	0.0	65.7	31.6
	F	47.4	0.3	0.2	82.3	24.8	54.9	83.1	83.1
2	S	30.0	30.0	30.0	30.0	30.0	30.0	48.3	49.3
	F	34.1	0.0	0.0	29.6	30.0	46.0	69.2	69.8
3	S	45.0	45.0	45.0	45.0	45.0	45.0	40.9	42.1
	F	45.0	45.0	45.0	45.0	45.0	45.0	40.9	42.1
4	S	45.0	45.0	0.0	0.0	45.0	45.0	53.5	55.0
	F	23.8	67.5	4.4	33.5	43.0	51.7	84.5	84.5
5	S	0.0	0.0	45.0	45.0	0.0	0.0	82.4	52.0
	F	55.5	1.6	82.0	−45.1	−35.3	131.8	86.0	86.0
6	S	90.0	90.0	90.0	90.0	90.0	90.0	65.7	31.6
	F	137.4	90.3	90.2	172.3	114.7	144.9	83.1	83.1
7	S	90.0	0.0	90.0	0.0	90.0	0.0	110.0	39.8
	F	130.5	9.1	92.7	7.0	126.2	40.2	85.3	85.3
8	S	90.0	90.0	90.0	0.0	0.0	0.0	76.4	30.2
	F	152.1	114.9	85.9	4.9	27.2	62.7	85.8	85.8
9	S	0.0	0.0	0.0	90.0	90.0	90.0	85.3	30.2
	F	45.3	5.3	0.6	91.2	180.3	124.5	82.3	82.3
10	S	10.0	20.0	30.0	40.0	50.0	60.0	91.6	50.6
	F	23.5	−10.2	16.0	−0.0	0.3	66.5	72.0	72.0

S: Starting values, F: Final optimum values.

Strength and stiffness are necessary for aircraft wings. For the present problem the strength of the wings is first satisfied by a modified, fully stressed design method. After that aeroelastic characteristics are satisfied by applying an indirect optimization technique. The technique used in the present example is the method that finds an active constraint first and treats the constraint as an equality constraint. An optimization problem with equality constraints can be solved by Lagrange's multiplier method. Iterative equations are derived

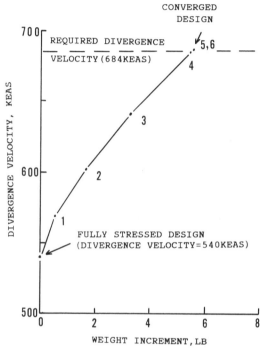

Fig. 14. Resizing steps for increased divergence velocity.[12]

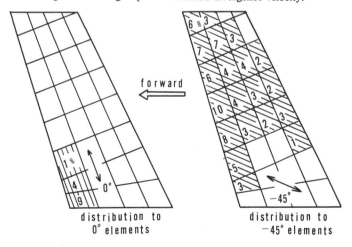

Fig. 15. Distribution of weight increment among elements resized for increased divergence velocity[12] (percentages shown are for both upper and lower covers).

from the method. For a forwardswept wing a divergence speed is now taken to be an active constraint. Upper and lower surfaces of the wing are made of composite laminates. Their fiber directions are selected to be 0°, +45°, −45°, and 90°. The thickness of the layers that corresponds to the above directions is taken as a design parameter. The process of iterative calculation is shown in Fig. 14. The initial point of the iteration is obtained from a modified, fully stressed design method. At the initial point a prescribed divergence speed is not satisfied. An iterative calculation is made to satisfy the divergence constraint with minimum mass increase. The added layer thicknesses to the fully stressed design are shown in Fig. 15. The figure shows that a −45° layer must be added to the upper and the lower surfaces of the wing to satisfy the divergence criterion. It can be understood that the addition of the −45° layer gives downward twist of the wing under bending and makes the divergence speed higher, and that the phenomenon is physically similar to the example shown in Fig. 7. For further information References 13 and 14 are useful.

V. CONCLUSIONS

Deformations and buckling characteristics of composite structures have been discussed. It is found that fiber directions have great influence upon these characteristics. It is concluded that the application of optimization techniques will be necessary for the design of composite structures. Several examples of optimum design of composite structures show the usefulness of optimization techniques.

REFERENCES

1) Iyama, H., Nomoto, M., and Hirano, Y., Tensile strength of orthotropic plates with loaded and unloaded holes. *J. Jpn. Soc. Composite Materials*, Vol. 13, No. 4 (1987), pp. 169–179.
2) Jones, R.M., *Mechanics of Composite Materials*, Scripta Book Co. (1975).
3) Hirano, Y., Iino, A., and Iyama, H., Extension-twisting coupling of two-layered angle-ply laminates. *Materials System*, Vol. 5 (1986), pp. 59–64.
4) NASA/Lockheed, L-1011 advanced composite fin.
5) Ashton, J.E. and Whitney, J.M., *Theory of Laminated Plates*, Technomic Publishing Co. (1970), p. 93.
6) Hirano, Y., Buckling of angle-ply laminated circular cylindrical shells. *J. Appl. Mech.*, Vol. 46, No. 1 (1979), pp. 233–234.
7) Hirano, Y., Optimization of laminated composite cylindrical shells for axial buckling. *Trans. Jpn. Soc. Aero. Space Sci.*, Vol. 26, No. 73 (1983), pp. 154–162.
8) Hirano, Y., Structural optimization. *J. Jpn. Soc. Aero. Space Sci.*, Vol. 32, No. 360 (1984), pp. 46–51.
9) Hirano, Y., Optimum design of laminated plates under axial compression. *AIAA J.*, Vol. 17, No. 9 (1979), pp. 1017–1019.
10) Hirano, Y., Optimum design of laminated plates under shear. *J. Composite Materials*, Vol. 13, Oct. (1979), pp. 329–334.
11) Stroud, W.J. and Agranoff, N., Minimum-mass design of filamentary composite panels under combined loads. *NASA TN D-8257* (1976).
12) Lerner, E. and Markowitz, J., An efficient structural resizing procedure for meeting static aeroelastic design objectives. *J. Aircraft*, Vol. 16, No. 2 (1979), pp. 65–71.
13) Herz, T.J., Shirk, M.H., Ricketts, R.H., and Weisshaar, T.A., On the track of practical forward-swept wings. *Astronautics Aeronautics*, Jan. (1982), pp. 40–52.
14) Shirk, M.H., Herz, T.J., and Weisshaar, T.A., Aeroelastic tailoring—theory, practice, and promise. *J. Aircraft*, Vol. 23, No. 1 (1986), pp. 6–18.

Calculation of the Two-Dimensional Effective Thermal Conductivity of Media with Regularly Dispersed Parallel Cylinders

Nozomu WATARI and Nobunori OSHIMA

College of Industrial Technology, Nihon University, Chiba

The present paper aims to estimate the effective thermal conductivity of two-dimensional media with regularly dispersed materials by applying the boundary element method. Media with parallel cylinders dispersed at square lattice points are considered. In the case of regularly arranged circular cylinders, the present authors have also developed an exact analytical method (the connection method).

The present method is compared with the previous one by carrying out numerical calculation, and the result is satisfactory. The method is also applied to analyzing the case of cylindrical materials with several other shapes of cross-section.

I. INTRODUCTION

In treating problems of thermal conduction in dispersed media, it is usually sufficient to calculate the effective thermal conductivity instead of analyzing the detailed thermal field. The present paper aims to estimate the effective thermal conductivity of media with parallel cylindrical bodies dispersed regularly by applying the boundary element method.[1,2] It is assumed that parallel cylinders are arranged in a square lattice and that two-dimensional thermal conduction takes place in the plane normal to these cylinders.

In the case of regularly arranged circular cylinders, we can carry out an exact theoretical analysis, and the present authors have also proposed an exact analytical method (the connection method).[3] Effective and accurate results have been obtained in the case of square arrangement.

However, in the case where materials with rectangular cross-section are dispersed, it is difficult to estimate theoretically the effective thermal conductivity. Therefore, it is necessary to develop as accurate an approximate method as possible. In the present paper, we compute directly the approximate values of heat flux across the boundary of an elementary domain by applying the boundary element method. The effective thermal conductivity is obtained from these values. The present method is extremely effective, since it is obtained without calculating the values of T (temperature) and q (heat flux) at interior points.

The boundary element method to be used will be based on constant element approximation. Considering the boundary conditions, simultaneous linear equations are obtained. The approximate values of heat flux q or temperature T are obtained by solving these equations. All the coefficients of these equations vary with the volume fraction of dispersed cylinders; however, they can be estimated analytically by calculating for the only value of the volume fraction.

The present method is compared with the connection method. In the case where a number of small materials are dispersed randomly in a continuous media, the cell-model method is efficient.[4] The result by this method is compared with the present one. We think that these examinations are useful in developing the approximate method to be applied to cylindrical materials with several other shapes of cross-section.

II. MODELLING OF ANALYTICAL DOMAIN AND BASIC RELATION

We shall consider the case of parallel cylinders in a square lattice (Fig. 1). Let a be the radius of a cylinder, $2R$ the distance between axes of two neighboring cylinders, k_d the thermal conductivity of a dispersed cylinder and k_c the thermal conductivity of continuous media.

We shall assume that the macroscopic temperature gradient is in the direction of the x-axis. From the symmetry and linearity of the problem, it is sufficient to obtain the solution for the temperature gradient in the x-direction and the y-direction. The latter is obtained by transforming the former symmetrically with respect to the line $x=y$. Then the solution to arbitrary directions of thermal conduction is obtained by combining the above two solutions linearly.

In this case, if we make a square domain with side $2R$ around a cylinder, it forms a unit of domain of whole fields. We shall call this unit domain $\bar{D}$ a "cell". $QQ_1Q_1'Q'Q$ in Fig. 2 denotes the cell. From the symmetry of temperature distribution, it is enough to study the domain $OAQBO$, which is one-fourth of the cell. OA and BQ are axes of symmetry and OB and AQ are axes of antisymmetry. Then, boundary conditions of the cell are given by

$$T = 0 \quad \text{on} \quad OB,$$
$$T = RX \quad \text{on} \quad AQ,$$
$$\partial T/\partial y = 0 \quad \text{on} \quad OA \text{ and } BQ \qquad (1)$$

where T is the temperature and X is a representative gradient of average temperature. In the case of two-dimensional steady thermal conduction, the temperature T is governed by the two-dimensional Laplace equation:

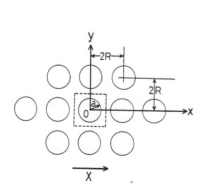

Fig. 1. Medium with dispersed parallel cylinders.

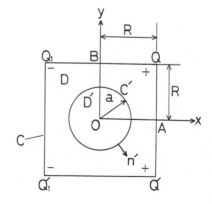

Fig. 2. Illustration of a cell $\bar{D}$.

$$\Delta T = 0. \tag{2}$$

Let us consider the cell $\bar{D}$, divided into cylinder D' (internal domain) and a continuous medium D (external domain). The interface between the two domains will be called C', which is a circle of radius a. On the interface C', across which thermal conductivity k is discontinuous, the following conditions are demanded for the connection of both sidefield solutions:

$$T = T', \tag{3}$$

$$q = \alpha q', \quad \alpha = k_d/k_c \tag{4}$$

where

$$q = \partial T/\partial n$$

and n' is the normal of C' outward from D'. q means the heat flux, except for multiplication of conductivity.

Let k_{eff} be effective thermal conductivity which is defined by

$$k_{eff}/k_c = \int_A^Q (\partial T/\partial x) dy \Big/ RX \tag{5}$$

(see Fig. 2). The effective thermal conductivity will be formulated as a function of the volume fraction z_d, which is given by

$$z_d = \pi a^2/4R^2 \tag{6}$$

From the above relation we have

$$a/R = 2(z_d/\pi)^{1/2}. \tag{7}$$

In the case of a square lattice arrangement, we have

$$z_d \leq \frac{\pi}{4} \doteq 0.78540. \tag{8}$$

III. NUMERICAL ANALYSIS

Assume that the body is two-dimensional and its boundary C is divided into N boundary elements $E_j (j = 1, 2, \ldots, N)$, as shown in Fig. 3. The node P_j is the middle point of each element E_j for the constant element formulation. The values of T and q are assumed to be constant on each element and equal to the values at the mid-node of the element.

For a given point P_i, the following discretized form is obtained:

$$\sum_{j=1}^{N} h_{ij} T_j = \sum_{j=1}^{N} g_{ij} q_j. \tag{9}$$

where

$$h_{ij} = \frac{\delta_{ij}}{2} + \frac{1}{2\pi} \int_{E_j} \frac{\partial}{\partial n}\left(\log \frac{1}{r}\right) ds, \tag{10}$$

$$g_{ij} = \frac{1}{2\pi} \int_{E_j} \log \frac{1}{r} ds \tag{11}$$

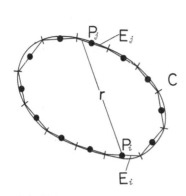
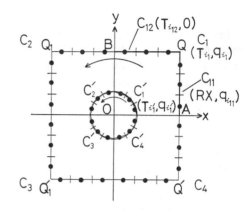

Fig. 3. General pattern of constant element discretization and the location of nodes for boundary element method.

Fig. 4. Discretization and nodes of the C and C' by the constant element method. Boundary conditions and direction of integration contour are shown on the figure.

and δ_{ij} denotes Kronecker's delta, r being the distance between the mid-node P_i and a point on the element E_j.

Let $C_k(k=1, \sim, 4)$ be the external surface (circumference of the cell) belonging to each quadrant and C'_k be the interface (circle C') to each quadrant, as shown in Fig. 4.

Let M be the number of boundary elements on C_{11} and C_{12} and M' be that on C'_1. The total number of mid-nodes is $8M+4M'$ ($M=M'=3$ in Fig. 4). These points are located in symmetry with respect to origin and numbered in anticlockwise order.

Let us analyze the two domains separately. The equations corresponding to domain D can be written as

$$\sum_j h_{ij}T_j + \sum_{j'} h_{ij'}T_{j'} = \sum_j g_{ij}q_j + \sum_{j'} g_{ij'}q_{j'} \tag{12}$$

$$\sum_j h_{i'j}T_j + \sum_{j'} h_{i'j'}T_{j'} = \sum_j g_{i'j}q_j + \sum_{j'} g_{i'j'}q_{j'} \tag{13}$$

where $1 \leq i,j \leq 8M$ (on C), $8M+1 \leq i',j' \leq 8M+4M'$ (on C') T_j and q_j are the temperature and heat flux on the external surface of domain D, and $T_{j'}$ and $q_{j'}$ the temperature and heat flux on the interface C', considering that it belongs to domain D.

For domain D' we have

$$\sum_{j'} h'_{i'j'}T'_{j'} = \sum_{j'} g'_{i'j'}q'_{j'} \tag{14}$$

where $T'_{j'}$ and $q'_{j'}$ are the temperature and heat flux on the interface C', considering that it belongs to domain D'.

Because of the symmetry of the problem, it is enough to consider only a quarter of all nodes, and we can reduce unknowns. From the symmetry or antisymmetry of temperature distribution, we have

$$T_{i_2} = -T_{i_1}, \ T_{i_3} = -T_{i_1}, \ T_{i_4} = T_{i_1}$$
$$T_{i'_2} = -T_{i'_1}, \ T_{i'_3} = -T_{i'_1}, \ T_{i'_4} = T_{i'_1} \tag{15}$$

where i_k and i'_k denote the corresponding mid-nodes belong to C_k and C'_k ($k=1, \sim, 4$) respectively. Similarly, Eq. (15) is also valid for q.

From the above relations, Eqs. (12), (13), and (14) can be written as

$$\sum_j H_{ij} T_j + \sum_{j'} H_{ij'} T_{j'} = \sum_j G_{ij} q_j + \sum_{j'} G_{ij'} q_{j'} \tag{16}$$

$$\sum_j H_{i'j} T_j + \sum_{j'} H_{i'j'} T_{j'} = \sum_j G_{i'j} q_j + \sum_{j'} G_{i'j'} q_{j'} \tag{17}$$

$$\sum_{j'} H'_{i'j'} T_{j'} = \sum_{j'} G'_{i'j'} q'_{j'} \tag{18}$$

where $1 \leq i, j \leq 2M$, $2M + 1 \leq i', j' \leq 2M + M'$.

and

$$H_{i'j} = h_{i'j_1} - h_{i'j_2} - h_{i'j_3} + h_{i'j_4}, \tag{19}$$

$$H_{i'j'} = h_{i'j'_1} - h_{i'j'_2} - h_{i'j'_3} + h_{i'j'_4}, \tag{20}$$

$$G_{i'j} = g_{i'j_1} - g_{i'j_2} - g_{i'j_3} + g_{i'j_4}, \tag{21}$$

$$G_{i'j'} = g_{i'j'_1} - g_{i'j'_2} - g_{i'j'_3} + g_{i'j'_4}. \tag{22}$$

Similarly, Eqs. (19) through (22) are also valid for i

and

$$H'_{i'j'} = H_{i'j'}, \tag{23}$$

$$G'_{i'j'} = G_{i'j'} \tag{24}$$

where

$$j_1 = j, \quad j_2 = 4M - j + 1, \quad j_3 = 4M + j, \quad j_4 = 8M - j + 1, \tag{25}$$

$$j'_1 = 6M + j', \quad j'_2 = 10M + 2M' - j' + 1,$$

$$j'_3 = 6M + 2M' + j', \quad j'_4 = 10M + 4M' - j' + 1. \tag{26}$$

From Eq. (18), it follows that

$$q'_{j'} = \sum_{l'} \sum_{k'} A_{j'k'} H'_{k'l'} T'_{l'}, \tag{27}$$

where $(A_{j'k'}) = (G'_{i'j'})^{-1}$. $(\ )^{-1}$ denotes the inverse matrix.

From the conditions in Eqs. (3) and (4), it follows that

$$q_{j'} = \sum_{l'} B_{j'l'} T_{l'}, \tag{28}$$

where

$$B_{j'l'} = \alpha \sum_{k'} A_{j'k'} H'_{k'l'} \quad (\alpha = k_d/k_c). \tag{29}$$

Substituting Eq. (28) into Eqs. (16) and (17),

$$-\sum_j G_{ij} q_j + \sum_j H_{ij} T_j + \sum_{l'} D_{il'} T_{l'} = 0, \tag{30}$$

$$-\sum_j G_{i'j} q_j + \sum_j H_{i'j} T_j + \sum_{l'} D_{i'l'} T_{l'} = 0. \tag{31}$$

Equations (30) and (31) can be written together as follows:

$$-\sum_j G_{Ij}q_j + \sum_j H_{Ij}T_j + \sum_{I'} D_{II'}T_{I'} = 0 \qquad (32)$$

where

$$D_{II'} = H_{II'} - \sum_{j'} G_{Ij'}B_{j'I'}. \qquad (33)$$

$$(1 \leq I \leq 2M + M')$$

From the boundary conditions (1),

$$T_{j_{11}} = RX \quad (\text{on } C_{11}),$$
$$q_{j_{12}} = 0 \quad (\text{on } C_{12}) \qquad (34)$$

where $1 \leq j_{11} \leq M$, $M + 1 \leq j_{12} \leq 2M$.

Considering the boundary conditions in Eq. (34) and transposing all unknowns to the left-hand side, the following system of equations is obtained:

$$-\sum_{j=1}^{M} G_{Ij}q_j + \sum_{j=M+1}^{2M} H_{Ij}T_j + \sum_{j'=2M+1}^{2M+M'} D_{Ij'}T_{j'} = -RX \cdot \sum_{j=1}^{M} H_{Ij} \qquad (35)$$

where $1 \leq I \leq 2M + M'$.

Equation (35) is a simultaneous linear equations with $2M + M'$ unknowns.

By solving Eq. (35), q_j/RX $(j=1, \ldots, M)$ are obtained. Then k_{eff}/k_c is calculated by Eq. (5).

However, the integration is replaced by the following summation:

$$k_{eff}/k_c = \sum_{j=1}^{M} q_j \bigg/ M, \qquad (36)$$

since the discretization is based on the constant element method.

IV. SIMPLIFICATION OF CALCULATION OF COEFFICIENTS

The coefficients in Eq. (35) vary with the volume fraction z_d. However, we can simplify part of the calculation of coefficients as follows. First, the coefficients corresponding to the external surface C depend only on M but are independent of the volume fraction z_d.

Secondly, the coefficients $h_{i'j'}$, $g_{i'j'}(2M+1 \leq i', j' \leq 2M+M')$ corresponding to C'(circle), if $z_d = K \times 0.1$ $(1 \leq K \leq 5\pi/2)$, vary with the increase of K.

However, let L_K be the length of elements on the circle C', r_K be the distance between mid-node and element and n'_K be the normal of C' outward from D'. Then, it is evident that

$$\partial r_K/\partial n'_K = \partial r_1/\partial n'_1.$$

From the above relations and Eq. (20), we have

$$(H_{i'j'})_K = (H_{i'j'})_{K=1} \qquad (37)$$

where $(\ )_{K=1}$ denotes the value for $K=1$.

From Eq. (10), we have

$$(g_{i'j'_1})_K = -\frac{1}{2\pi}(\log \sqrt{K}) \cdot L_K + \sqrt{K} \cdot (g_{i'j'_1})_{K=1}. \qquad (38)$$

Equation (38) is also valid for other j'_k ($k=2, 3, 4$).
From Eqs. (38) and (22) we obtain

$$(G_{i'j'})_K = \sqrt{K} \cdot (G_{i'j'})_{K=1}. \tag{39}$$

Therefore, it is sufficient for both coefficients corresponding to C' (circle) to calculate the values for $K=1$ ($z_d=0.1$). Equations (37) and (39) are also applied to the case where the shape of the cross-section of the dispersed cylindrical material has symmetry with respect to both x- and y-axis. The cross-coefficients between C and C' have to be calculated for every K.

V. NUMERICAL CALCULATION AND EXAMINATION

V-1. Circular Cylinder

We first give the volume fraction z_d, from which the size of the cylinder is determined. Next, we take $M = M'$ in the present paper. The integrals h_{ij} and g_{ij} (Eqs. (10) and (11)) are calculated using a 4-point Gauss integration formula, and the coefficients in Eq. (35) are formed by the method mentioned above (section IV). Finally, solving Eq. (35), the effective thermal conductivity k_{eff}/k_c is determined by Eq. (36). The result by the present method is plotted in Fig. 5.

For $z_d > \pi/4$, we can take the pattern as shown in Fig. 6. The phase considered to be continuous so far is enclosed by crushed cylinders and becomes the discrete phase for $z_d > \pi/4$. Though the shape of the discrete phase is not a circle, the scheme of numerical calculation mentioned in the preceding sections can be applied, except that the coordinate values of nodes on the interface are modified. The result is plotted as line 4) of Fig. 5 ($0.8 \leq z_d \leq 1.0$).

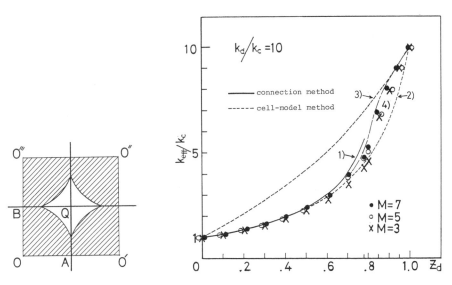

Fig. 5. The effective thermal conductivity k_{eff}/k_c vs. the volume fraction z_d. M denotes the number of boundary elements on half of one side of the cell. (circular cylinders)

Fig. 6. Illustration of a cell for $z_d > \pi/4$.

The solid line 1) is the result obtained by an exact analytical method (the connection method). The broken lines 2) and 3) are the results obtained by the cell model method. 2) is the result in the case where a number of cylinders are dispersed randomly, and the effective thermal conductivity is easily calculated by Eq. (40).

In the case of Fig. 6, assume that the central domain of the thermal conductivity k_c is replaced by a cylinder with the same cross-sectional area. Then the pattern becomes the one shown in Fig. 1, where k_c and k_d are exchanged, and z_d and $z_c = 1 - z_d$ are likewise exchanged.

The result is shown in line 3) of (Fig. 5) and can be calculated by Eq. (41).

$$k_{eff}/k_c = (1 + K_1)/(1 - K_1), \qquad (40)$$

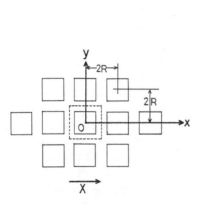

Fig. 7. Medum with dispersed parallel cylinders with square cross section.

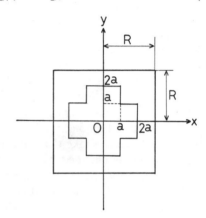

Fig. 8. Cross shape.

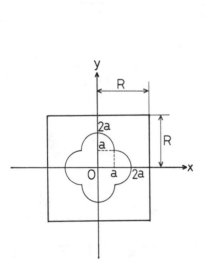

Fig. 9. Cross shaped blossom.

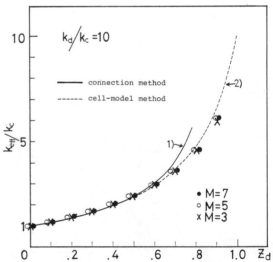

Fig. 10. The effective thermal conductivity k_{eff}/k_c vs. the volume fraction z_d. M denotes the number of boundary elements on half of one side of the cell. (square shape)

Table 1. Effective thermal conductivity k_{eff}/k_c vs. the volume fraction z_d for $M=7$.
M denotes the number of boundary elements on half of one side of the cell.

z_d	⊕	⊞	✠	✿	Random (cell-model-2))
0.1	1.1789	1.1976	1.2020	1.1900	1.1782
0.2	1.3894	1.4294	1.4465	1.4196	1.3913
0.3	1.6460	1.7052	1.7556	1.7131	1.6506
0.4	1.9681	2.0369	2.1645	2.1148	1.9730
0.5	2.3917	2.4445	2.7457	2.7351	2.3846
0.6	2.9911	2.9623	3.6752		2.9286
0.7	3.9559	3.6499	5.5439		3.6809
0.78	4.8469				
0.8	5.2794	4.6199			4.7895
0.85	6.9177				
0.9	8.0606	6.1227			6.5862
0.95	9.0682				
1.0	10.026				10.000

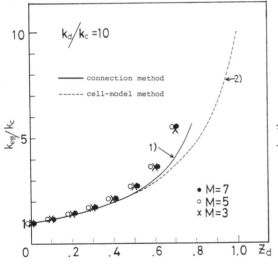

Fig. 11. The effective thermal conductivity k_{eff}/k_c vs. the volume fraction z_d. M denotes the number of boundary elements on half of one side of the cell. (cross shape)

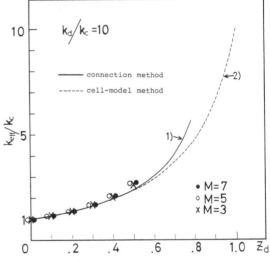

Fig. 12. The effective thermal conductivity k_{eff}/k_c vs. the volume fraction z_d. M denotes the number of boundary elements on half of one side of the cell. (cross shaped blossom)

where
$$K_1 = \frac{k_d - k_c}{k_d + k_c} \cdot z_d.$$

$$k_{eff}/k_c = (k_d/k_c) \cdot (1 + K'_1)/(1 - K'_1), \tag{41}$$

where
$$K'_1 = \frac{k_c - k_d}{k_c + k_d}(1 - z_d).$$

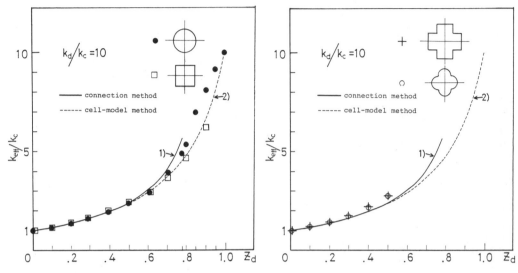

Fig. 13. Comparison of Fig. 6 and Fig. 10 for $M=7$. M denotes the number of boundary elements on half of one side of the cell.

Fig. 14. Comparison of Fig.11 and Fig. 12 for $M=7$. M denotes the number of boundary elements on half of one side of the cell.

It is enough for convergence of the present method to take $M=7$ (M is the number of boundary elements on C_{11}), and it is noted that the result by the present method almost coincides with an exact analytical method (the connection method) and a cell model method up to $z_d = 0.7$, as shown in Fig. 5.

V-2. Cylindrical Materials with Several other Shapes of Cross-Section

In this section, we shall calculate the effective thermal conductivity in the case where a number of cylindrical materials with shapes of cross-section as shown in Figs. 7, 8, and 9 are dispersed in a square lattice. We can carry out the elimination of k_{eff}/k_c using the same program as in the case of a circular cylinder, except that the coordinate values and the number (Eq. (26)) of nodes on the interface and the volume fraction z_d (Eq. (6)) are modified. The results are shown in Table 1 and Figs. 10, 11, and 12.

Note that the symbols ×, ○, and ● are marked in Fig. 5 and Figs. 10 through 12 on locations shifted slightly horizontally in order to avoid closely overlapping illustration.

In the case where a number of small materials are dispersed randomly, Eq. (40) is obtained easily by the simple cell model method where the dispersed materials are approximated by cylinders.[5]

The result is dependent only on the volume fraction z_d. We can realize the effectiveness of the above-mentioned method from Fig. 5 and Fig. 10.

VI. CONCLUSION

In the present paper, heat fluxes going through the cell of composite material with parallel cylinders arranged in a square lattice are calculated approximately by applying the boundary element method, and the effective thermal conductivity is obtained easily.

Making use of symmetry, the reduction of unknowns is carried out. The coefficients in the simultaneous linear equations vary with the increase of the volume fraction z_d. However, we can get a part of coefficients without calculation from those of a standard case for any other z_d. This simplification is also applicable to the case where the dispersed cylindrical material has symmetry with respect to both x- and y-axis.

In the case of a circular cylinder, it is enough for convergence of the present method to take $M = 7$, and the result by the present method almost coincides with an exact analytical method (the connection method) up to $z_d = 0.7$. The present method is also applied to analyze the case of cylindrical materials with several other shapes of cross-section.

REFERENCES

1) Brebbia, C.A., *The Boundary Element Method for Engineers* (Pentech Press., 1984), pp. 46–103.
2) Brebbia, C.A., and Walker, S., *Boundary Element Techniques in Engineering* (Newnes-Butterworth, 1980), pp. 180–183.
3) Oshima, N. and Watari, N., Thermal Conductivity of Two-Dimensional Dispersive Media with Regular Arrangement. *Theoretical and Applied Mechanics*, 34 (Univ. Tokyo Press, 1986), pp. 377–386.
4) Oshima, N., Theoretical estimation of thermal conductivity of media with cylinders dispersed in square array. *Trans. Jpn. Soc. Mech. Eng.*, 52–475 (1986), 1095. (in Japanese)
5) Oshima, N., Calculation of mechanical coefficients of multiphase fluids (I). *Trans. Jpn. Soc. Mech. Eng.*, 44–384 (1978), 2759. (in Japanese)

Anisotropic Theory of Rapidly Deforming Granular Assembly

Masami NAKAGAWA

Department of Civil Engineering, Tohoku University, Sendai

A rapid deformation of an aggregate of identical, rough, inelastic, circular disks is considered. In accordance with the results from both theoretical and numerical simulations, we assume a sequence of highly dissipative binary interactions among disks and introduce the so-called anisotropic velocity distribution function based on the second moment of velocity fluctuations. Constitutive relations are then found for a dilute rectilinear flow.

I. INTRODUCTION

We are interested in the theory that enables us to describe rapid deformations of granular materials and to predict the mechanism for developing and sustaining them. Such deformations seem to be important in snow avalanches, rock debris slides, sediment transport, and pack-ice flows.

In attempting to describe such phenomena it seems natural to exploit the similarities between the colliding grains and the agitated gas molecules. However, because collisions between the grains of a granular material inevitably dissipate energy, standard kinetic theory must be extended to incorporate this. The advantage of this extension of the kinetic theory is that it provides definite forms for the rate of energy dissipation and fluxes of momentum and energy. They depend on measurable particle properties alone.

However, kinetic theoretical calculations[1-3] and numerical simulations[4,5] that incorporate energy dissipation suggest the inadequacy of the theory with small energy dissipation and the use of the isotropic Maxwellian velocity distribution. In particular, Walton and Braun[5] show the pronounced anisotropy in the velocity distribution in dilute flows. They conclude that the highly dissipative interactions among the disks is the cause of this anisotropy. Nakagawa[6] also analytically verified the above observation by investigating the mechanism of binary collision in detail. He found that when a collection of frictional disks is rapidly sheared, then a majority of interactions between disks take place in such a way that the energy dissipation is most pronounced.

In accordance with these observations, Jenkins and Richman[7] have recently introduced an anisotropic Maxwellian velocity distribution function for identical, smooth, inelastic disks. Here the anisotropic Maxwellian function is extended in order to develop a two-dimensional theory that incorporates the roughness of the disks.

II. STATISTICAL PRELIMINARIES

A dilute plane flow of identical, frictional, inelastic, circular disks of mass m and diameter σ is considered. Kinetic theory will be employed to develop balance laws for mean values of disk properties and to derive constitutive relations that relate the sources and fluxes to the mean fields.

The disk velocity is c, spin is ω, and spin velocity is s such that $s \equiv \sqrt{I/m}\,\omega$. Spin is always oriented perpendicular to the plane where a two-dimensional granular flow is defined. Then the mean values of a disk property, ψ, are defined as

$$n\langle\psi\rangle \equiv \int \psi(c, s) f^1(c, s, r, t)\,dc \qquad (1)$$

in which n is the number density, r is the position vector of the center of a disk, and $f^1(c, s, r, t)$ is the single velocity distribution function. Then mean velocity and mean spin velocity are defined as $u = \langle c \rangle$, $w = \langle s \rangle$, and fluctuations in disk velocity C and spin velocity S are defined as $\langle C \rangle \equiv \langle c - u \rangle = 0$ and $\langle S \rangle \equiv \langle s - w \rangle = 0$, respectively.

For disks resuming binary collisions, the complete pair distribution function at contact is assumed to be the product of the two single velocity distribution functions as shown below:

$$f^2(c_1, s_1, r_1, c_2, s_2, r_2, t) \equiv f^1(c_1, s_1, r, t) f^1(c_2, s_2, r, t). \qquad (2)$$

Subscripts 1 and 2 denote disk 1 and 2 or their respective properties.

It should be noted that the single velocity distribution functions in expression (2) are evaluated at the same position. This is justified by observing that, in a dilute flow of disks, each disk travels long distances between collisions; when treating collisions between pairs of disks, the presence of neighboring disks is completely ignored. When the system involved is fairly dense, however, the complete pair distribution function is the product of two single velocity distribution functions, each being evaluated at its center, and the radial distribution function g. This factor g incorporates the influence of the area excluded by the disks on their collision frequency. When no correlation between positions is assumed, as in a dilute system, g is unity.

With Eq. (2), the collisional rate of change of any disk property in a dilute system may be expressed as

$$\mathfrak{C}(\psi) \equiv \frac{1}{2}\int \Delta\psi f^1(c_1, s_1 r, , t) f^1(c_2, s_2, r, t)$$
$$\sigma(g \cdot k)\,dk\,dc_1\,ds_1\,dc_2\,ds_2 \qquad (3)$$

where $\Delta\psi \equiv \psi_1' + \psi_2' - \psi_1 - \psi_2$ and the integration is to be carried out over all possible values of k, c_1, s_1, c_2, s_2 for impending collisions, $g \cdot k > 0$. Here, the vector $g = c_1 - c_2$ is the relative velocity. When flow is dilute, the rate of change of a disk property is thoroghly due to the collisional source.

III. VELOCITY DISTRIBUTION FUNCTION

Based on the second moment $K_{\alpha\beta} \equiv \langle C_\alpha C_\beta \rangle$, the anisotropic Maxwellian single velocity distribution function is defined:

$$F_0(c, s, r, t) \equiv n/2\pi\,(\det K)^{1/2}(2\pi\tilde{T})^{1/2}\exp[-C_\alpha C_\beta K_{\alpha\beta}^{-1}/2]\exp[-S^2/2\tilde{T}], \qquad (4)$$

in which $\det K$ is the determinant of K and $\tilde{T} \equiv \langle S^2 \rangle$ is the granular spin temperature. When the difference in the energies of the velocity fluctuations among perpendicular axes can be ignored, the usual isotropic Maxwellian velocity distribution function

$$f_0(C, S, r, t) \equiv n/(2\pi \bar{T})(2\pi \tilde{T})^{1/2} \exp[-C_\xi C_\xi/2\bar{T}] \exp[-S^2/2\tilde{T}] \tag{5}$$

is recovered. In expression (5) the granular temperature associated with the fluctuation in disk velocity is defined as $\bar{T} \equiv 1/2\langle C^2 \rangle$. When expression (4) is used in place of f^1 in (1) and (3), various means and collisional rates of change of disk properties are derived based on the proposed velocity distribution function.

IV. BALANCE LAWS

For a disk property ψ that is a function of disk velocity and spin velocity, the balance law for the rate of change of its mean has the form

$$\partial \langle n\psi \rangle/\partial t = n\langle F/m \cdot \partial \psi/\partial c \rangle - \nabla \cdot \langle nc\,\psi \rangle + \mathfrak{C}(\psi), \tag{6}$$

where F represents the units of force. When ψ is expressed in terms of fluctuations of the velocity and spin, then Eq. (6) must be transformed to

$$D\langle \rho\psi \rangle/Dt + \langle \rho\psi \rangle \partial u_\xi/\partial r_\xi + \partial \langle \rho C_\xi \psi \rangle/\partial r_\xi$$
$$+ [Du_\xi/Dt - F_\xi/m]\partial \psi/\partial C_\xi + \rho[Dw/Dt - F_3/m]\langle \partial \psi/\partial S \rangle$$
$$+ \langle \rho C_\xi \partial \psi/\partial C_i \rangle \partial u_i/\partial r_\xi = \mathfrak{C}(m\psi) \tag{7}$$

where F_ξ is the external force and F_3 is $\sqrt{m/I}$ times the external couple acting on a disk. The usual summation convention is followed. Greek subscripts take the values of 1 and 2.

When ψ is identified with 1, C_α, S, S^2, and $C_\alpha C_\beta$ in Eq. (7), the balance laws for mass, linear momentum, mean spin, spin energy and second moment are respectively obtained; and they are
Balance of mass:

$$D\rho/Dt + \rho \partial u_\xi/\partial r_\xi = 0, \tag{8}$$

Balance of linear momentum:

$$\rho[Du_\alpha/Dt - F_\alpha/m] + \partial \langle \rho C_\xi C_\alpha \rangle/\partial r_\xi = 0, \tag{9}$$

Balance of mean spin:

$$\rho[Dw/Dt - F_3/m] + \partial \langle \rho C_\xi S \rangle/\partial r_\xi = \mathfrak{C}(mS), \tag{10}$$

Balance of spin energy:

$$\rho D\tilde{T}/Dt + \partial \langle \rho C_\xi S^2 \rangle/\partial r_\xi + 2\langle \rho C_\xi S \rangle \partial w/\partial r_\xi = \mathfrak{C}(mS^2) \tag{11}$$

and

Balance of second moment:

$$\rho DK_{\alpha\beta}/Dt + \partial \langle \rho C_\xi C_\alpha C_\beta \rangle/\partial r_\xi + \langle \rho C_\beta C_\xi \rangle \partial u_\alpha/\partial r_\xi$$
$$+ \langle \rho C_\alpha C_\xi \rangle \partial u_\beta/\partial r_\xi = \mathfrak{C}(m C_\alpha C_\beta). \tag{12}$$

In Eq. (9) the second term is the divergence of the transport contribution to the stress tensor; the collisional rate of change of linear momentum vanishes because the linear momentum is conserved in a collision. In Eq. (10) the second term is the divergence of the transport contribution to the couple stress. In Eq. (11), $\langle \rho C_\xi S^2 \rangle$ is the transport contribution to the flux of spin energy.

In Eq. (12) the tensor $K_{\alpha\beta} \equiv \langle C_\alpha C_\beta \rangle$ is the second moment, and when $\alpha = \beta$, the balance of translational energy is obtained.

Balance of translational energy:

$$\rho D\bar{T}/Dt + \partial[1/2\langle \rho C_\xi C^2 \rangle]/\partial r_\xi + \langle \rho C_\alpha C_\xi \rangle \partial u_\alpha/\partial r_\xi = \mathfrak{C}(1/2 m C^2) \quad (13)$$

where $2\bar{T} \equiv K_{\xi\xi} = \langle C_\xi C_\xi \rangle$ as defined earlier.

V. CONSTITUTIVE RELATIONS

In obtaining constitutive relations, a steady homogeneous rectilinear flow is considered. The flow between two parallel plates is sheared with constant strain rate of 2λ. The physical x and y axes are taken in directions parallel and perpendicular to the flow, as shown in Fig. 1. Then the x component of the velocity depends only upon y, and it is $u = 2\lambda y$. The velocity gradient can be decomposed into the symmetric and antisymmetric parts, D and W, respectively:

$$D = \begin{bmatrix} 0 & \lambda \\ \lambda & 0 \end{bmatrix} \text{ and } W = \begin{bmatrix} 0 & \lambda \\ -\lambda & 0 \end{bmatrix} \quad (14)$$

where D is the stretching and W is the spin.

The direction of the eigenvector of D corresponding to the eigenvalue λ can be obtained by rotating the x axis counterclockwise by an angle of $\pi/4$ as shown in Fig. 2. The same figure also shows that the direction of the eigenvector of K corresponding to the eigenvalue K_1 is related to the x axis by a counterclockwise rotation through an angle $\phi + \pi/4$. The eigenvector of K corresponding to its second eigenvalue K_2 forms an angle of $\phi + \pi/4$ with the y axis. A parameter α which measures the anisotropy of K is defined by

$$\alpha \equiv (K_2 - K_1)/2\bar{T}. \quad (15)$$

The results of the integration for the collisional rate of changes of translational energy,

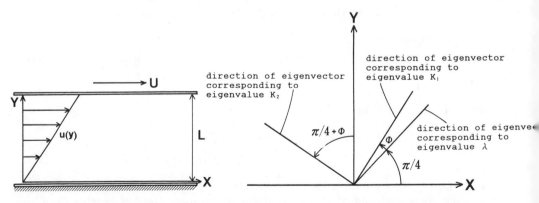

Fig. 1. Two-dimensional simple shear flow with strain rate of 2λ.

Fig. 2. Eigenvectors of K.

spin energy, and second moment can be expressed in terms of the following two basic integrals:

$$\Omega(\alpha) \equiv \int (1 - \alpha \cos 2\theta)^{1/2} \, d\theta, \tag{16}$$

and

$$\Phi(\alpha) \equiv \int (1 - \alpha \cos 2\theta)^{-1/2} \, d\theta. \tag{17}$$

Their expressions in terms of these integrals are shown in the Appendix.

In the simple shearing flow n, ∇u, $\bar{T}$, $\tilde{T}$, and $K_{\alpha\beta}$ are all constant and the contracted third moment vanishes. The mean spin is zero and the spin Eq. (10) is identically satisfied.

Under these conditions Eqs. (11), (12), and (13) will be simplified. The functions of α in them, I (α) through IX (α), are also defined in the Appendix. The balance of spin energy (11) simplifies in response to the requirement that $\mathfrak{C}(mS^2) = 0$. This then determines the temperature ratio as

$$\frac{\tilde{T}}{\bar{T}} = \frac{\text{II}(\alpha) + 2\text{III}(\alpha)}{[1/a - 1/\kappa] \, \text{IV}(\alpha)}. \tag{18}$$

The temperature ratio vanishes for smooth disks. In contrast to the isotropic case, the temperature ratio here depends upon the coefficient of restitution, e, through the parameter α. This weak dependence on e is an anisotropic correction to the temperature ratio. Later the dependence of $\tilde{T}/\bar{T}$ upon various parameters will be explored.

The balance of the translational energy (13) simplifies to

$$\langle \rho C_\alpha C_\xi \rangle \partial u_\alpha / \partial r_\xi = \mathfrak{C}(1/2 m \, C^2). \tag{19}$$

When the rate of strain tensor $\mathbf{D}$ is deviatoric, this further simplifies to

$$\rho \hat{K}_{\alpha\xi} D_{\xi\alpha} = \mathfrak{C}(1/2m \, C^2), \tag{20}$$

where $\wedge$ indicates the deviatoric quantity.

After much algebra this can be written as

$$2\pi^{3/2} \, \alpha(R/\nu)\cos 2\phi = \tilde{R}(\alpha), \tag{21}$$

where ν is the area fraction and the parameter $R \equiv \sigma\lambda/4\bar{T}^{1/2}$ is the measure of the strength of the mean shear relative to that of the velocity fluctuations and

$$\tilde{R}(\alpha) \equiv 2[a(1-a) - (b^2 + 2ab - b)] \, \text{I}(\alpha)$$
$$+ a(1 - a) \, [\text{II}(\alpha) + 2\text{III}(\alpha)] - \frac{a^2}{\kappa} \, \text{IV}(\alpha) \tilde{T}/\bar{T}. \tag{22}$$

The balance of the second moment (12) reduces to

$$\rho \, K_{\xi\beta}(D_{\alpha\xi} + W_{\alpha\xi}) + \rho K_{\xi\alpha}(D_{\beta\xi} + W_{\beta\xi}) = \mathfrak{C}(m C_\alpha C_\beta). \tag{23}$$

The constants a and b in Eq. (22) are defined in the Appendix. When the roughness coefficient $\beta = -1$, $a = 0$. β is assumed to range between -1 and 1. For values of β between 0 and 1, β may be considered to be the coefficient of restitution for the tangential deformation of rough disks. Values of β between -1 and 0 can be interpreted in terms of the coefficient of friction between rigid rough disks that slip during a collision.

From this, the equation governing the deviatoric part of the second moment is obtained:

$$2\rho \bar{T}\hat{D}_{\alpha\beta} + \rho \hat{K}_{\xi\alpha}(D_{\beta\xi} + W_{\beta\xi}) + \rho \hat{K}_{\xi\beta}(D_{\alpha\xi} + W_{\alpha\xi})$$
$$- \rho \hat{K}_{\xi\nu} D_{\nu\xi} \delta_{\alpha\beta} = \mathfrak{C}(m\, C_\alpha\, C_\beta) \qquad (24)$$

This equation is next written with respect to an orthonormal basis composed of the eigenvectors of K. Then the two independent components are extracted. The first is that associated with $\alpha = \beta = 1$,

$$2\pi^{3/2}\,(R/\nu)\cos 2\phi = \alpha\, \tilde{Q}(\alpha), \qquad (25)$$

where

$$\tilde{Q}(\alpha) \equiv -\{2[a(1-a) - (b^2 + 2ab - b)]\,\mathrm{V}(\alpha) + \frac{a^2}{\kappa}\,\mathrm{VI}(\alpha)\tilde{T}/\bar{T}$$
$$- a(1-a)\,\mathrm{VII}(\alpha) - 2a(1-a)\,\mathrm{VIII}(\alpha) - 2[2a(1-a)$$
$$+ b(1-2a)]\,\mathrm{IX}(\alpha)\}. \qquad (26)$$

The off-diagonal component is simply

$$\sin 2\phi - \alpha = 0. \qquad (27)$$

Equations (18), (21), (25), and (27) are then solved for $\tilde{T}/\bar{T}$, ϕ, α, and R. The term $(R/\nu)\cos 2\phi$ can be eliminated between Eqs. (21) and (25) to yield a single equation for the determination of α:

$$\tilde{R}(\alpha) = \alpha^2\, \tilde{Q}(\alpha), \qquad (28)$$

in which the expression for the temperature ratio (18) is to be used in $\tilde{R}(\alpha)$ and $\tilde{Q}(\alpha)$. There are two ways of solving Eq. (28). One is to solve it numerically; another is to obtain an approximate analytical solution. Here the latter approach is employed since it makes possible to write the intermediate expressions of interest in terms of the material properties of the disks.

In Eq. (28) the integrand of each integral in $\tilde{R}(\alpha)$ and $\tilde{Q}(\alpha)$ is first expanded in the Taylor series and then integrated term by term. The resulting expression for α^2 is, up to an error of the order of α^4,

$$\alpha^2 = \Lambda/\Gamma \qquad (29)$$

where

$$\Lambda \equiv \pi\left[(1+e)(1-e) + \frac{\kappa(1+\beta)(1-\beta)}{2\kappa + (1-\beta)}\right] \qquad (30)$$

and

$$\Gamma \equiv \pi\{9/16(1+e)(1-e) + 1/2(1+\kappa)\,[\kappa(1+\beta)(1-e)$$
$$+ (1+e)(2 + \kappa(1-\beta))]$$
$$+ \kappa(1+\beta)/32(1+\kappa)^2\,[(2 + \kappa(1-\beta)$$
$$+ 7\kappa(1+\beta)^2/(2\kappa + (1-\beta))]\} \qquad (31)$$

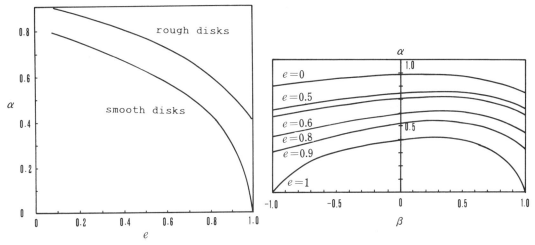

Fig. 3. The variation of α with coefficient of restitution, e. Here $\beta=0$ and $\kappa=1/2$.

Fig. 4. The variation of α with roughness coefficient, β, for various values of coefficient of restitution, e. Here $\kappa=1/2$.

When $1-\beta$ is small, Eq. (29) is reduced to the smooth solution given by Jenkins and Richman.[7] In Fig. 3 the variation of the α with the coefficient of restitution e for $\kappa=0.5$ and $\beta=0$ is given. This is the variation in α when the collisional dissipation associated with the tangential component of the relative contact velocity is most pronounced. On the same figure, Jenkins and Richman's prediction for smooth disks is also given.

The degree of anisotropy decreases as the coefficient of restitution increases from zero to one. The vanishing anisotropy can only be realized when disks are perfectly elastic and perfectly smooth. Even if disks are perfectly elastic, a significant amount of anisotropy remains owing to the roughness of the disks.

In Fig. 4, the variation in α with β is shown for various values of e. It is noted again that the isotropy is only achieved when $e=1$ and $\beta=\pm 1$. The peaks of the variation of α over β are skewed toward the positive region. However, most of the physical materials possess β ranging between -1 and 0, and such disks do exhibit the maximum anisotropy at $\beta=0$ regardless of the values of e.

Based on the approximate values of α, the temperature ratio (18) may be written as

$$\tilde{T}/\bar{T} = [1 + 3\Lambda/16\Gamma]/[(1/a - 1/\kappa)(1 - \Lambda/16\Gamma)]. \tag{32}$$

In Fig. 5 the variation of the temperature ratio with the roughness coefficient β for various values of e is shown.

There is a weak dependence of the temperature ratio on the coefficient of restitution. The increase in $\tilde{T}/\bar{T}$ with β is rather small for negative values of β compared to those for positive β. This is because as β increases from 0 to 1, the degree of the reversals of the tangential component of the contact velocity increases. This results in an increase of spin energy. The equipartition of the total energy among disks' degree of freedom can only be achieved when disks are perfectly rough ($\beta=1$) and perfectly elastic ($e=1$).

The determination of ϕ in terms of α is very simple, once the values of α are known

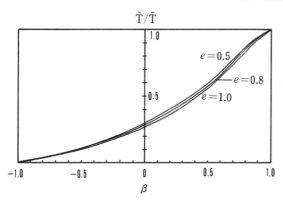

Fig. 5. The variation of the temperature ratio, $\tilde{T}/\bar{T}$, with roughness coefficient, β, for various values of coefficient of restitution, e. Here $\kappa = 1/2$.

Fig. 6. The variation of the angle, ϕ, with coefficient of restitution, e. Here $\beta = 0$ and $\kappa = 1/2$.

from Eq. (29). The variations of ϕ with the coefficient of restitution and with the roughness coefficient are shown in Figs. 6 and 7, respectively.

Once α, ϕ, and $\tilde{T}/\bar{T}$ are known, the variation of R/ν with α may be expressed as

$$2R/\nu = \pi^{-3/2} \alpha \tilde{Q}(\alpha)/(1 - \alpha^4)^{1/2}. \tag{33}$$

In Fig. 8 the variation of $R^* = 8\nu R$ with roughness coefficient β for various values of coefficient of restitution e is plotted. Here ν is assumed to be 0.2. The values of R^* increases with roughness until a critical value of β is reached, after which R^* decreases.

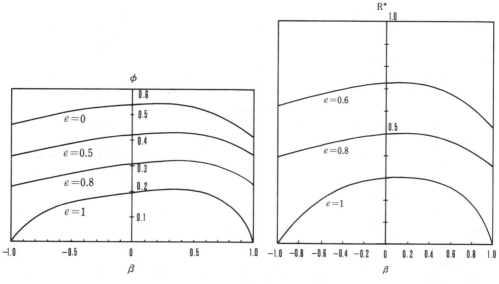

Fig. 7. The variation of the angle, ϕ, with roughness coefficient, β, for various values of coefficient of restitution, e. Here $\kappa = 1/2$.

Fig. 8. The variation of $R^* \equiv 8\nu R$ for various values of coefficient of restitution, e. Here $\beta = 0$ and $\kappa = 1/2$.

When a flow is dilute, the only contribution to the stress tensor is that of transport. Thus the shear stress, $-Pxy$, can be found by expressing the $x-y$ components of deviatoric second moment in terms of α and ϕ:

$$-Pxy = \rho\, \alpha\, \bar{T} \cos 2\phi. \tag{34}$$

When the relationships $R = \sigma\, \lambda/4\bar{T}^{1/2}$, $\rho = 4m\,\nu/\pi\,\sigma^2$, and $\cos 2\phi = (1 - \alpha^2)^{1/2}$ are used, this can be expressed as

$$-Pxy = 2\eta\, \lambda \tag{35}$$

where

$$\eta \equiv m\nu\, \lambda\, \alpha(1 - \alpha^2)^{1/2}/8\pi\, R^2 \tag{36}$$

is the viscosity. The variation of the viscosity with the area fraction ν is shown in Fig. 9 for $0 < \nu < 0.2$, $\beta = 0$, and various values of e. There is an excellent agreement with the results obtained by the computer simulation by Campbell and Gong[4] and Walton and Braun[5]. As shown in the figure, η appears to asymptote to infinity as ν approaches 0. When the flow densities are low, interparticle collisions are infrequent and the inelastic collisions cannot effectively damp out the random velocities. This appears to contribute to large random velocities and corresponding large transport contributions to the stress tensor. This also results in a large shear viscosity.

It can also be seen in the same figure that the reduction of elasticity of the disks signifi-

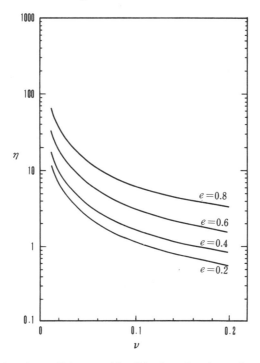

Fig. 9. The variation of viscosity coefficient, η, with solid volume fraction, ν, for various values of coefficient of restitution, e. Here $\beta=0$, $\kappa=1/2$ and $\lambda=5$.

cantly reduces the viscosity at low densities where the stress is primarily due to the momentum carried by the particles.

VI. CONCLUSION

Anisotropic Maxwellian velocity distribution function is introduced and constitutive relations are found based upon it. The anisotropy of the second moment of velocity fluctuations is found to decrease with the coefficient of restitution and to reach its peak when energy dissipation due to collisions is most pronounced. The temperature ratio shows a weak dependence on the coefficient of restitution and also shows a relatively slow increase for $-1 < \beta < 0$ where the roughness coefficient, β, is interpreted in terms of the coefficient of friction between rigid rough disks that slip during a collision. The viscosity is found to decrease with increased solid volume fraction and with decreased coefficient of restitution. The result shows good agreement with those by the numerical simulations.

Acknowledgment

The author wish to thank Professors James T. Jenkins at Cornell University and Mark W. Richman at Worcester Polytechnical Institute for their valuable advice.

REFERENCES

1) Lun, C.K.K., Savage, S.B., Jeffrey, D.J., and Shepurniy, J., Kinetic theories for granular flow: inelastic particles in Couette flow and slightly inelastic particles in a general flow field. *J. Fluid Mech.*, **130** (1984), pp. 187–202.
2) Jenkins, J.T. and Richman, M.W., Grad's 13-moment system for a dense gas of inelastic spheres. *Arch. Rat'l. Mech. Anal.*, **87** (1985), pp. 355–377.
3) Jenkins, J.T. and Richman, M.W., Kinetic theory for plane flows of a dense gas of identical, rough inelastic, circular disks. *Physics of Fluids*, **28** (1985), pp. 3485–3494.
4) Campbell, C.S. and Gong, A., The stress tensor in a two-dimensional granular shear flow. *J. Fluid Mech.*, **164** (1986), pp. 107–125.
5) Walton, O.R. and Braun, R.L., Viscosity, granular-temperature, and stress calculations for shearing assemblies of inelastic, frictional disks. *J. Rheol.*, **30** (1986), pp. 949–980.
6) Nakagawa, M., Kinetic theory for plane flows of rough, inelastic, circular disks. Ph.D. Thesis (1987), Cornell University.
7) Jenkins, J.T. and Richman, M.W., Plane simple shear of smooth inelastic, circular disks. The anisotropy of the fluctuation energy in the dilute and dense limits. *J. Fluid Mech.* (in press).

APPENDIX

The results of the integration for the collisional rate of changes of translational energy, spin energy, and second moment can be expressed in terms of two basic integrals (16) and (17), and they are:

Translational Energy

$$\mathfrak{C}\left(\frac{m}{2}C^2\right) = \frac{2\sigma n^2}{\sqrt{\pi}} \bar{T}^{3/2}\left\{A[\mathrm{I}(\alpha) + \frac{1}{2}\mathrm{II}(\alpha) + \mathrm{III}(\alpha)] + B\mathrm{I}(\alpha) + \frac{C}{8}\mathrm{IV}(\alpha)\tilde{T}/\bar{T}\right\}, \quad (A1)$$

Spin Energy

$$\mathfrak{C}\left(\frac{m}{2}S^2\right) = \frac{\sigma n^2}{4\sqrt{\pi}} C\bar{T}^{3/2}[\mathrm{II}(\alpha) + 2\mathrm{III}(\alpha) - \left(\frac{1}{a} - \frac{1}{\kappa}\right)\mathrm{IV}(\alpha)\tilde{T}/\bar{T}], \quad (A2)$$

and

Second Moment

$$\mathfrak{C}(mC_\alpha C_\beta) = \frac{\sigma n^2}{\sqrt{\pi}}[P(\alpha)\delta_{\alpha\beta} + Q(\alpha)\hat{K}_{\alpha\beta}], \quad (A3)$$

where the hat denotes the deviatoric part and $P(\alpha)$ and $Q(\alpha)$ are defined by

$$P(\alpha) = 2\bar{T}^{3/2}\left\{A[\mathrm{I}(\alpha) + \frac{1}{2}\mathrm{II}(\alpha) + \mathrm{III}(\alpha)] + B\mathrm{I}(\alpha) + \frac{C}{8}\mathrm{IV}(\alpha)\tilde{T}/\bar{T}\right\},$$

and

$$Q(\alpha) \equiv 2\bar{T}^{1/2}\left\{A[-V(\alpha) + \frac{1}{2}\mathrm{VII}(\alpha) + \mathrm{VIII}(\alpha) + 2\mathrm{IX}(\alpha)] \right.$$
$$\left. + \frac{C}{8}\mathrm{VI}(\alpha)\tilde{T}/T + D[-V(\alpha) + \mathrm{IX}(\alpha)] - \frac{E}{2}V(\alpha)\right\},$$

respectively.

In expressions (A1), (A2) and (A3), the coefficients A, B, C, D and E are defined as

$$A \equiv -ma(1-a), \qquad B \equiv m(b^2 + 2ab - b)$$

$$C \equiv \frac{4ma^2}{\kappa}, \qquad D \equiv -mb(1-2a)$$

and

$$E \equiv 2mb^2$$

and the functions of α are defined as

$$\mathrm{I}(\alpha) = \frac{4}{3}\Omega(\alpha) + \frac{\alpha^2 - 1}{3}\Phi(\alpha),$$

$$\mathrm{II}(\alpha) = (1 - \alpha^2)\Phi(\alpha),$$

$$\mathrm{III}(\alpha) = \frac{2(2\alpha^2 + 1)}{15} \Omega(\alpha) + \frac{2\alpha^2 - 1}{15} \Phi(\alpha),$$

$$\mathrm{IV}(\alpha) = \Omega(\alpha),$$

$$\mathrm{V}(\alpha) = \frac{3\alpha^2 + 1}{5\alpha^2} \Omega(\alpha) - \frac{\alpha^2 - 1}{5\alpha^2} \Phi(\alpha),$$

$$\mathrm{VI}(\alpha) = -\frac{1}{3\alpha^2} \Omega(\alpha) - \frac{\alpha^2 - 1}{3\alpha^2} \Phi(\alpha),$$

$$\mathrm{VII}(\alpha) = \frac{\alpha^2 - 1}{\alpha^2} \Omega(\alpha) - \frac{\alpha^2 - 1}{\alpha^2} \Phi(\alpha),$$

$$\mathrm{VIII}(\alpha) = -\frac{2(3\alpha^2 - 4)}{15\alpha^2} \Omega(\alpha) + \frac{8(\alpha^2 - 1)}{15\alpha^2} \Phi(\alpha),$$

and

$$\mathrm{IX}(\alpha) = \frac{2(3\alpha^2 + 1)}{15\alpha^2} \Omega(\alpha) + \frac{2(\alpha^2 - 1)}{15\alpha^2} \Phi(\alpha).$$

In these equations the constants a and b are defined as

$$a \equiv \kappa(1 + \beta)/2(1 + \kappa),$$
$$b \equiv r - a,$$
$$r \equiv (1 + e)/2$$

and

$$\kappa \equiv 4I/m\sigma^2.$$

Application of the Method of Continuous Distribution of Dislocations to an Interface Crack Problem

Hideaki KASANO, Kazuhiro SHIMOYAMA,* and Hiroyuki MATSUMOTO

*Department of Mechanical Engineering, Tokyo Institute of Technology, * Graduate School, Tokyo Institute of Technology, Tokyo*

The singular stress field at the tips of an interface crack between fiber and matrix is investigated using a micromechanical model of a unidirectional carbon fiber-reinforced composite material. Application of the method of continuous distribution of dislocations (CDD) to this problem leads to a Cauchy type singular integral equation of the second kind, which is numerically solved by use of Lobatto-Jacobi integration rule with a finite number of collocation points. Numerical results are presented for the stress intensity factors and the crack surface displacements.

I. INTRODUCTION

The method of continuous distribution of dislocations (CDD) provides a powerful tool to analyze many kinds of crack problems. Its application to a crack problem leads to a direct formulation in terms of an integral equation if the fundamental stress field due to one or two dislocations is known. The integral equation is generally a singular integral equation, which is very convenient to evaluate the singularity of the stress field peculiar to such crack problems. We have therefore employed this method to investigate some crack propagation problems in carbon fiber-reinforced composite materials.[1-4]

Following a series of these investigations we consider here an interface crack problem which models a debonding between fiber and matrix. It is shown that a suitable limiting process is necessary to apply this method to such interface crack problems and that the singular integral equation can be solved without any tedious manipulation by use of a numerical integration rule with a finite number of collocation points.

Numerical results are presented for the stress intensity factors at the crack tips and the crack surface displacements.

II. MATHEMATICAL FORMULATIONS

The micromechanical model used for a unidirectional carbon fiber-reinforced composite is a bonded plane consisting of an isotropic and an orthotropic half-plane which simulate the isotropic matrix and the carbon fiber with elastic anisotropy, respectively.[1-4] First we consider the fundamental stress fields due to two edge dislocations in the bonded plane model as shown in Fig. 1 within the framework of the two-dimensional elasticity theory. The stress functions for the two edge dislocations with their Burgers vectors parallel to the x-axis and symmetrically located at points (ξ, η) and $(\xi, -\eta)$ in the isotropic half-plane of the model as shown in Fig. 1(a) are expressed in the following forms:

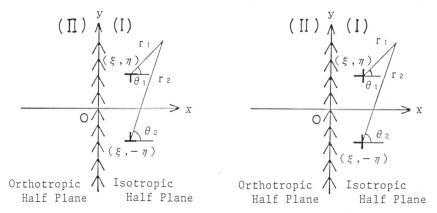

Fig. 1. A pair of edge dislocations existing in the neighbourhood of bimaterial interface.
(a) Burgers vectors b_x parallel to x axis,
(b) Burgers vectors b_y parallel to y axis.

$$\chi^{\mathrm{I}} = \frac{2Gb_x}{\pi(\kappa+1)}(r_1 \cdot \log r_1 \cdot \sin\theta_1 - r_2 \cdot \log r_2 \cdot \sin\theta_2)$$

$$+ \int_0^\infty \{g_1(\lambda) \cdot x + g_2(\lambda)\} e^{-\lambda x} \cos\lambda y \, d\lambda \tag{1}$$

$$\chi^{\mathrm{II}} = \int_0^\infty \{g_3(\lambda) \cdot \chi_s + g_4(\lambda) \cdot \chi_d\} d\lambda \tag{2}$$

and $\kappa = 3 - \nu/1 + \nu$ for plane stress and $3 - 4\nu$ for plane strain with the superscripts I and II referring to the isotropic and the orthotropic half-planes, respectively, where b_x represents the magnitude of the Burgers vector, G the modulus of rigidity, and $g_1(\lambda) \sim g_4(\lambda)$ are unknown functions which are to be determined from the boundary conditions. The polar coordinates (r_1, θ_1) and (r_2, θ_2) with their origins located at the dislocation points are related to the Cartesian coordinates (x, y) as follows:

$$r_1^2 = (x - \xi)^2 + (y - \eta)^2, \quad r_2^2 = (x - \xi)^2 + (y + \eta)^2$$

$$\theta_1 = \tan^{-1}\left(\frac{x-\xi}{y-\eta}\right), \quad \theta_2 = \tan^{-1}\left(\frac{x-\xi}{y+\eta}\right). \tag{3}$$

The stress functions χ_s and χ_d in Eq. (2) are given by[5]

$$\chi_s = \chi_1 + \chi_2 \quad \chi_d = \frac{\chi_1 - \chi_2}{\sqrt{\mu_1} - \sqrt{\mu_2}}$$

$$\chi_i = e^{\sqrt{\mu_i}\lambda x} \cos\lambda y \quad (i = 1, 2) \tag{4}$$

where μ_i's are constants depending upon the elastic constants of the orthotropic half-plane. The boundary conditions for this case, where the perfect bond at the interface ($x=0$) between both planes is assumed, can be stated as follows:

$$(\sigma_x^{\mathrm{I}})_{x=0} = (\sigma_x^{\mathrm{II}})_{x=0}, \quad (\tau_{xy}^{\mathrm{I}})_{x=0} = (\tau_{xy}^{\mathrm{II}})_{x=0}$$

$$\left(\frac{\partial u_x^{\mathrm{I}}}{\partial y}\right)_{x=0} = \left(\frac{\partial u_x^{\mathrm{II}}}{\partial y}\right)_{x=0}, \quad \left(\frac{\partial v_y^{\mathrm{I}}}{\partial y}\right)_{x=0} = \left(\frac{\partial v_y^{\mathrm{II}}}{\partial y}\right)_{x=0}. \tag{5}$$

By substituting the stress and displacement components corresponding to Eqs. (1) and (2) into Eq. (5) and using the method of Fourier transforms we have a system of linear algebraic equations for $g_1(\lambda) \sim g_4(\lambda)$, from which the stress field for Fig. 1(a) is determined. The stress field for the two edge dislocations located at the interface $(0, \eta)$ and $(0, -\eta)$ is derived from the above results by taking the limit of ξ approaching zero. For later analysis, only the normal stress σ_x^I and the shear stress τ_{xy}^I will be written here:

$$\sigma_x^I = \frac{2G\, b_x}{\pi(\kappa+1)}\left[\frac{(y-\eta)\{3x^2+(y-\eta)^2\}}{\{x^2+(y-\eta)^2\}^2} - \frac{(y+\eta)\{3x^2+(y+\eta)^2\}}{\{x^2+(y+\eta)^2\}^2}\right.$$

$$- \frac{b_{01}}{2}\left\{\frac{y-\eta}{x^2+(y-\eta)^2} - \frac{y+\eta}{x^2+(y+\eta)^2}\right\}$$

$$\left. - a_{01}\left\{\frac{y-\eta}{\{x^2+(y-\eta)^2\}^2} - \frac{y+\eta}{\{x^2+(y+\eta)^2\}^2}\right\}x^2\right] \tag{6}$$

$$\tau_{xy}^I = -\frac{2G\, b_x}{\pi(\kappa+1)}\left[\frac{x\{x^2-(y-\eta)^2\}}{\{x^2+(y-\eta)^2\}^2} - \frac{x\{x^2-(y+\eta)^2\}}{\{x^2+(y+\eta)^2\}^2}\right.$$

$$+ \frac{a_{01}-b_{01}}{2}\left\{\frac{x}{x^2+(y-\eta)^2} - \frac{x}{x^2+(y+\eta)^2}\right\}$$

$$\left. - \frac{a_{01}}{2}\left\{\frac{x^2-(y-\eta)^2}{\{x^2+(y-\eta)^2\}^2} - \frac{x^2-(y+\eta)^2}{\{x^2+(y+\eta)^2\}^2}\right\}x\right] \tag{7}$$

where a_{01} and b_{01} are constants depending upon the elastic constants of both half-planes.

On the other hand, the stress functions for the case of Fig. 1(b) are given as follows:

$$\chi^I = \frac{2G\, b_y}{\pi(\kappa+1)}(r_1 \cdot \log r_1 \cdot \cos\theta_1 + r_2 \cdot \log r_2 \cdot \cos\theta_2)$$

$$+ \int_0^\infty \{g_1'(\lambda)\cdot x + g_2'(\lambda)\}e^{-\lambda x}\cos\lambda y\, d\lambda \tag{8}$$

$$\chi^{II} = \int_0^\infty \{g_3'(\lambda)\cdot \chi_s + g_4'(\lambda)\cdot \chi_d\}\, d\lambda \tag{9}$$

where b_y represents the magnitude of the Burgers vector parallel to the y-axis and $g_1'(\lambda) \sim g_4'(\lambda)$ are unknown functions to be determined from the boundary conditions. The stress field corresponding to Eqs. (8) and (9) can be obtained in the same way as in Fig. 1(a).

The normal stress σ_x^I and the shear stress τ_{xy}^I, when the two edge dislocations shown in Fig. 1(b) are located at the interface, are given by

$$\sigma_x^I = \frac{2G\, b_y}{\pi(\kappa+1)}\left[\frac{x\{x^2-(y-\eta)^2\}}{\{x^2+(y-\eta)^2\}^2} + \frac{x\{x^2-(y+\eta)^2\}}{\{x^2+(y+\eta)^2\}^2}\right.$$

$$- \frac{b_{01}'}{2}\left\{\frac{x}{x^2+(y-\eta)^2} + \frac{x}{x^2+(y+\eta)^2}\right\}$$

$$\left. - \frac{a_{01}'}{2}\left\{\frac{x^2-(y-\eta)^2}{\{x^2+(y-\eta)^2\}^2} + \frac{x^2-(y+\eta)^2}{\{x^2+(y+\eta)^2\}^2}\right\}x\right] \tag{10}$$

$$\tau^{\text{I}}_{xy} = \frac{2G\,b_y}{\pi(\kappa+1)}\left[\frac{(y-\eta)\{x^2-(y-\eta)^2\}}{\{x^2+(y-\eta)^2\}^2}+\frac{(y+\eta)\{x^2-(y+\eta)^2\}}{\{x^2+(y+\eta)^2\}^2}\right.$$
$$+\frac{a'_{01}-b'_{01}}{2}\left\{\frac{y-\eta}{x^2+(y-\eta)^2}+\frac{y+\eta}{x^2+(y+\eta)^2}\right\}$$
$$\left.-a'_{01}\left\{\frac{y-\eta}{\{x^2+(y-\eta)^2\}^2}+\frac{y+\eta}{\{x^2+(y+\eta)^2\}^2}\right\}x^2\right] \tag{11}$$

where a'_{01} and b'_{01} are constants depending upon the elastic constants of both isotropic and orthotropic half-planes. With these results we next consider an interface crack problem as shown in Fig. 2, where the crack surfaces are subjected to the normal stress $P(y)$ and the shear stress $Q(y)$.

By continuously distributing the infinitesimal dislocations along the assumed crack surfaces as shown in Fig. 3(a) and (b), and by superposing the resulting stress fields and taking the limit of x approaching zero together with a boundary condition on the crack surfaces, we have the following integral equations to determine the dislocation density:

$$\lim_{x\to 0}\left\{-\frac{2G}{\pi(\kappa+1)}\int_0^{c_1}\left[\frac{(y-\eta)\{3x^2+(y-\eta)^2\}}{\{x^2+(y-\eta)^2\}^2}-\frac{(y+\eta)\{3x^2+(y+\eta)^2\}}{\{x^2+(y+\eta)^2\}^2}\right.\right.$$
$$-\frac{b_{01}}{2}\left\{\frac{y-\eta}{x^2+(y-\eta)^2}-\frac{y+\eta}{x^2+(y+\eta)^2}\right\}$$
$$\left.-a_{01}x^2\left\{\frac{y-\eta}{\{x^2+(y-\eta)^2\}^2}-\frac{y+\eta}{\{x^2+(y+\eta)^2\}^2}\right\}\right]f(\eta)d\eta$$
$$+\frac{2G}{\pi(\kappa+1)}\int_0^{c_1}\left[\frac{x\{x^2-(y-\eta)^2\}}{\{x^2+(y-\eta)^2\}^2}+\frac{x\{x^2-(y+\eta)^2\}}{\{x^2+(y+\eta)^2\}^2}\right.$$
$$-\frac{b'_{01}}{2}\left\{\frac{x}{x^2+(y-\eta)^2}+\frac{x}{x^2+(y+\eta)^2}\right\}$$
$$\left.\left.-\frac{a'_{01}}{2}x\left\{\frac{x^2-(y-\eta)^2}{\{x^2+(y-\eta)^2\}^2}+\frac{x^2-(y+\eta)^2}{\{x^2+(y+\eta)^2\}^2}\right\}g(\eta)d\eta\right\}\right.$$
$$=-P(y),\quad (|y|<c_1) \tag{12-1}$$

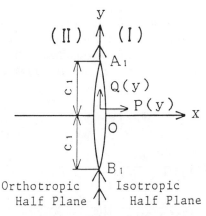

Fig. 2. Interface crack.

$$\lim_{x\to 0}\left\{\frac{2G}{\pi(\kappa+1)}\int_0^{c_1}\left[\frac{x\{x^2-(y-\eta)^2\}}{\{x^2+(y-\eta)^2\}^2}-\frac{x\{x^2-(y+\eta)^2\}}{\{x^2+(y+\eta)^2\}^2}\right.\right.$$

$$+\frac{a_{01}-b_{01}}{2}\left\{\frac{x}{x^2+(y-\eta)^2}-\frac{x}{x^2+(y+\eta)^2}\right\}$$

$$\left.-\frac{a_{01}}{2}x\left\{\frac{x^2-(y-\eta)^2}{\{x^2+(y-\eta)^2\}^2}-\frac{x^2-(y+\eta)^2}{\{x^2+(y+\eta)^2\}^2}\right\}\right]f(\eta)d\eta$$

$$+\frac{2G}{\pi(\kappa+1)}\int_0^{c_1}\left[\frac{(y-\eta)\{x^2-(y-\eta)^2\}}{\{x^2+(y-\eta)^2\}^2}+\frac{(y+\eta)\{x^2-(y+\eta)^2\}}{\{x^2+(y+\eta)^2\}^2}\right.$$

$$+\frac{a'_{01}-b'_{01}}{2}\left\{\frac{y-\eta}{x^2+(y-\eta)^2}+\frac{y+\eta}{x^2+(y+\eta)^2}\right\}$$

$$\left.\left.-a'_{01}x^2\left\{\frac{y-\eta}{\{x^2+(y-\eta)^2\}^2}+\frac{y+\eta}{\{x^2+(y+\eta)^2\}^2}\right\}\right]g(\eta)d\eta\right\}$$

$$=-Q(y),\quad(|y|<c_1) \tag{12-2}$$

where $f(\eta)$ and $g(\eta)$ represent the dislocation density of which the Burgers vectors are parallel to the x- and the y-axes, respectively. It should be noted here that the stress field for Fig. 3(b) is the same as that of Fig. 3(c).

Here, by noting that $f(\eta)$ is an odd function and $g(\eta)$ an even function, and by expanding the integral interval from $(0, c_1)$ to $(-c_1, c_1)$ we have

$$\lim_{x\to 0}\left\{-\frac{2G}{\pi(\kappa+1)}\right.$$

$$\int_{-c_1}^{c_1}\left[\frac{(y-\eta)\{3x^2+(y-\eta)^2\}}{\{x^2+(y-\eta)^2\}^2}-\frac{b_{01}}{2}\cdot\frac{y-\eta}{x^2+(y-\eta)^2}-a_{01}x^2\frac{y-\eta}{\{x^2+(y-\eta)^2\}^2}\right]f(\eta)d\eta+\frac{2G}{\pi(\kappa+1)}$$

$$\left.\int_{-c_1}^{c_1}\left[\frac{x\{x^2-(y-\eta)^2\}}{\{x^2+(y-\eta)^2\}^2}-\frac{b'_{01}}{2}\cdot\frac{x}{x^2+(y-\eta)^2}-\frac{a'_{01}}{2}x\cdot\frac{x^2-(y-\eta)^2}{\{x^2+(y-\eta)^2\}^2}\right]g(\eta)d\eta\right\}$$

$$=-P(y),\quad(|y|<c_1) \tag{13-1}$$

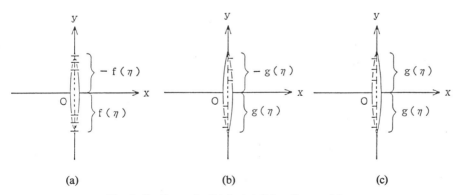

Fig. 3. Continuously distributed dislocation model.

$$\lim_{x\to 0}\left\{\frac{2G}{\pi(\kappa+1)}\right.$$

$$\int_{-c_1}^{c_1}\left[\frac{x\{x^2-(y-\eta)^2\}}{\{x^2+(y-\eta)^2\}}+\frac{a_{01}-b_{01}}{2}\cdot\frac{x}{x^2+(y-\eta)^2}-\frac{a_{01}}{2}x\frac{x^2-(y-\eta)^2}{\{x^2+(y-\eta)^2\}^2}\right]f(\eta)d\eta+\frac{2G}{\pi(\kappa+1)}$$

$$\left.\int_{-c_1}^{c_1}\left[\frac{(y-\eta)\{x^2-(y-\eta)^2\}}{\{x^2+(y-\eta)^2\}^2}+\frac{a'_{01}+b'_{01}}{2}\frac{y-\eta}{x^2+(y-\eta)^2}-a'_{01}x^2\frac{y-\eta}{\{x^2+(y-\eta)^2\}^2}\right]g(\eta)d\eta\right\}$$

$$=-Q(y),\ (|y|<c_1). \tag{13-2}$$

The above equations reduce to the following singular integral equations of the second kind through a suitable limiting process:[6]

$$\alpha_2 g(y)+\frac{1}{\pi}\int_{-c_1}^{c_1}\frac{f(\eta)}{\eta-y}d\eta=\alpha_{22}P(y),\quad (|y|<c_1) \tag{14-1}$$

$$-\alpha_1 f(y)+\frac{1}{\pi}\int_{-c_1}^{c_1}\frac{g(\eta)}{\eta-y}d\eta=\alpha_{11}Q(y),\quad (|y|<c_1) \tag{14-2}$$

where

$$\alpha_1=\frac{a_{01}-b_{01}}{a'_{01}-b'_{01}-2},\quad \alpha_2=\frac{b'_{01}}{b_{01}-2},$$

$$\alpha_{11}=\frac{\kappa+1}{G}\cdot\frac{1}{a'_{01}-b'_{01}-2},\quad \alpha_{22}=\frac{\kappa+1}{G}\cdot\frac{1}{b_{01}-2}. \tag{15}$$

In addition, the equations

$$\int_{-c_1}^{c_1}f(\eta)d\eta=0,\quad \int_{-c_1}^{c_1}g(\eta)d\eta=0 \tag{16}$$

are used to ensure the single-valuedness of the displacements. For analytical convenience, we transform Eqs. (14–1) and (14–2) into a single integral equation of complex form. That is, multiplying both sides of (14–1) by $-i\sqrt{\alpha_1/\alpha_2}$ with imaginary unit $i=\sqrt{-1}$ and adding the resulting equation to Eq. (14–2) we get

$$-\gamma\phi(y)+\frac{1}{\pi i}\int_{-c_1}^{c_1}\frac{\phi(\eta)}{\eta-y}d\eta=\alpha_{11}Q(y)-i\alpha_{22}\sqrt{\alpha_1/\alpha_2}P(y),\quad (|y|<c_1) \tag{17}$$

where

$$\gamma=\alpha_2\sqrt{\alpha_1/\alpha_2},\quad \phi(y)=\sqrt{\alpha_1/\alpha_2}f(y)+ig(y). \tag{18}$$

Then, Eq. (16) becomes as follows:

$$\int_{-c_1}^{c_1}\phi(\eta)d\eta=0. \tag{19}$$

By making a change of variables such as

$$s_1=\frac{c_1}{\eta},\quad t_1=\frac{y}{c_1},\quad \phi(t_1)\equiv\phi(c_1 t_1),\quad P(t_1)\equiv P(c_1 t_1),\quad Q(t_1)\equiv Q(c_1 t_1) \tag{20}$$

Eqs. (17) and (19) reduce to

$$-\gamma\phi(t_1) + \frac{1}{\pi i}\int_{-1}^{1}\frac{\phi(s_1)}{s_1 - t_1}ds_1 = \alpha_{11}Q(t_1) - i\alpha_{22}\sqrt{\alpha_1/\alpha_2}\,P(t_1), \quad (|t_1| < 1) \quad (21)$$

$$\int_{-1}^{1}\phi(s_1)ds_1 = 0. \quad (22)$$

The stress intensity factors are obtained in the following. When its right-hand side terms $P(t_1)$ and $Q(t_1)$ are replaced by $-(\sigma_x)_{x=0}$ and $-(\tau_{xy})_{x=0}$, Eq. (21) holds for $t_1 \geq 1$ as well as $|t_1| < 1$. Then, since $\phi(t_1) = 0$ for $t_1 \geq 1$, it becomes as follows:

$$\alpha_{22}\sqrt{\alpha_1/\alpha_2}(\sigma_x)_{x=0} + i\alpha_{11}(\tau_{xy})_{x=0} = -\frac{1}{\pi}\int_{-1}^{1}\frac{\phi(s_1)}{s_1 - t_1}ds_1, \quad (t_1 \geq 1). \quad (23)$$

By noting here the dominant part of Eq. (21) and considering that $\phi(t_1)$ has singularities at $t_1 = \pm 1$ we assume the unknown function to be

$$\phi(t_1) = (1 - t_1)^\alpha (1 + t_1)^\beta F_1(t_1) \quad (24)$$

where, in general, α and β are complex numbers given by

$$\alpha = -\frac{1}{2} - i\omega, \quad \beta = -\frac{1}{2} + i\omega, \quad \omega = \frac{1}{2\pi}\log\left(\frac{1+\gamma}{1-\gamma}\right) \quad (25)$$

and the function $F_1(t_1)$ is continuous and bounded in $|t_1| \leq 1$. By substituting Eq. (24) into the right-hand side of Eq. (23) and by using the mathematical formulae we have the following approximate expressions for $(\sigma_x)_{x=0}$ and $(\tau_{xy})_{x=0}$ in the close neighborhood of the crack tips:

$$\alpha_{22}\sqrt{\alpha_1/\alpha_2}(\sigma_x)_{x=0} + i\alpha_{11}(\tau_{xy})_{x=0} = 2^\beta(t_1 - 1)^\alpha \sqrt{1 - \gamma^2}\,F_1(1). \quad (26)$$

Here, we define the stress intensity factors by

$$K_I^* + iK_{II}^* = \lim_{t_1 \to 1}\sqrt{\pi c_1}\,(t_1 - 1)^{-\alpha}(t_1 + 1)^{-\beta}\{\alpha_{22}\sqrt{\alpha_1/\alpha_2}(\sigma_x)_{x=0} + i\alpha_{11}(\tau_{xy})_{x=0}\} \quad (27)$$

and then substituting Eq. (26) into Eq. (27) we get

$$K_I^* + iK_{II}^* = \sqrt{\pi c_1}\sqrt{1 - \gamma^2}\,F_1(1). \quad (28)$$

In the numerical calculations, we use the following stress intensity factors instead of K_I^* and K_{II}^* to compare the present results with those of Erdogan's paper,[7] where the half-planes are both isotropic except for their different elastic constants:

$$K_I = \frac{1}{\alpha_{22}\sqrt{\alpha_1/\alpha_2}}K_I^*, \quad K_{II} = \frac{1}{\alpha_{11}}K_{II}^*. \quad (29)$$

On the other hand, the relative displacements on the interface crack surfaces are expressed in terms of $\phi(\eta)$ as follows:

$$U_x = \frac{1}{\alpha_{22}\sqrt{\alpha_1/\alpha_2}} \cdot \frac{1}{P_0 c_1}\,\text{Re}\int_{c_1}^{y}\phi(\eta)d\eta \quad (30\text{-}1)$$

$$V_y = \frac{1}{\alpha_{11}} \cdot \frac{1}{P_0 c_1}\,\text{Im}\int_{c_1}^{y}\phi(\eta)d\eta \quad (30\text{-}2)$$

where P_0 is a constant pressure $P(y) = P_0$.

III. NUMERICAL CALCULATIONS

The singular integral equation of the second kind given by Eq. (21) together with Eq. (22) can be solved by reducing it to an equivalent Hilbert problem, transforming it into a Fredholm integral equation with a regular kernel, or by expanding the solution in the form of orthogonal polynomials.

The method used here is one proposed by Theocaris et al.[8] where the singular integral equation is discretized by using the Lobatto-Jacobi integration rule with a finite number of collocation points. That is, substitution of Eq. (24) into Eqs. (21) and (22) and use of the Lobatto-Jacobi integration rule lead to a system of n linear algebraic equations for n unknown functions $F_1(s_{11})$, $F_1(s_{12})$, ..., $F_1(s_{1n})$ as follows:

$$\sum_{k=1}^{n} W_{1k} \frac{1}{s_{1k} - t_{1r}} F_1(s_{1k}) = \alpha_{11} Q(t_{1r}) - i \alpha_{22} \sqrt{\alpha_1/\alpha_2} P(t_{1r}) \quad (r=1, 2, \ldots, n-1)$$

$$\sum_{k=1}^{n} W_{1k} F_1(s_{1k}) = 0 \tag{31}$$

where W_{1k} is a weight, and s_{1k} and t_{1r} are collocation points[8] which are generally complex numbers except for $s_{11}=1$ and $s_{1n}=-1$. The merit of this method is that numerical calculation is straightforward and that $F_1(1)$ related to the stress intensity factors as in Eq. (28) can be obtained directly.

Figure 4 shows an example of the collocation points plotted on the complex plane. We find that s_{1k} and t_{1r} tend to approach the points on the real axis $[-1, 1]$ with an increased number of n. Table 1 compares the present solution for the stress intensity factors with the exact one obtained by Erdogan,[7] where the half-planes are both isotropic materials with the elastic constants of epoxy resin ($G=1.16$ GPa, $\nu=0.35$) and glass fiber ($G=28.73$ GPa, $\nu=0.2$), and a constant pressure $P(y)=P_0$ and $Q(y)=0$ is assumed on the crack surfaces. It is confirmed from this table that the results obtained by the present method have good convergency and precision.

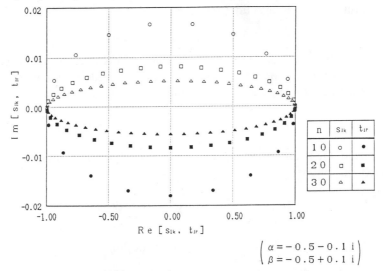

Fig. 4. Collocation points s_{1k} and t_{1r} on complex plane.

Table 1. Results by present method.

n	$K_I/P_0\sqrt{\pi c_1}$	$K_{II}/P_0\sqrt{\pi c_1}$
2	1.00004861	0.19251962
3	1.00000002	0.19251026
4	1.00000000	0.19251026
10	1.00000000	0.19251026
Exact solution:	(1.00000000)	(0.19251026)

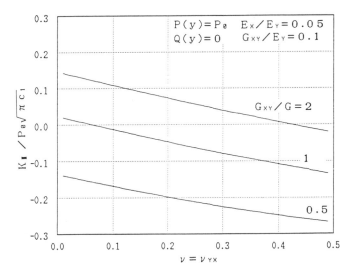

Fig. 5. Mode II stress intensity factor as a function of Poisson's ratio.

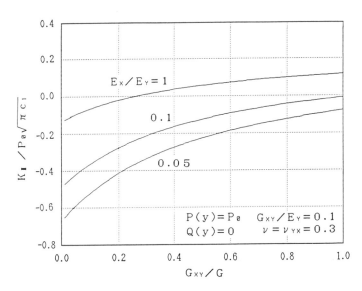

Fig. 6. Mode II stress intensity factor as a function of G_{xy}/G.

IV. NUMERICAL RESULTS

It is assumed that the crack surfaces are subjected to a constant pressure $P(y)=P_0$ and $Q(y)=0$. In this case the mode I stress intensity factor is $K_I=p_0\sqrt{\pi c_1}$, which is independent of the elastic constants of both half-planes. Therefore, in the following, only the mode II stress intensity factor K_{II} is considered. Figure 5 shows the normalized stress intensity factor $K_{II}/p_0\sqrt{\pi c_1}$ as a function of Poisson's ratio $\nu(=\nu_{yx})$ with a fixed value of $E_x/E_y=0.05$ and $G_{xy}/E_y=0.1$. The stress intensity factor decreases almost linearly with an increased value of $\nu=\nu_{yx}$, and the value is larger for a large value of G_{xy}/G.

Figure 6 shows the same factor as a function of G_{xy}/G with a fixed value of $\nu=\nu_{yx}=0.3$

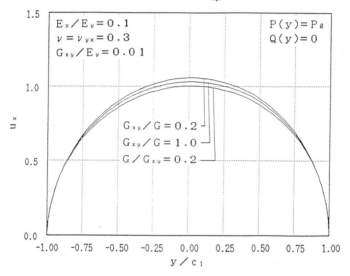

Fig. 7. Relative displacement u_x on crack surfaces.

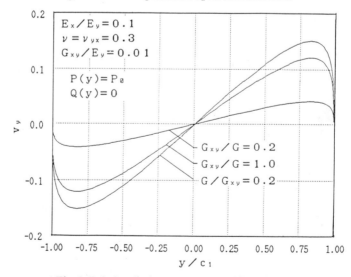

Fig. 8. Relative displacement v_y on crack surfaces.

and $G_{xy}/E_y=0.1$. It increases with an increased value of G_{xy}/G and is larger for a large value of E_x/E_y. Figures 7 and 8 show the relative displacements u_x and v_y on the crack surfaces as a function of positions with a fixed value of $E_x/E_y=0.1$, $\nu=\nu_{yx}=0.3$, and $G_{xy}/E_y=0.01$. Although the values of u_x are almost the same in the neighborhood of the crack tips irrespective of the different value of G_{xy}/G, there appears a slight difference between them in the middle part of the crack where the relative displacement in the x-direction is the largest. On the other hand, v_y yields the maximum value in the neighborhood of the crack tips and its maximum value is larger for a large value of G_{xy}/G.

V. CONCLUSIONS

The CDD method along with a suitable limiting process enables us to formulate an interface crack problem in terms of a Cauchy type singular integral equation of the second kind. This equation is solved straightforwardly using a numerical integration rule with a finite number of collocation points. The validity of the present method is confirmed from some numerical examples.

REFERENCES

1) Kasano, H., Watanabe, T., Matsumoto, H., and Nakahara, I., Singular stress fields at the tips of a crack normal to the bimaterial interface of isotropic and anisotropic half planes. *Bull. JSME*, Vol. 29, No. 258 (1986), pp. 4043–4049.
2) Kasano, H., Watanabe, T., Matsumoto, H., and Nakahara, I., Stress intensity factors and order of stress singularities at the tips of a crack normal to the bimaterial interface. *JSME Int. J.*, Vol. 30, No. 259 (1987) pp. 44–50.
3) Kasano, H., Watanabe, T., Matsumoto, H., and Nakahara, I., Influence of material inhomogeneity on the singular crack tip stress fields in composite materials. *Role of Fracture Mechanics in Modern Technology* (North-Holland, 1987), p. 695.
4) Kasano, H., Watanabe, T., Matsumoto, H., and Nakahara, H., Plane elastostatic analysis of a crack passing through a bimaterial interface. *JSME Int. J.*, Vol. 30, No. 267 (1987), pp. 1375–1382.
5) Kasano, H., Matsumoto, H., and Nakahara, I., Stress function approach in anisotropic elasticity theory and its application. *Preprint of 36th National Congress on Theoretical and Applied Mechanics* (1986), pp. 83–84 (in Japanese).
6) Erdogan, F., Simultaneous dual integral equations with trigonometric and Bessel kernels. *ZAMM.*, Vol. 48, No. 4 (1968), pp. 217–225.
7) Erdogan, F., et al., *Mechanics of Fracture*, Vol. 1 (Nodhoff Publishing Co., (1973), p. 369.
8) Theocaris, P.S. and Ioakimidis, N.I., A method of numerical solution of Cauchy type singular integral equation with generalized kernels and arbitrary complex singularities. *J. Computational Phys.*, Vol. 30 (1979), pp. 309–323.

Scattering of Elastic Waves in a Particulate Composite with Interfacial Layers (Phase Velocity and Attenuation)

Yasuhide SHINDO*, Subhendu K. DATTA,[2]* and Hassel M. LEDBETTER[3]*

Department of Mechanical Engineering II, Tohoku University, Sendai, [2] Department of Mechanical Engineering, University of Colorado, Boulder, [3]* Fracture and Deformation Division, National Bureau of Standards, Boulder

> The scattering of time harmonic compressional and shear waves by a spherical inclusion with an imperfect interface is considered. The interface is assumed to be a thin layer with nonhomogenous elastic properties. We apply the results of the single scattering problem to the propagation of coherent plane waves in a particle-reinforced composite medium. The composite medium contains a random distribution of inclusions of the same size with layers of the same thickness. Numerical results are obtained for a moderately wide range of frequencies, and the effects of interface properties on scattering cross sections, attenuation coefficients of energy, and phase velocities and attenuations of coherent plane waves are discussed in detail.

I. INTRODUCTION

The propagation of harmonic waves in a particle-reinforced composite material has been a subject of recent interest.[1-6] One often determines the effective elastic moduli and damping in such a composite material using ultrasound and nondestructively characterizing the macroscopic mechanical properties of the material. Mal and Bose[7] considered an isotropic elastic material containing a random distribution of identical and imperfectly bonded spherical particles of another elastic material and determined the dynamic elastic moduli of the composite material in the Rayleigh or low-frequency limit. At long wavelengths the wave speeds thus calculated are nondispersive and hence provide the values for the static effective elastic properties. Recently, Vardan et al.[8] developed a multiple scattering theory of elastic waves by randomly distributed spherical inclusions in an elastic medium and calculated phase velocity and coherent attenuation for longitudinal wave incidence. The scattering of an ultrasonic wave in the composite medium results in a frequency-dependent velocity and attenuation of the wave. The propagation of low frequency waves in composites containing dilute concentrations of spherical inclusions was considered by Norris,[9] and relations for the dispersive wave speeds and attenuations at low frequency, when the inclusions are voids, were derived.

For a composite medium with imperfect interfaces between the matrix and the second phase, the actual details of the calculation are complicated. In this paper we consider the scattering of time harmonic plane compressional and shear waves in a particulate-reinforced composite with interfacial layers. The problem of scattering of elastic waves by a spherical inclusion with an interface layer is analyzed and the results for the single scattering problem are then used to consider the elastic wave propagation in the composite. The composite contains a random distribution of spherical inclusions surrounded by thin boundary layers of variable elastic properties. The effective complex wave numbers follow from the coherent wave equations which depend only upon the scattering amplitude of the single scattering problem. Dispersive plane wave velocities and attenuations are studied as a function of concentrations of scatterers and frequency, and the effect of the interface properties on the coherent wave propagation is discussed in detail. The method of solution is such that the numerical results can be obtained at any desired finite frequency.

II. STATEMENT OF THE PROBLEM AND THE SINGLE SCATTERING FIELD

We consider a random distribution of identical spherical inclusions of radius a in an infinite matrix. Let λ_2, μ_2, ρ_2, ν_2 be the Lamé constants, the mass density, the Poisson's ratio of the matrix, and λ_1, μ_1, ρ_1, ν_1 those of the inclusions. We assume that a thin layer of uniform thickness h but variable material properties λ, μ is present at the interface separating the matrix from each sphere.

In order to study the plane-wave propagation in a particle-reinforced composite medium with interfacial layers, we first consider the scattered field due to a single spherical inclusion with an interface layer. Consider the composite sphere shown in Fig. 1. Let (x, y, z) be the Cartesian coordinate system with origin at the center of the sphere and (r, θ, ϕ) be the corresponding spherical polar coordinates. Let λ, μ be expressed as

$$\lambda + 2\mu = (\lambda_0 + 2\mu_0)f(r) \quad (a < r < a + h)$$
$$\mu = \mu_0 g(r) \quad (a < r < a + h) \quad (1)$$

where $f(r)$ and $g(r)$ are general functions of r, and λ_0 and μ_0 are Lamé constants of the interface material at some value of $r(a < r < a + h)$.

We consider a plane longitudinal (P) wave propagating in the positive z-direction or a plane shear (S) wave polarized in the x-direction and propagating in the positive z-direction. Thus,

$$e^{-i\omega t} \boldsymbol{u}^i = w_0 e^{i(k_1 z - \omega t)} \boldsymbol{e}_z + u_0 e^{i(k_2 z - \omega t)} \boldsymbol{e}_x \quad (2)$$

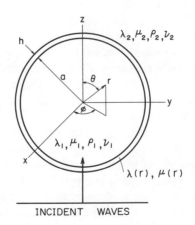

Fig. 1. A spherical inclusion with an interfacial layer and incident waves.

where ω is the circular frequency of the wave, t is the time, e_x, e_z are unit vectors in the axial directions, and w_0, u_0 are the amplitudes of the incident P and S waves. k_1, k_2 are the wave numbers of the P and S waves in the matrix,

$$k_1 = \omega/c_1$$
$$k_2 = \omega/c_2 \tag{3}$$

where $c_1 = \{(\lambda_2 + 2\mu_2)/\rho_2\}^{1/2}$, $c_2 = (\mu_2/\rho_2)^{1/2}$ denote the longitudinal and shear wave speeds in the matrix. The time factor $\exp(-i\omega t)$ will be omitted throughout the analysis.

We make the assumption that h is very much smaller than the wavelength of the propagating wave. The boundary conditions for this problem are

$$\sigma_{rr}^t = \sigma_{rr}^s + \sigma_{rr}^i$$
$$\sigma_{r\theta}^t = \sigma_{r\theta}^s + \sigma_{r\theta}^i$$
$$\sigma_{r\phi}^t = \sigma_{r\phi}^s + \sigma_{r\phi}^i \qquad (r = a) \tag{4}$$

$$u_r^s + u_r^i - u_r^t = \frac{hk_{1i}}{\lambda_0 + 2\mu_0}\sigma_{rr}^t$$

$$u_\theta^s + u_\theta^i - u_\theta^t = \frac{hK_{2i}}{\mu_0}\sigma_{r\theta}^t$$

$$u_\phi^s + u_\phi^i - u_\phi^t = \frac{hK_{2i}}{\mu_0}\sigma_{r\phi}^t \qquad (r = a) \tag{5}$$

where $\sigma_{ij}(\sigma_{rr}, \sigma_{r\theta}, \sigma_{r\phi})$ is the elastic stress tensor, $u_i(u_r, u_\theta, u_\phi)$ is the displacement vector, superscripts t, s, and i stand for the transmitted, scattered, and incident field quantities, respectively, and K_{1i} and K_{2i} are

$$K_{1i} = \int_0^1 \frac{dx}{f(a + hx)}$$

$$K_{2i} = \int_0^1 \frac{dx}{g(a + hx)}. \tag{6}$$

The scattered and transmitted displacement fields may be expressed in the forms[10]

$$\boldsymbol{u}^s = \sum_{n=0}^{\infty}\sum_{m=-1}^{1}[A_{mn}^s \boldsymbol{L}_{mn}^{(3)} + B_{mn}^s \boldsymbol{M}_{mn}^{(3)} + C_{mn}^s \boldsymbol{N}_{mn}^{(3)}] \tag{7}$$

$$\boldsymbol{u}^t = \sum_{n=0}^{\infty}\sum_{m=-1}^{1}[A_{mn}^t \boldsymbol{L}_{mn}^{(1)\prime} + B_{mn}^t \boldsymbol{M}_{mn}^{(1)\prime} + C_{mn}^t \boldsymbol{N}_{mn}^{(1)\prime}] \tag{8}$$

where A_{mn}^s, B_{mn}^s, C_{mn}^s, and A_{mn}^t, B_{mn}^t, C_{mn}^t are the unknown constants to be solved. Spherical vector wave functions $\boldsymbol{L}_{mn}^{(3)}$, $\boldsymbol{M}_{mn}^{(3)}$, and $\boldsymbol{N}_{mn}^{(3)}$ appearing above are given by

$$\boldsymbol{L}_{mn}^{(3)} = \left[\frac{\partial}{\partial r}h_n^{(1)}(k_1 r)P_n^m(\cos\theta)\boldsymbol{e}_r + h_n^{(1)}(k_1 r)\frac{1}{r}\frac{\partial}{\partial \theta}P_n^m(\cos\theta)\boldsymbol{e}_\theta \right.$$
$$\left. + \frac{im}{r\sin\theta}h_n^{(1)}(k_1 r)P_n^m(\cos\theta)\boldsymbol{e}_\phi\right]e^{im\phi}$$

$$\boldsymbol{M}_{mn}^{(3)} = \left[\frac{im}{\sin\theta} h_n^{(1)}(k_2 r) P_n^m(\cos\theta) \boldsymbol{e}_\theta - h_n^{(1)}(k_2 r) \frac{\partial}{\partial\theta} P_n^m(\cos\theta) \boldsymbol{e}_\phi\right] e^{im\phi}$$

$$\boldsymbol{N}_{mn}^{(3)} = \left[\frac{n(n+1)}{r} h_n^{(1)}(k_2 r) P_n^m(\cos\theta) \boldsymbol{e}_r + \frac{1}{r}\frac{\partial}{\partial r}\{rh_n^{(1)}(k_2 r)\}\frac{\partial}{\partial\theta} P_n^m(\cos\theta) \boldsymbol{e}_\theta\right.$$

$$\left. + \frac{im}{r\sin\theta}\frac{\partial}{\partial r}\{rh_n^{(1)}(k_2 r)\} P_n^m(\cos\theta) \boldsymbol{e}_\phi\right] e^{im\phi} \tag{9}$$

where $\boldsymbol{e}_r$, $\boldsymbol{e}_\theta$, $\boldsymbol{e}_\phi$ are the unit vectors along the r-, θ-, and ϕ-directions, $h_n^{(1)}$ is the spherical Hankel function of the first kind, and P_n^m is the associated Legendre function of the first kind. $\boldsymbol{L}_{mn}^{(1)\prime}$, $\boldsymbol{M}_{mn}^{(1)\prime}$, and $\boldsymbol{N}_{mn}^{(1)\prime}$ are obtained by replacing $h_n^{(1)}$, k_1, k_2 by the spherical Bessel function of the first kind j_n, $k_1' = \omega/c_1'$, $k_2' = \omega/c_2'$, respectively, in Eq. (9).

From boundary conditions (4) and (5), equations for the determination of A_{mn}^s, B_{mn}^s, and C_{mn}^s are found to be

$$A_{mn}^s/a = i(k_1 a)^3 [P_n \Phi_{mn} w_0 + Q_n X_{mn} u_0]$$
$$C_{mn}^s/a = i(k_2 a)^3 [R_n \Phi_{mn} w_0 + S_n X_{mn} u_0] \tag{10}$$
$$Q_b B_{mn}^s = -S_b Y_{mn} u_0 \tag{11}$$

where

$$\begin{bmatrix} i(k_1 a)^3 P_n & i(k_1 a)^3 Q_n \\ i(k_2 a)^3 R_n & i(k_2 a)^3 S_n \end{bmatrix} = -P_q^{-1} R_s$$

$$P_q = (2h/a)K_i L_{ns} - M_{ns} + (\mu_2/\mu_1)M_{nt} L_{nt}^{-1} L_{ns}$$

$$R_s = (2h/a)K_i L_{ni} - M_{ni} + (\mu_2/\mu_1)M_{nt} L_{nt}^{-1} L_{ni} \tag{12}$$

$$\Phi_{mn} = \frac{i^{n-1}}{k_1 a}(2n+1)\delta_{m0}$$

$$X_{mn} = \frac{i^{n-1}}{2k_2 a}\frac{2n+1}{n(n+1)}\{\delta_{m1} - n(n+1)\delta_{m,-1}\} \tag{13}$$

$$Q_b = (h/a)k_{22}^i\{(n-1)h_n^{(1)}(k_2 a) - k_2 a h_{n+1}^{(1)}(k_2 a)\} - h_n^{(1)}(k_2 a)$$
$$+ (\mu_2/\mu_1)\Delta(k_2' a)\{(n-1)h_n^{(1)}(k_2 a) - k_2 a h_{n+1}^{(1)}(k_2 a)\}$$

$$\Delta(k_2' a) = j_n(k_2' a)/\{(n-1)j_n(k_2' a) - k_2' a j_{n+1}(k_2' a)\} \tag{14}$$

$$Y_{mn} = \frac{i^{n-1}}{2}\frac{2n+1}{n(n+1)}\{\delta_{m1} + n(n+1)\delta_{m,-1}\}. \tag{15}$$

S_b is obtained by replacing $h_n^{(1)}$ by j_n in Eq. (14) and δ_{mn} is the Kronecker delta. In Eq. (12), the matrices $\boldsymbol{L}_{ni}$, $\boldsymbol{M}_{ni}$ are given by

$$\boldsymbol{L}_{ni} = \begin{bmatrix} L_{11}^{ni} & L_{12}^{ni} \\ L_{21}^{ni} & L_{22}^{ni} \end{bmatrix} \tag{16}$$

$$L_{11}^{ni} = \{n^2 - n - (k_2 a)^2/2\}j_n(k_1 a) + 2k_1 a j_{n+1}(k_1 a)$$
$$L_{12}^{ni} = n(n+1)\{(n-1)j_n(k_2 a) - k_2 a j_{n+1}(k_2 a)\}$$
$$L_{21}^{ni} = (n-1)j_n(k_1 a) - k_1 a j_{n+1}(k_1 a)$$
$$L_{22}^{ni} = \{n^2 - 1 - (k_2 a)^2/2\}j_n(k_2 a) + k_2 a j_{n+1}(k_2 a) \tag{17}$$

$$\boldsymbol{M}_{ni} = \begin{bmatrix} M_{11}^{ni} & M_{12}^{ni} \\ M_{21}^{ni} & M_{22}^{ni} \end{bmatrix} \tag{18}$$

$M_{11}^{ni} = nj_n(k_1 a) - k_1 a j_{n+1}(k_1 a)$

$M_{12}^{ni} = n(n+1)j_n(k_2 a)$

$M_{21}^{ni} = j_n(k_1 a)$

$$M_{22}^{ni} = (n+1)j_n(k_2 a) - k_2 a j_{n+1}(k_2 a). \tag{19}$$

$\boldsymbol{L}_{ns}$, $\boldsymbol{M}_{ns}$ are obtained from $\boldsymbol{L}_{ni}$, $\boldsymbol{M}_{ni}$, respectively, by replacing j_n and j_{n+1} by $h_n^{(1)}$ and $h_{n+1}^{(1)}$, respectively, in Eqs. (16) and (18). $\boldsymbol{L}_{nt}$, $\boldsymbol{M}_{nt}$ are also obtained from $\boldsymbol{L}_{ni}$, $\boldsymbol{M}_{ni}$, respectively, by replacing k_1 and k_2 by k_1' and k_2', respectively. The matrix $\boldsymbol{K}_i$ is given by

$$\boldsymbol{K}_i = \begin{bmatrix} k_{11}^i & k_{12}^i \\ k_{21}^i & k_{22}^i \end{bmatrix} \tag{20}$$

$k_{11}^i = \{\mu_2/(\lambda_0 + 2\mu_0)\} K_{1i}$

$k_{12}^i = k_{21}^i = 0$

$$k_{22}^i = (\mu_2/\mu_0) K_{2i}. \tag{21}$$

The scattered field at a large distance from the sphere follows from Eq. (7) letting r tend to ∞. This yields

$$u_r^s \sim \frac{e^{ik_2 r}}{r} g(\theta, \phi) \tag{22}$$

$$u_\theta^s \sim \frac{e^{ik_2 r}}{r} h_1(\theta, \phi)$$

$$u_\phi^s \sim \frac{e^{ik_2 r}}{r} h_2(\theta, \phi) \tag{23}$$

where

$$g(\theta, \phi) = \sum_{n=0}^{\infty} A_{0n}^s (-i)^n P_n(\cos\theta) \tag{24}$$

$$h_1(\theta, \phi) = \sum_{n=0}^{\infty} \sum_{m=-1}^{1} (-i)^n \left\{ C_{mn}^s \frac{\partial P_n^m}{\partial \theta}(\cos\theta) + \frac{B_{mn}^s}{k_2} \frac{m}{\sin\theta} P_n^m(\cos\theta) \right\} e^{im\phi}$$

$$h_2(\theta, \phi) = - \sum_{n=0}^{\infty} \sum_{m=-1}^{1} (-i)^{n+1} \left\{ C_{mn}^s \frac{m}{\sin\theta} P_n^m(\cos\theta) + \frac{B_{mn}^s}{k_2} \frac{\partial P_n^m}{\partial \theta}(\cos\theta) \right\} e^{im\phi}. \tag{25}$$

The function $g(\theta, \phi)$ is termed the far-field scattering amplitude for the scattering P waves and the functions $h_1(\theta, \phi)$ and $h_2(\theta, \phi)$ the far-field scattering amplitudes in the θ- and ϕ-directions, respectively, for the scattered S waves. The scattering cross sections for incident P and S waves are then[11]

$$\Sigma_p = \frac{4\pi}{k_1} I_m[g(0, \phi)]$$

$$= \frac{4\pi}{k_1} I_m\left[\sum_{n=0}^{\infty} (-i)^n A_{0n}^s\right] \quad (26)$$

$$\Sigma_s = \frac{4\pi}{k_2} I_m[b_\theta(0, \phi)h_1(0, \phi) + b_\phi(0, \phi)h_2(0, \phi)]$$

$$= \frac{4\pi}{k_2} I_m\left[\sum_{n=1}^{\infty} \frac{(-i)^n}{2}\left\{n(n+1)C_{1n}^s - C_{-1n}^s + \frac{n(n+1)}{k_2} B_{1n}^s + \frac{1}{k_2} B_{-1n}^s\right\}\right] \quad (27)$$

where

$$b_\theta = \cos\theta \cos\phi$$
$$b_\phi = -\sin\theta. \quad (28)$$

III. PLANE-WAVE PROPAGATION IN THE RANDOM PARTICULATE COMPOSITE

Once the scattered field due to a single inclusion is known, the effective wave speed and attenuation of the coherent waves through the composite can easily be calculated. We find that the attenuation coefficients of energy α_p and α_s are given by[12]

$$\alpha_p = (3c/4\pi\, a^3) \Sigma_p \quad (29)$$

$$\alpha_s = (3c/4\pi\, a^3) \Sigma_s \quad (30)$$

where c is the volume concentration of randomly distributed inclusions in the matrix.

It can be shown that for long wavelengths the effective wave speeds of plane longitudinal and shear waves are given by[6,7]

$$(k_{10}^*/k_1)^2 = \frac{(1 + 9cP_1)(1 + 3cP_0)\{1 + (3cP_2/2)(2 + 3\tau^2)\}}{1 - 15cP_2(1 + 3cP_0) + (3cP_2/2)(2 + 3\tau^2)} \quad (31)$$

$$(k_{20}^*/k_2)^2 = \frac{(1 + 9cP_1)\{1 + (3cP_2/2)(2 + 3\tau^2)\}}{1 + (3cP_2/4)(4 - 9\tau^4)}. \quad (32)$$

Using Eqs. (31) and (32), the static effective elastic properties λ^*, μ^*, ρ^*, and ν^* are then given by

$$\lambda^* + 2\mu^* = (\lambda_2 + 2\mu_2)\{-c(1 - \rho_1/\rho_2) + 1\}(k_1/k_{10}^*)^2$$

$$\mu^* = \mu_2\{-c(1 - \rho_1/\rho_2) + 1\}(k_2/k_{20}^*)^2$$

$$\rho^* = c\,\rho_1 + (1 - c)\,\rho_2$$

$$\nu^* = \frac{2 - (k_{20}^*/k_{10}^*)^2}{2\{1 - (k_{20}^*/k_{10}^*)^2\}}. \quad (33)$$

At low concentrations of inclusions we can use the following dispersion relations[9,13]

$$(K_1/k_1)^2 = 1 + (3c/k_1^2 a^3) \sum_{n=0}^{\infty} (-i)^n A_{0n}^s \quad (34)$$

$$(K_2/k_2)^2 = 1 + (3c/k_2^2 a^3) \sum_{n=1}^{\infty} \frac{(-i)^n}{2} \left\{ n(n+1)C_{1n}^s - C_{-1n}^s + \frac{n(n+1)}{k_2} B_{1n}^s + \frac{1}{k_2} B_{-1n}^s \right\}. \quad (35)$$

IV. NUMERICAL RESULTS AND DISCUSSION

To examine the effect of interface properties on the phase velocities and attenuation of coherent plane waves through the composite medium, the far-field scattering amplitudes (24) and (25) have been computed numerically. The considered composite was a lead-epoxy composite. The constituent properties are given in Table 1. Two special cases of the interface material are considered. Case I refers to the case of the interface material through which the elastic properties vary linearly from those of the inclusions to those of the matrix, and case II refers to the case when the interface material possesses constant properties. Then we have

Table 1. Material properties of lead-epoxy composite.

		ρ_1 (kg/m³)	μ_1 (GPa)	$\lambda_1 + 2\mu_1$ (GPa)	ν_1
Lead		11300	8.35	55.46	0.411
Epoxy		ρ_2 (kg/m³)	μ_2 (GPa)	$\lambda_2 + 2\mu_2$ (GPa)	ν_2
		1202	1.71	8.36	0.372

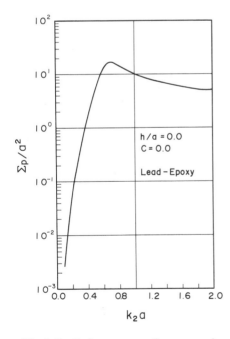

Fig. 2. Scattering cross section versus frequency for P waves.

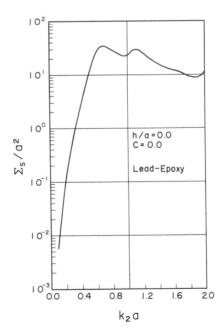

Fig. 3. Scattering cross section versus frequency for S waves.

(I) $$f(r) = \frac{\lambda_1 + 2\mu_1}{\lambda_0 + 2\mu_0} + \frac{\lambda_2 + 2\mu_2 - (\lambda_1 + 2\mu_1)}{\lambda_0 + 2\mu_0}(r-a)/h$$

$$g(r) = \frac{\mu_1}{\mu_0} + \frac{\mu_2 - \mu_1}{\mu_0}(r-a)/h$$

$$K_{1i} = \frac{\lambda_0 + 2\mu_0}{\lambda_2 + 2\mu_2 - (\lambda_1 + 2\mu_1)} \ln\left\{1 + \frac{\lambda_2 + 2\mu_2 - (\lambda_1 + 2\mu_1)}{\lambda_1 + 2\mu_1}\right\}$$

$$K_{2i} = \frac{\mu_0}{\mu_0 - \mu_1} \ln\left(1 + \frac{\mu_2 - \mu_1}{\mu_1}\right) \tag{36}$$

(II) $$f(r) = g(r) = 1$$

$$K_{1i} = K_{2i} = 1$$

$$\lambda_0 + 2\mu_0 = \{(\lambda_1 + 2\mu_1) + (\lambda_2 + 2\mu_2)\}/2$$

$$\mu_0 = (\mu_1 + \mu_2)/2. \tag{37}$$

In Figs. 2 and 3, the normalized scattering cross sections Σ_p/a^2 and Σ_s/a^2 are plotted as functions of frequency k_2a for $h/a=0.0$ and $c=0.0$. The effect of a thin interface layer on Σ_p/a^2 and Σ_s/a^2 at different frequencies for $c=0.0$ is shown in Table 2. Figure 4 and 5 show the variations of the attenuations of energy α_p/k_1 and α_s/k_2 with the frequency at different values of the volume fraction c for $h/a=0.0$. The attenuations α_p/k_1 and α_s/k_2 are calculated after replacing the matrix properties by the effective composite properties at low frequencies given by Eq. (33). The attenuation curves for α_p/k_1 have increasing peaks at smaller frequencies with decreasing concentration. The peaks of the attenuation curves

Table 2. Effect of interface on scattering cross sections.

Lead-epoxy	k_2a / h/a	0.1	0.2	0.5	1.0	1.5	2.0
Σ_p/a^2	0.0	0.325×10^{-2}	0.588×10^{-1}	0.537×10^1	0.102×10^2	0.634×10^1	0.515×10^1
	0.1 (I)	0.326×10^{-2}	0.592×10^{-1}	0.574×10^1	0.988×10^1	0.609×10^1	0.505×10^1
	0.1 (II)	0.325×10^{-2}	0.591×10^{-1}	0.568×10^1	0.993×10^1	0.613×10^1	0.506×10^1
Σ_s/a^2	0.0	0.725×10^{-2}	0.130×10^0	0.112×10^2	0.234×10^2	0.137×10^2	0.110×10^2
	0.1 (I)	0.724×10^{-2}	0.131×10^0	0.120×10^2	0.269×10^2	0.126×10^2	0.109×10^2
	0.1 (II)	0.724×10^{-2}	0.130×10^0	0.119×10^2	0.261×10^2	0.128×10^2	0.109×10^2

Table 3. Effect of interface on attenuations of energy.

Lead-epoxy	k_2a / h/a	0.1	0.2	0.5	1.0	1.5	2.0
α_p/k_1 ($c=0.1$)	0.0	0.123×10^{-2}	0.115×10^{-1}	0.434×10^0	0.377×10^0	0.209×10^0	0.103×10^0
	0.1 (I)	0.125×10^{-2}	0.118×10^{-1}	0.481×10^0	0.380×10^0	0.208×10^0	0.103×10^0
	0.1 (II)	0.124×10^{-2}	0.117×10^{-1}	0.472×10^0	0.381×10^0	0.208×10^0	0.103×10^0
α_s/k_2 ($c=0.1$)	0.0	0.126×10^{-2}	0.116×10^{-1}	0.379×10^0	0.336×10^0	0.259×10^0	0.110×10^0
	0.1 (I)	0.127×10^{-2}	0.118×10^{-1}	0.421×10^0	0.355×10^0	0.292×10^0	0.107×10^0
	0.1 (II)	0.127×10^{-2}	0.118×10^{-1}	0.413×10^0	0.354×10^0	0.291×10^0	0.107×10^0

Scattering of Elastic Waves in a Particulate Composite 215

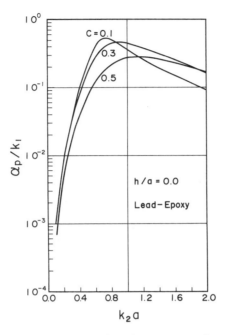

Fig. 4. Effect of concentration on attenuation of energy versus frequency for P waves.

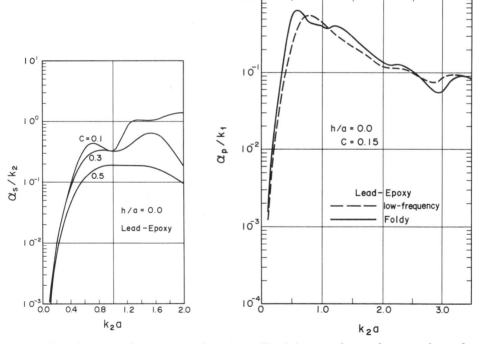

Fig. 5. Effect of concentration on attenuation of energy versus frequency for S waves.

Fig. 6. A comparison of attenuation of energy versus frequency for P waves between low frequency and Foldy.

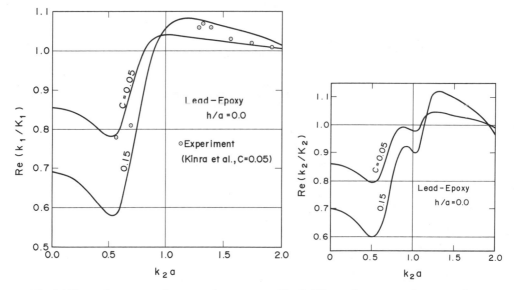

Fig. 7. Effect of concentration on phase velocity versus frequency for P waves.

Fig. 8. Effect of concentration on phase velocity versus frequency for S waves.

Table 4. Effect of interface on phase velocities.

Lead-epoxy	k_2a h/a	0.1	0.2	0.5	1.0	1.5	2.0
	0.0	1.318	1.340	1.496	0.951	0.958	0.988
$Re(k_1/K_1)$	0.1 (I)	1.323	1.346	1.513	0.941	0.958	0.990
($c=0.1$)	0.1 (II)	1.322	1.345	1.510	0.943	0.957	0.990
	0.0	1.310	1.331	1.473	1.057	0.937	1.018
$Re(k_2/K_2)$	0.1 (I)	1.316	1.338	1.492	1.071	0.936	1.019
($c=0.1$)	0.1 (II)	1.315	1.337	1.489	1.069	0.936	1.019

for α_s/k_1 also increase in magnitude with decreasing concentration. In Fig. 6, a comparison of the attenuation α_p/k_1 is made between low frequency and Foldy for $h/a=0.0$ and $c=0.15$. The attenuation α_p/k_1 for Foldy is calculated with matrix properties replaced by composite properties given by Eqs. (34) and (35). The agreement is good for the frequency considered. The interface effect on α_p/k_1 and α_s/k_2 at different frequencies for $c=0.1$ is shown in Table 3. The interface effect is pronounced in the neighborhood of k_2a reaching the peaks, and increases the attenuations α_p/k_1 at $k_2a=0.1$, 0.2, 0.5, and 1.0 and α_s/k_1 at $k_2a=0.1$, 0.2, 0.5, 1.0, and 1.5.

In Fig. 7, the normalized phase velocity $Re(k_1/K_1)$ for P waves is plotted as a function of frequency k_2a for $h/a=0.0$ and $c=0.05$ and 0.15. A comparison of the phase velocity made between theory and experiment[14] for $c=0.05$. The agreement is good for the frequency considered. The normalized phase velocity $Re(k_2/K_2)$ for S waves is also shown in Fig. 8. The phase velocities are sensitive to the frequency. In Table 4, we have shown the interface effect on $Re(k_1/K_1)$ and $Re(k_2/K_2)$ at different frequencies for $c=0.1$. The effect of the interface depends on the frequency. Figures 9 and 10 show the variations of the

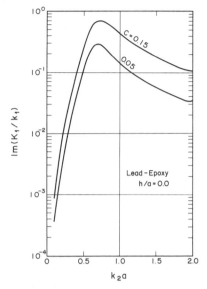

Fig. 9. Effect of concentration on coherent attenuation versus frequency for P waves.

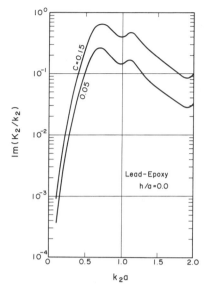

Fig. 10. Effect of concentration on coherent attenuation versus frequency for S waves.

Table 5. Effect of interface on coherent attenuations.

Lead-epoxy	k_2a \ h/a	0.1	0.2	0.5	1.0	1.5	2.0
$I_m(K_1/k_1)$ ($c=0.1$)	0.0	0.651×10^{-3}	0.579×10^{-2}	0.190×10^0	0.283×10^0	0.117×10^0	0.687×10^{-1}
	0.1 (I)	0.649×10^{-3}	0.580×10^{-2}	0.200×10^0	0.277×10^0	0.112×10^0	0.673×10^{-1}
	0.1 (II)	0.650×10^{-3}	0.580×10^{-2}	0.198×10^0	0.278×10^0	0.113×10^0	0.675×10^{-1}
$I_m(K_2/k_2)$ ($c=0.1$)	0.0	0.660×10^{-3}	0.583×10^{-2}	0.182×10^0	0.264×10^0	0.117×10^0	0.643×10^{-1}
	0.1 (I)	0.656×10^{-3}	0.582×10^{-2}	0.193×10^0	0.299×10^0	0.107×10^0	0.636×10^{-1}
	0.1 (II)	0.657×10^{-3}	0.582×10^{-2}	0.191×10^0	0.292×10^0	0.109×10^0	0.637×10^{-1}

normalized attenuations of $I_m(K_1/k_1)$ and $I_m(K_2/k_2)$ with the frequency k_2a for $h/a=0.0$ and $c=0.05$ and 0.15. The interface effect on the attenuations is also shown in Table 5.

In conclusion, scattering of compressional and shear waves by a spherical inclusion with an interface was analyzed and the results of the single scattering problem were applied to plane-wave propagation in a particulate composite with interfacial layers. The interface effect can increase or decrease scattering cross sections, attenuation coefficients of energy, and effective phase velocities and attenuations, and depend on the frequency. The numerical results were obtained for any given finite frequencies. The interface effect for case I is more pronounced than that for case II.

REFERENCES

1) Kinra, V.K., Petraitis, M.S., and Datta, S.K., Ultrasonic wave propagation in a random particulate composite. *Int. J. Solids Struct.*, Vol. **16**, No. 4 (1980), pp. 301–312.

2) Kinra, V.K. and Anand, A., Wave propagation in a random particulate composite at long and short wavelengths. *Int. J. Solids Struct.*, Vol. **18**, No. 5 (1982), pp. 367–380.
3) Selvadula, A.P.S. and Voyiadjis, G. (eds.), *Mechanics of Material Interfaces*, Amsterdam: Elsevier (1986), p. 131.
4) Sayers, C.M. and Smith, R.L., Ultrasonic velocity and attenuation in an epoxy matrix containing lead inclusions. *J. Phys. D. Appl. Phys.*, Vol. **16**, No. 7 (1983), pp. 1189–1194.
5) Sayers, C.M., Scattering of ultrasound by minority phases in polycrystalline metals. *Wave Motion*, Vol. **7**, No. 1 (1985), pp. 95–104.
6) Ledbetter, H.M. and Datta, S.K., Effective wave speeds in an SiC-particle-reinforced Al composite. *J. Acoust. Soc. Am.*, Vol. **79**, No. 2 (1986), pp. 239–248.
7) Mal, A.K. and Bose, S.K., Dynamic elastic moduli of a suspension of imperfectly bonded spheres. *Proc. Camb. Phil. Soc.*, Vol. **76**, No. 3 (1974), pp. 587–600.
8) Varadan, V.K., Ma, Y., and Varadan, V.V., A multiple scattering theory for elastic wave propagation in discrete random media. *J. Acoust. Soc. Am.*, Vol. **77**, No. 2 (1985), pp. 375–385.
9) Norris, A.N., Scattering of elastic waves by spherical inclusions with applications to low frequency wave propagation in composites. *Intl. J. Engng. Sci.*, Vol. **24**, No. 8 (1986), pp. 1271–1282.
10) Stratton, J.A., *Electromagnetic Theory*, McGraw-Hill Book Co. (1941), p. 414.
11) Barratt, P.J. and Collins, W.D., The scattering cross-section of an obstacle in an elastic solid for plane harmonic waves. *Proc. Camb. Phil. Soc.*, Vol. **61**, No. 4 (1965), pp. 969–981.
12) Yamakawa, Y., Scattering and attenuation of elastic waves. *Geophys. Mag.*, Vol. **31**, No. 1 (1962), pp. 97–103.
13) Foldy, L.L., The multiple scattering of waves. *Phys. Rev.*, Vol. **67**, No. 3,4 (1945), pp. 107–119.
14) Kinra, V.K., Ker, E., and Datta, S.K., Influence of particle resonance on wave propagation in a random particulate composite. *Mech. Res. Comm.*, Vol. **9**, No. 2 (1982), pp. 109–114.

Whole-Field Strain Measurements by Moiré and their Application to Composite Materials

Kazuo Kunoo, Hiroichi Ohira, Kousei Ono, and Naoki Sato

Department of Aeronautical Engineering, Kyushu University, Fukuoka

In this paper we develop a method to interpolate moiré fringe data by a piecewise linear interpolation function in a triangular region. The experimental data interpolated are so flexible that results can easily be expressed in desired form, such as displacement, three components of strains, principal strains, and strains in directions parallel or perpendicular to a fiber direction in the case of composite materials.

Comparisons between digital moiré analyses and strain gauge analyses showed good agreement and confirmed the validity of digital moiré analyses in practical use. We performed an experiment on a composite laminate with a hole and showed the results in strain contour lines.

I. INTRODUCTION

Composite materials with their high strength per density ratio are used for primary aircraft structures. However, it is difficult to understand their behavior completely because composite materials have strong anisotropy. Experimental strain analysis is usually conducted using a conventional strain gauge method. It is hard, however, to measure whole-field strains with a strain gauge method which gives point-to-point information and is therefore not suitable for understanding the complex mechanisms of composite materials.

On the other hand, conventional moiré analysis yields whole-field displacement information and can be useful to apply for composite materials. However moiré analysis of strains requires such tedious and time-consuming work that the amount of generated data that must be reduced can cause disenchantment with the whole-field method. To cope with these difficulty, we introduce digital image processing technique.

In a previous study,[1] we developed a digital image processing system for moiré analysis of strains, where a digital image processing technique was developed and applied to a simple extension test. A comparison between the strain gauge method and moiré method showed good agreement.

In this report, we study a method to interpolate extracted moiré fringe data by piecewise linear functions. This method is very useful for the experimental analysis of a specimen with irregular boundaries such as a hole, a notch, or a cutout. Next we apply the digital image processing system to a composite laminate specimen with a hole subjected to uniaxial tension.

II. INTERPOLATION OF MOIRÉ FRINGES

Moiré fringes processed by the digital image processing system express contour lines of specimen displacement. We will get two sets of specimen displacement in the x- and y-

directions for each loading. In the experiment a whole region of moiré image is divided into triangular subregions. Introduction of a piecewise linear function to interpolate moiré fringe displacement in a triangular region makes it possible to treat a specimen with complicated boundaries such as a hole, a notch, or a cutout. Then the procedure to process moiré fringe data is almost the same as the finite element analysis used for structures.

We apply the least squares approximation to get interpolation functions of measured moiré fringe data. Least squares error between measured displacement u_m, v_m, and interpolation function $u(x, y)$, $v(x, y)$ is given by Eq. (1)

$$E = \frac{1}{2} \sum_m^I w_m[u(x_m, y_m) - u_m]^2 + \frac{1}{2} \sum_m^J w_m[v(x_m, y_m) - v_m]^2 + qU \qquad (1)$$

where w, q is a weighting factor, I, J are the number of data u, v, respectively, given in a triangular region. U is the strain energy of a plate for the case of plane stress. If strain energy of a plate is not added to least squares error, a matrix for an element that has no data becomes singular.

By the assumption of linear displacement within a triangular element as shown in Fig. 1, the following equations are obtained

$$u = \alpha_1 + \alpha_2 x + \alpha_3 y, \qquad v = \beta_1 + \beta_2 x + \beta_3 y. \qquad (2)$$

We define nodal point displacement as

$$\{\delta\}^T = [u_i v_i u_j v_j u_k v_k]. \qquad (3)$$

Displacement in a triangular element is expressed by nodal point displacement as

$$u = [N_i\ 0\ N_j\ 0\ N_k\ 0]\{\delta\}, \qquad v = [0\ N_i\ 0\ N_j\ 0\ N_k]\{\delta\} \qquad (4)$$

where

$$N_1 = (a_1 + b_1 x + c_1 y)/A, \quad 1 = i, j, k. \qquad (5)$$

$$a_i = x_j y_k - x_k y_j, \quad b_i = y_j - y_k, \quad c_i = x_k - x_j. \qquad (6)$$

Substituting Eq. (4) into Eq. (1), and taking the derivative with respect to nodal point displacement, we get

$$\partial E/\partial\{\delta\} = \sum_m^I w_m\{[N_x]_m^T[N_x]_m\{\delta\} - u_m[N_x]_m^T\} + \sum_m^J w_m\{[N_y]_m^T[N_y]_m\{\delta\} - v_m[N_y]_m^T\} + q[K_E]\{\delta\} = 0. \qquad (7)$$

Therefore

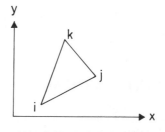

Fig. 1. Triangular finite element.

$$([K_x] + [K_y] + q[K_E])\{\partial\} = \{P_x\} + \{P_y\} \tag{8}$$

where

$$[K_x] = \sum_m^I w_m [N_x]_m^T [N_x]_m, \quad [K_y] = \sum_m^J w_m [N_y]_m^T [N_y],$$

$$\{P_x\} = \sum_m^I w_m u_m [N_x]_m^T, \quad \{P_y\} = \sum_m^J w_m v_m [N_y]_m^T \tag{9}$$

$$[N_x] = \begin{bmatrix} N_i^2 & 0 & N_i N_j & 0 & N_i N_k & 0 \\ 0 & 0 & 0 & 0 & 0 & 0 \\ N_i N_j & 0 & N_j^2 & 0 & N_j N_k & 0 \\ 0 & 0 & 0 & 0 & 0 & 0 \\ N_i N_k & 0 & N_j N_k & 0 & N_k^2 & 0 \\ 0 & 0 & 0 & 0 & 0 & 0 \end{bmatrix}$$

$$[N_y] = \begin{bmatrix} 0 & 0 & 0 & 0 & 0 & 0 \\ 0 & N_i^2 & 0 & N_i N_j & 0 & N_i N_k \\ 0 & 0 & 0 & 0 & 0 & 0 \\ 0 & N_i N_j & 0 & N_j^2 & 0 & N_j N_k \\ 0 & 0 & 0 & 0 & 0 & 0 \\ 0 & N_i N_k & 0 & N_j N_k & 0 & N_k^2 \end{bmatrix} \tag{10}$$

where A is an area of a triangular element, and $[K_E]$ is a stiffness matrix of a triangular element for constant strains.

In Eq. (8), matrices are symmetric, which means that we may use a conventional finite element program to solve Eq. (8) for nodal displacement. Strains within an element are constant and given by Eq. (11)

$$\begin{Bmatrix} \varepsilon_x \\ \varepsilon_y \\ \gamma_{xy} \end{Bmatrix} = \frac{1}{2A} \begin{bmatrix} b_i & 0 & b_j & 0 & b_k & 0 \\ 0 & c_i & 0 & c_j & 0 & c_k \\ c_i & b_i & c_j & b_j & c_k & b_k \end{bmatrix} \{\partial\}. \tag{11}$$

III. EXPERIMENTS ON COMPOSITE MATERIALS

The specimen was a 16-ply carbon fiber reinforced laminate 40 mm in width, 2.7 mm in thickness of 45° degree symmetric construction with a 6-mm diameter hole and was loaded in uniaxial tension. Use of moiré analysis requires printing of a 40-line/mm grating to the composite specimen. Direct printing of the grating is impossible because carbon fiber composites are black. To manage this problem, we first printed the grating on an aluminum vaporized polystylene film and then stuck it on the surface of a carbon fiber composite specimen.

Figure 2 shows the original image of moiré fringes loaded 10016N. The finite element mesh layout used in this experiment is shown in Fig. 3. In each triangular region, a piece-

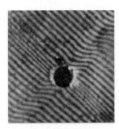 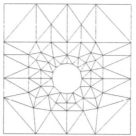 

Fig. 2. Original imupage at 10016N load.   Fig. 3. Finite element mesh layout.   Fig. 4. Extracted moiré fringe and interpolated fringe.

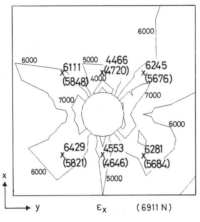 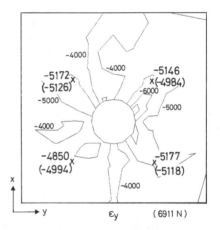

Fig. 5. Normal strain distribution in x-direction at 6911N load.

Fig. 6. Normal strain distribution in y-direction at 6911N load.

wise linear interpolation function was calculated. Figure 4 shows the moiré fringes extracted from the original image by means of digital image processing and contour dotted lines calculated by the method given above. There are some discrepancies between extracted and interpolated moiré fringes because a rather coarse mesh layout was used.

Figure 5 shows the contour lines for normal strains in the direction of applied load 6911N. The bracketed numerals indicate the strain gauge results at the point shown by x; the numerals without parentheses at those points show the results of moiré analyses. Comparisons of the moiré and strain gauge method shows fair agreement. Figure 6 shows the contour lines for normal strains perpendicular to the applied load. The same notations given in Fig. 5 are used. Again the results of the moiré method agree well with those of the strain gauge method.

Figure 7 shows the contour lines for shear strains. In this case, shear strains calculated by classical lamination theory is zero. The strains displayed by Fig. 7 are very small compared with normal strains shown in Fig. 5 or Fig. 6, confirming the validity of moire analysis. The strain concentration is not too high at this stage of loading.

Figure 8 shows the contour lines for normal strains in the direction of applied load 10016N. Strain concentrations near the hole, especially in the 45° direction are prominent.

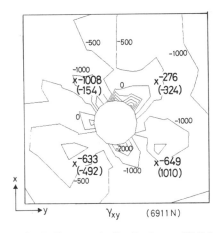

Fig. 7. Shear strain distribution at 6911N

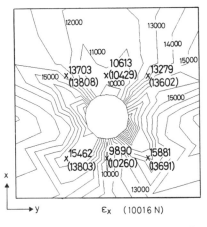

Fig. 8. Normal strain distribution in

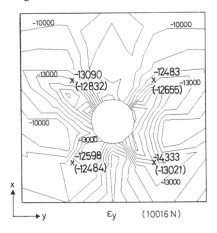

Fig. 9. Normal strain distribution in
y-direction at 10016N load.

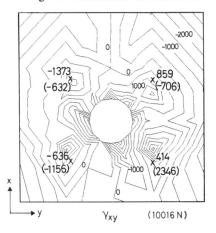

Fig. 10. Shear strain distribution at
10016N load.

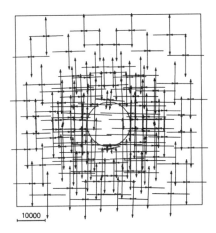

Fig. 11. Principal strain distribution at 10016N load.

Figure 9 shows the contour lines for normal strains perpendicular to the applied load. There is strain concentration around the hole. Figure 10 shows the contour lines for shear strains. In this case, strain concentrations can be seen around the hole, but their magnitude is rather small compared with normal strains. The principal strain distribution is shown in Fig. 11, where the direction of the principal strains is almost parallel to the x, y coordinate. This is one of the features of symmetric laminates. Comparison between moiré analysis and strain gauge analysis of strains in Figs. 4–10 shows fair agreement.

IV. DISCUSSIONS AND CONCLUSIONS

In this report, we developed a digital image processing system for moiré analysis of strains and applied it to a composite laminate with a hole. Comparison between moiré and strain gauge analyses results showed good agreement and confirmed the validity of moiré analyses in practical use.

The use of digital techniques for fringe analysis greatly increases the potential of a moiré analysis of strains. Our digital image processing system is so flexible that experimental data can easily be converted into desired form, such as displacement, 3 components of strains, principal strain, and strains in a fiber direction or a direction perpendicular to a fiber of a composite laminate.

Acknowledgments

The authors would like to thank Mr. Tadasi Nagayasu, technical assistant, Kyushu University, for his contribution to our computer software. We would also like to express our gratitude to Mr. Masahiro Ogawa and Mr. Toshihiro Mae, Kyushu University graduate students, for cooperating in the experiments.

This study was supported in part by a grant-in-aid for Scientific Research from the Ministry of Education, Japan.

REFERENCE

1) Kunoo, K., Ohira, H., Ono, K., and Sato, N., Digital image processing for moiré analysis of strain. *Theoretical and Applied Mechanics*, Vol. 36 (Univ. Tokyo Press, 1988), pp. 213–217.

VI

MECHANICS IN NEW ENVIRONMENTS

Space Structures and Applied Mechanics

——Morphology in Space——

Koryo MIURA

The Institute of Space and Astronautical Science, Sagamihara, Kanagawa

I. INTRODUCTION

New types of structural forms are emerging in space technology. Recognition of the need for such new forms became a reality when astronauts returning to earth after a long stay in space discovered that they could not walk in a normal upright position after they left the spacecraft. Our body structures were created under the gravitational vectors which restrict various structures on earth. When the astronauts returned, we learned that the morphological development of all structures on earth, including minerals and architectural and biological structures, are dependent on the gravitational principle.

It is interesting to consider what type of scientific principles in space (mechanical and geometrical) are imposed on various structures to be developed in space. What are the morphological structures produced under the new principles and what are the structural concepts for them? The answers to these questions will fill in the blank pages on space morphology in the near future, but at present, the following is a compilation of bits and pieces of information on these topics gathered from the limited experiences of the author.

II. PRINCIPLE OF MORPHOLOGICAL DEVELOPMENT IN SPACE

There are two major conditions: the mechanical condition and the geometrical condition. Let us use a solar sail as an example of a basic space structure. A solar sail is a space ship that is used in interplanetary navigation and is propelled by the flux of solar light pressure. The major components of the structure are, as shown in Fig. 1, a sail consisting of four triangular reflecting membranes with frames and cables to support the sail. A large antenna and a solar-power satellite have similar large planes and frames. Various types of loading pressure exert on these space structures are shown in Fig. 2.[1]

The two major environmental loads are the aerodynamic resistance at lower altitudes and the solar pressure which is constant near the earth. Other forces, such as solar wind and electromagnetic force due to the charging of the spaceship, are fairly low in terms of their order of magnitude when compared with the former two factors. The other type of force is inertial and depends on the distribution of mass. As can be seen in the figure, whatever the factors may be, there is no single dominant force comparable to gravity. This is the major environmental characteristic in space.

Next we must determine the factors of the geometrical condition. Space structures are

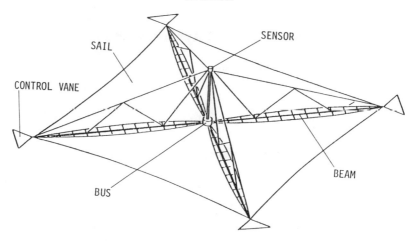

Fig. 1. A solar sail.

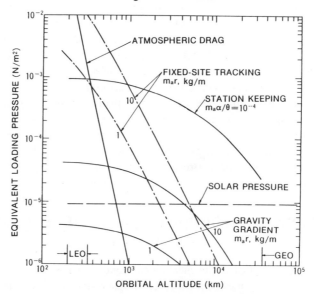

Fig. 2. Loads in space environment.

generally large. This is a result of the fact that the electromagnetic waves and sunlight being dealt with are very weak. Certainly there is no spatial limitation in space or on the earth, although the means of transportation by rockets and shuttles is stringently restricted. There are two types of transportation and construction methods of space structures: the erectable type in which structural materials are transported to space and constructed, and the deployable type in which the structures are built and packed on the earth and then deployed in space. In particular, the latter geometrical transformation is not seen in structures on the earth and has a special characteristic not described here. As a natural extension of the transformation, there exists very active transformation which can be called metamorphosis.

III. FOLDING ONE-DIMENSIONAL STRUCTURES

The most basic items in a folding one-dimensional structure are beams and columns and are found in the solar sail. The primary condition is the geometrical condition. The issue here is how to fold easily these giant one-dimensional structures. The answer is found not in the telescopic method of inserting cylinders within one another and folding them down, but rather, in forming the material into the shape of a coil.

Figure 3 shows the structure, called the coilable longeron mast, to be attached to the Aurora Observation Satellite, EXOS-D. The main feature is that three longerons are packed into a coil. The surprising phenomenon is that when a stretched mast is pulled in the shrinking direction, i.e., when an axial force is applied, the structure automatically forms a coil structure. Superficially, there is no external force used to produce the coil structure in this case. The question remains as to how this coil structure is produced. A similar phenomenon

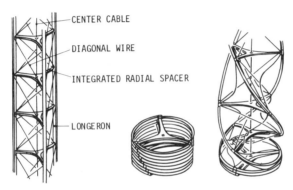

Fig. 3. A coilable longeron mast.

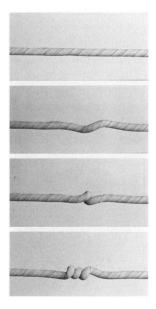

Fig. 4. Knotting formation of rubber bands.

is observed in the case of model airplanes powered by rubber bands. As one winds the band, it becomes tight and twists. When it reaches a certain (limiting) point, the first knot develops. As the winding continues, more knots are produced, and thus the number of knots increases (Fig. 4). The development of the knotting formation is discrete and never continuous. Therefore, no half-knots are produced. A knot is a single coil, and thus the above phenomenon is a jumping phenomenon from a torsion to a bending mode.

In terms of the resulting actions, the folding phenomenon has a close relationship to the three-dimensional elastica.[2] When solving Kirchhoff's equation by imposing various boundary conditions on a thin stick, coil-shaped solutions can be obtained. Thus, in the method of folding a longeron, the longeron buckles by the axial forces and a coil is formed. Then the coil can be folded into a completely flat ring. This process is repeated successively. In theory, the phenomenon can be outlined as above; however, as a practical engineering problem, this is merely a starting point. The three-dimensi onal ela sticahas many other solutions. Even if desired solutions are obtained, there is no definite rule as to when and where one knot appears and the subsequent knots develop. In the designing process, a knot must start at a designated end and the subsequent knot must follow immediately after the first one. That is, elastic instability must be under complete control.[3]

IV. FOLDING OF TWO-DIMENSIONAL STRUCTURES (MEMBRANES)

Giant plane membranes will frequently be used in the near future in space missions. As with the solar sail mentioned above, the solar-powered satellite, space radar, and lens antennas are typical examples of these two-dimensional type structures. Therefore, the technology necessary for the construction and packing of these giant membranes on the earth and their deployment in space is needed.

When the geometrical problem of folding a plane is formally (superficially) presented, it can take the form of the figure outlined in Fig. 5. At this stage, let us find the method needed to uniformly fold an arbitrary plane into a point. Here, the solution must be periodical in two orthogonal directions. This is obvious when one considers the huge size of the plane; in this case, a nonperiodic solution is impractical.

The first idea in solving this geometrical problem involves the application of the theory

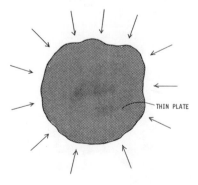

Fig. 5. Geometric problem of folding a plane.

of elasticity which consists of the following syllogism.[4,5] Let us consider it hypothetically, as if it is an elastic deformation problem of a plate. Then the problem can be solved by the finite deformation differential equation for a plate presented by von Kármán. By making the thickness of the plate unlimitedly close to zero, the first geometrical problem can be solved. In the unlimitedly thin plate, the ratio of the bending to the in-plane component of distortion energy is unlimitedly close to zero. The deformation, as can be represented by a piece of paper, must be the solution of inextensional deformation of infinite in-plane stiffness, that is, the solution of folding a plane.

Figure 6 shows 10 periodic solutions arbitrarily selected for the solution calculated using the procedure described above. It shows the counter lines of their out-of-plane deformation. When the least energy solutions were selected from these solutions, the final relief of the solution is as shown in Fig. 7. It may be too exaggerated to say that the solution can be obtained within the application limit (finite deformation) of the von Karman equations. However, it can be confirmed that by combining the solution of finite deformation with a solution of geometric approach, the solution is definitely true.

Aside from the original purpose of the problem, it is important to note that there has

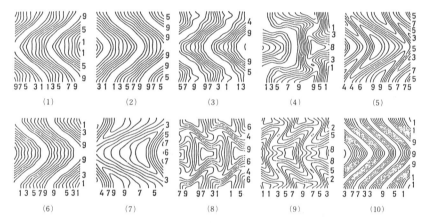

Fig. 6. Various solutions for deformation of a biaxially shortened infinite plate.

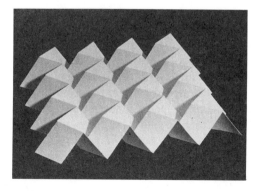

Fig. 7. Relief of a solution of folding a plane.

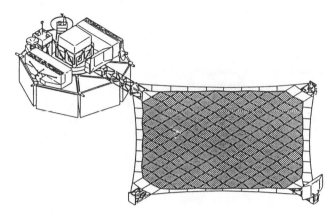

Fig. 8. A space flyer and two-dimensional array.

been a great deal of discussion regarding the theoretical meaning of the solution. That is, its relationship to elastica. When the classification of elastica by Konopasek[6] is utilized, the solution corresponds to the three-dimensional elestica of a two-dimensional continuum. To use a simple expression, it is the elastica of a plate (plate-elastica). As the elastica of a column represents the elastic deformation of the column, the plate elastica must symbolize the elastic deformation of a plate. We know that elastica of a column is used for curve fitting because of the smooth characteristics of their curves. On the other hand, in the case of plates, as shown in Fig. 6, the solution does not seem to be an elastically curved surface since it has sharp ridge lines with different curvatures. This result demonstrates that the characteristics of deformation of a plate are not a deduction of the column solution.

Now, let us return to the original issue. Plans are underway to launch a two-dimensional deployable array on board the Space Flyer scheduled for 1992 (Fig. 8). The array utilizes the same principle as described above. It is a thin-membrane solar cell array which will be simultaneously deployed in two orthogonal directions, thus, it is called the "Two-dimensional Deployable Array Experiment."[7]

V. METAMORPHOSIS

If folding and deployment belong to one form of geometrical transformation, metamorphosis is an active form of transformation. In the deployable structure, the essential issues are two extreme conditions: folded and deployed states. Nonetheless, when other conditions are considered, the idea that a structure with a morphology which can freely change, i.e., the idea of metamorphosis, will emerge. If the structural change occurs in response to the requirement of the environment, it is very natural that the idea of adaptive structures emerge. In space, structures must respond to a variety of different requirements of payloads. Adaptive structures have often been developed from such requirements.

Let us consider a mechanism used to alter the shape of the simplest and most stable macroscopically one-dimensional spatial truss. First we must consider the condition in which the truss may alter its shape freely. In order to satisfy such conditions, each member of the truss must be able to alter its size independently while the length of the other members remains unaffected. Therefore, a truss with variable shapes is a statically determinate truss.

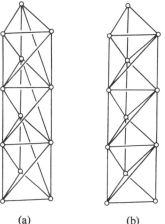
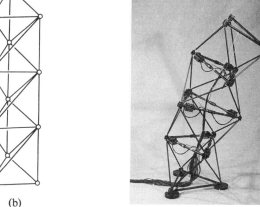

Fig. 9. Simplest statically determinate trusses. Fig. 10. The variable geometry truss.

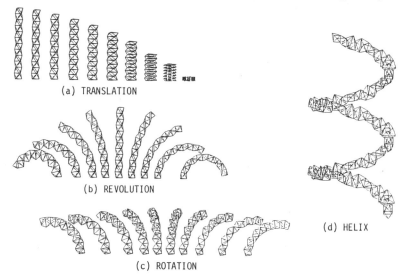

Fig. 11. Geometric adaptability of a VG truss.

A useful equation regarding the statistically determinate truss system is Maxwell's equation. In the equation, for a spatial truss, $M-3J=6$. Here, M and J represent the number of membranes and the number of joints, respectively. The number 6 on the right-hand side represents the reaction force. Since space structures are long and the number of modules is large, the system will be one in which we can ignore the 6 on the right-hand side. From the conditions that it is long and it is one-dimensionally repetitive, only two systems (Fig. 9) are available as basic systems. In Fig. 9 (a) and (b), the module is octagonal and it is in the form of a tetrahedron, respectively. The next selection is to choose the member to be variable. Here, all lateral members are made variable. The lateral members contain actuators and their length can independently be controlled from their initial unit length to $\sqrt{3}$ times their unit length (Fig. 10). We call this structure the variable geometry truss.[8]

Figure 11 shows a variety of spatial shapes to be created using these variable geometry trusses. To create an arbitrary curve in space, the truss only has to produce an arbitrary curvature and torsion. The truss satisfies these requirements and can smoothly change its shape. Of course, high stiffness as a stereoscopic truss is maintained.

REFERENCES

1) Hedgpeth, J.M., Critical requirements for the design of large space structures. *NASA CR-3484* (1981).
2) Love, A.E.H., *Mathematical Theory of Elasticity*, 4th edition, Cambridge University Press (1952).
3) Miura, K. et al., Simplex mast: an extendable mast for space applications. *14th International Symposium on Space Technology and Science*, Tokyo (1984).
4) Miura, K., Proposition of pseudo-cylindrical concave polyhedral shells. *IASS Symposium on Folded Plates and Prismatic Structures*, International Association for Shell Structures, Vienna (1970).
5) Tanizawa, K. et al., Large displacement configurations of bi-axially compressed infinite plate. *Trans. Jpn. Soc. Aeronautics Space Sciences*, Vol. 20 (1978).
6) Konopasek, M., Classical elastica theory and its generalizations. *NATO Adv. Study Inst. Ser. E*, 38 (1980).
7) Miura, K., et al., 2-D array experiment on board space flyer unit. *Space Solar Power Review*, Vol. 5 (1985).
8) Miura, K., et al., Variable geometry truss and its application to deployable truss and space crane arm. *Acta Astronautica*, Vol. 12 (1985).

Magnetomechanical Behavior of Solids

Kenzo MIYA and Toshiyuki TAKAGI

Nuclear Engineering Research Laboratory, The Faculty of Engineering, University of Tokyo, Tokai, Ibaraki

The basic theory needed for the study of magnetomechanical behavior of solids is briefly reviewed. In order to clarify this behavior, magnetic field, electromagnetic forces, and magnetic stresses must be evaluated. The analytical methods for some magnetomechanical problems in superconducting magnets and ferromagnetic materials are also discussed.

I. INTRODUCTION

Since the late 1960's a new engineering field has been growing related to the interaction between the elastic and electromagnetic fields. One motivation for its development has been the possibility of its applications to problems in geophysics and acoustics.

Recently, magnetomechanical problems have been recognized as important for engineering feasibility of fusion reactors, of which many components are in a strong magnetic field. Lorentz force, which is a result of the interaction between magnetic field and induced current, will load the components.[1] If ferromagnetic material is used in the magnetic field, it will deform due to the magnetic stress resulting from interaction between magnetic induction B and induced magnetized moment.[2]

Magneto-solid mechanics or magnetomechanics in solids is the study of the interaction between the magnetic and elastic fields described above. The need for magneto-solid mechanics arises from the increasing application and devices that employ high magnetic fields.[3] For example, recent advances in superconductivity have extended the list of these devices.

Maxwell equations are essential for the study of magnetomechanics in solids, and it is easy to treat the equations as diffusion equations for those familiar with numerical solution techniques such as the finite element method. The governing equations of electromagnetic fields are the Maxwell equations and Ohm's law, which have been studied for many years. Many engineering problems remain to be solved, however, including 1) superconducting magnets of a fusion reactor; 2) magnetic levitation; 3) electromagnetic forming; 4) magnetohydrodynamic generators; 5) hypervelocity projectiles; 6) superconducting rockets; 7) superconducting ships; and 8) fracture mechanics by electromagnetic force.

In this paper we first review the basic theory for the study of magnetomechanical behavior of solids. Since we do not have enough space to discuss all the above problems, magnetomechanical problems for superconducting coils and a ferromagnetic material are discussed here.

II. BASIC THEORY FOR MAGNETOMECHANICAL BEHAVIOR OF SOLIDS

Electrodynamics is related to the mechanical behavior in the studies of electromagnetics and treats a conductor as a rigid body. On the other hand magnetomechanics treats a conductor as a deformable solid. Therefore the basic theory for magnetomechanics consists of theories of electromagnetic field, electromagnetic force, elastic deformation, and their coupling.

II-1. Maxwell Equations

The Maxwell equations have differential and integral forms.[3] The former are expressed as follows:

$$\nabla \cdot J + q = 0 \quad \text{(conservation of charge)} \tag{1}$$

$$\nabla \cdot B = 0 \quad \text{(conservation of flux)} \tag{2}$$

$$\nabla \cdot D = q \quad \text{(Gauss's law)} \tag{3}$$

$$\nabla \times H = J + \dot{D} \quad \text{(Ampere's law)} \tag{4}$$

$$\nabla \times E + \dot{B} = 0 \quad \text{(Faraday's law of induction).} \tag{5}$$

The constitutive equations are

$$J = \sigma E \quad \text{(Ohm's law),} \tag{6}$$

$$B = \mu H \tag{7}$$

where J is current density, B magnetic induction, D electric flux density, q electric charge, H magnetic field, E electric field, μ permeability, and σ electric conductivity.

For the magnetomechanics in a conductor, we can consider charge and electric force to be zero.[3] The Maxwell equations are not all independent. The full set of the equations leads to hyperbolic differential equations which give the wave type solution. We do not need the wave type solution because for frequencies of importance to structural problems (less than 10^7 Hz) the wavelengths are much longer than conventional structures. Therefore the term $\dot{D}$ in Eq. (4) can be neglected. When this term is dropped, the equations become diffusion equations or elliptic equations.[3]

II-2. Electromagnetic Forces

We here discuss the forces between closed circuits C_1 and C_2 with steady currents I_1 and I_2. Total force on a circuit C_1 can be expressed by the integral

$$F = -B \times \oint_{C_1} I_1 \, ds_1. \tag{8}$$

Current I_2 on a circuit C_2 generates magnetic induction B_2. Then B_2 is given by Biot-Savart's law.

$$B = \frac{\mu_0}{4\pi} \oint_{C_2} \frac{I_2 ds_2 \times R}{R^3}. \tag{9}$$

Substitution of Eq. (9) into (8) gives the following double integral

$$F = \frac{\mu_0 I_1 I_2}{4\pi} \oint_{C_1} \oint_{C_2} \frac{ds_1 \times (ds_2 \times R)}{R^3}, \tag{10}$$

where R is a position vector, and ds_1, ds_2 are differential vectors directed along the direction of currents I_1 and I_2.

The body force per unit volume is given by the product of current density in an elastic body and the magnetic induction.

$$f = J \times B. \qquad (11)$$

Total force on the circuit is given by the following form

$$F = \oint_{C_1} I \times Bo \, ds.$$

If Bo is uniform, line integral of current gives zero force. But even in the case of zero force, the net moment or torque might not necessarily be zero. The moment about a point is given by the formula

$$C = \oint_{C_1} r \times (Idr \times Bo) \qquad (12)$$

where r is a position vector to the current element Idr. Using the vector identity,[3]

$$2(r \cdot B)dr = (r \times dr) \times B + d[r(r \cdot B)],$$

the moment can be written as

$$C = m \times B \qquad (13)$$

where m is an equivalent magnetic moment of the circuit and is defined by

$$m = \frac{1}{2} I \oint_{C_1} r \times dr. \qquad (14)$$

If a circuit is located under small gradient fields, the circuit receives magnetic force in addition to the moment. Magnetic induction is expanded in a Taylor series:

$$Bo(r) = Bo(0) + (r \cdot \nabla)Bo(0) + \cdots. \qquad (15)$$

Uniform magnetic induction $Bo(0)$ generates the net zero force. Then the force can be expressed by

$$F = I \int_{C_1} dr \times (r \cdot \nabla)Bo(0). \qquad (16)$$

Using the vector identity,[3]

$$2dr \times (r \cdot \nabla) = (r \times dr) \times \nabla + d[r(r \cdot \nabla)],$$

the force can be rewritten as

$$F = \frac{1}{2} I \left(\int (r \times dr) \times \nabla \right) \times Bo(0)$$
$$= (m \times \nabla) \times Bo(0). \qquad (17)$$

If m is constant with respect to the operator ∇, the force is expressed by the form

$$F = \nabla(m \cdot Bo) = (m \cdot \nabla)Bo. \qquad (18)$$

Equation (18) means that a circuit is accelerated toward the direction of the magnetic induction gradient. This suggests the possibility of acceleration of a circuit with current or a magnetic material by the gradient field.

III. MAGNETOELASTIC INSTABILITY OF SUPERCONDUCTING COILS

In a Tokamak fusion reactor a set of discrete superconducting coils, as shown in Fig. 1, are used to generate a toroidal magnetic field. Several patterns of magnetic loads such as centering force, circumferential tension, and overturning torque act on the coils. The mutual attractions are caused by the rest of the magnets and are balanced to result in zero force because of a symmetric array. This symmetric array causes a magnetoelastic instability of a set of toroidal coils if coil current is sufficiently high to overcome the bending stiffness of the coil. Moon[4] has demonstrated the magnetoelastic instability of the toroidal coils and also developed a modal analysis method for the instability. Miya[5,6] has applied a finite element method to this phenomenon and discussed the magnetomechanical behaviors of full-scale superconducting toroidal coils. Below the theoretical background for the magnetomechanics of toroidal coils is described.

III-1. Derivation of Basic Equations

In order to derive the basic equations, the energy method based on Lagrangian theory is now used. We assume the superconducting coils to be thin shell structures. The Lagrangian L is given by the form

$$L = T - U + W \qquad (19)$$

where T is kinetic energy, U elastic strain energy, and W magnetic energy. They are expressed by

$$T = \frac{1}{2} \sum_i \iint \rho \, \dot{q}^2 \, h \, dx dy \qquad (20)$$

$$U = \frac{1}{2} \sum_i \iint \{\varepsilon\}^T \{\sigma\} \, dx dy \qquad (21)$$

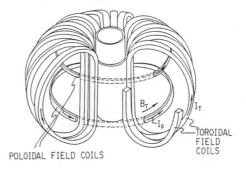

Fig. 1. Magnets in a tokamak reactor.

$$W = \sum_{i \neq m} \sum \iint A \cdot J h dx dy \tag{22}$$

where h is thickness of coil, ρ equivalent density of coil, q deflection of coil, $\{\varepsilon\}$ generalized strain, $\{\sigma\}$ generalized stress, A vector potential, and J current density vector. The following Lagrange equation must be satisfied for an electromechanical system in general,[7]

$$\frac{d}{dt}\left(\frac{\partial L}{\partial \dot{q}}\right) - \frac{\partial L}{\partial q} = F, \tag{23}$$

$$\frac{d}{dt}\left(\frac{\partial L}{\partial \dot{Q}}\right) - \frac{\partial L}{\partial Q} = E \tag{24}$$

where $\dot{q} = \partial q/\partial t$, and F is an external force acting on the coil. Q is the electric charge and its change $\dot{Q} = \partial Q/\partial t$ is a so-called current flowing in the coil. Substitution of Eqs. (20), (21), and (22) into Eq. (19) and then Eq. (19) into Eq. (23) yields

$$\sum_i \iint \rho \ddot{q} q h dx dy + \sum_i \frac{\partial}{\partial q}\left[\frac{1}{2}\iint \{\varepsilon\}^T \{\sigma\} dx dy\right]$$

$$- \sum_i \frac{\partial}{\partial q}\left[\sum_m \iint A \cdot J h dx dy\right] = \sum F_i. \tag{25}$$

The third term is coupled with movements of the rest of the coils and explains the magnetic stiffness. A substitution of Eq. (19) into Eq. (24) yields

$$\sum_i \frac{\partial}{\partial t}\left[\frac{\partial}{\partial J}\sum_m \iint A \cdot J h dx dy\right] + RJ = V, \tag{26}$$

where R is an electric resistance and V is the voltage of electrical source. This equation represents a conservation of electrical energy in the coil system including a power supply. The power supply provides electric energy to the coil system and Eq. (26) is satisfied. Usually we solve Eq. (25) only in order to know the magnetomechanical behavior of coils.

III-2. Magnetoelastic Instability of a Current Carrying Beam

In order to clarify characteristic features of magnetic stiffness, we now consider three elastic beams carrying the same currents as shown in Fig. 2(a). Dimensions and elastic constants of the two beams are the same. The magnetic energy stored after deformation is represented by

$$W = W_{12} + W_{13} + W_{23} \tag{27}$$

where W_{ij} is the magnetic energy stored between the beams i and j. The third term is not related to beam deformation. Finally we get the magnetic energy,[6]

$$W = \frac{\mu_0 J^2}{4} \frac{2}{d^2} \int_0^l q^2 \, dx. \tag{28}$$

In this problem the Lagrangian L is constructed as

$$L = \int_0^l \left[\frac{\rho A}{2}(\dot{q})^2 - \frac{EI}{2}\left(\frac{\partial^2 q}{\partial x^2}\right)^2 + \frac{P}{2}\left(\frac{\partial q}{\partial x}\right)^2 + \frac{\mu_0 J}{2\pi}\frac{q^2}{d^2}\right] dx. \tag{29}$$

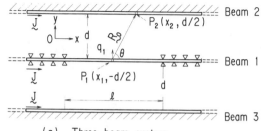

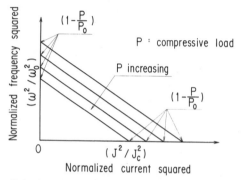

(b) Relation between beam current and its frequency

Fig. 2. Three-current carrying-beam system and linear relation between current squared and frequency squared.

Application of Hamilton's principle to Eq. (29) yields an equilibrium equation:

$$\rho A \frac{\partial^2 q}{\partial t^2} + EI \frac{\partial^4 q}{\partial x^4} + P \frac{\partial^2 q}{\partial x^2} - \frac{\mu_0 J^2}{\pi d^2} q = 0 \tag{30}$$

where P is compressive load, A sectional area, ρ density, and I the second moment of area. This equation is the well-known vibration equation of a beam. The fourth term represents a magnetoelastic coupling term.

Eigenvalues of Eq. (30) can be obtained easily and the solution is given by

$$\frac{J^2}{J_c^2} + \frac{\omega^2}{\omega_0^2} = \left(1 - \frac{P}{P_0}\right) \tag{31}$$

$$J_c = \frac{\pi^2 d}{l^2} \sqrt{\frac{\pi EI}{\mu_0}}, \tag{32}$$

$$P_c = \frac{\pi^2}{l^2} EI, \tag{33}$$

$$\omega_0 = \frac{\pi^2}{l^2} \sqrt{\frac{EI}{\rho A}} \tag{34}$$

where ω_0 is the natural frequency of beam 1 when J and P are zero, J_c the buckling current of beam 1 for $P=0$, and P_c the buckling load of beam 1 for $J=0$. Equation (31) also

III-3. Magnetoelastic Instability of Superconducting Coils

A similar relation as that given in section 3.2 holds true for the buckling and the magnetoelastic vibration of superconducting toroidal coils. We have analyzed the phenomena based on the finite element method.[5,6] For the structural analysis of superconducting coils, the finite element method is widely used. We will discuss the treatment of magnetic energy using the finite element method and the comparison between numerical and experimental results.

1) Finite Element Implementation

The perturbation of magnetic engergy is generally given by

$$\Delta W = \iint \Delta A \cdot I dxdy + \iint A \cdot \Delta I dxdy \tag{35}$$

where I is the current density vector per unit width. The first term on the left-hand side of Eq. (35) depends on the relative displacements between coils. The second term depends on the directional change of the current density vector. Here ΔW_A and ΔW_I denote the former and the latter, respectively. ΔW_A is finally obtained after lengthy calculation:

$$\Delta W_A = \frac{\mu_0}{8\pi} I^{(1)} I^{(2)} C_{11} \times \left[\frac{1}{R_0^3} \iint (A_i^{(2)} u_i^{(2)} - A_i^{(1)} u_i^{(1)}) dS^{(1)} dS^{(2)} \right.$$

$$+ \frac{3}{2R_0^5} \iint (A_i^{(2)} A_j^{(2)} u_i^{(2)} u_j^{(2)} + A_i^{(1)} A_j^{(1)} u_i^{(1)} u_j^{(1)}$$

$$- 2A_i^{(1)} A_j^{(2)} u_i^{(1)} u_j^{(2)}) dS^{(1)} dS^{(2)}$$

$$- \frac{1}{2R_0^3} (B_{ij}^{(2)} u_i^{(2)} u_j^{(2)} + B_{ij}^{(1)} u_i^{(1)} u_j^{(1)}$$

$$\left. - 2C_{ij} u_i^{(2)} u_j^{(1)}) dS^{(1)} dS^{(2)} \right] \tag{36}$$

where

$$A_i^{(k)} = R_0 \cdot a_i^{(k)} \quad (k = 1,2)$$

$$B_{ij}^{(k)} = a_i^{(k)} a \cdot a_j^{(k)} \quad (k = 1,2)$$

$$C_{ij}^{(k)} = a_i^{(k)} \cdot a_j^{(k)} \quad (k = 1,2)$$

and a is the unit vector defined in each element and expressed using the components of transformation matrix, T_{ij}, from global to local coordinates.

$$a_i^{(k)} = (T_{1i}^{(k)} \; T_{2i}^{(k)} \; T_{3i}^{(k)})$$

Some of the other variables are shown in Fig. 3. ΔW_I is also obtained as:

$$\Delta W_I = \frac{\mu_0}{8\pi} I^{(1)} I^{(2)} \frac{1}{R_0} \iint a_i^{(2)} \cdot \frac{\Delta I^{(1)}}{I^{(1)}} dS^{(1)} dS^{(2)} \tag{37}$$

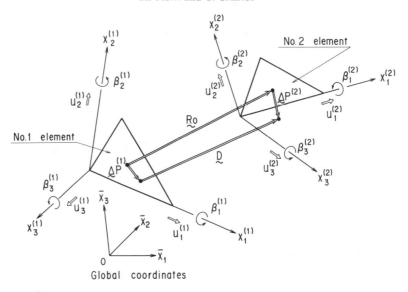

Fig. 3. Definition of displacements and coordinates.

where
$$\Delta I_1^{(1)} = -\frac{1}{2} I^{(1)} T_{11}^{(1)} \beta_y^2 + I^{(1)} T_{31}^{(1)} \beta_y$$

$$\Delta I_2^{(1)} = -\frac{1}{2} I^{(1)} T_{21}^{(1)} \beta_x^2 - I^{(1)} T_{31}^{(1)} \beta_x$$

$$\Delta I_3^{(1)} = -\frac{1}{2} I^{(1)} T_{31}^{(1)} \{\beta_x^2 + \beta_y^2\} - I^{(1)} T_{11}^{(1)} \beta_y$$
$$+ I^{(1)} T_{21}^{(1)} \beta_x$$

$$\beta_x = \frac{\partial u_3}{\partial x_2}, \quad \beta_y = -\frac{\partial u_3}{\partial x_1}.$$

The first term of Eq. (25) is expressed as:

$$\sum_i \iint \rho\{\ddot{q}\} h dx dy = [M_i]\{\ddot{q}\} + [M_0]\{\ddot{q}\} \tag{38}$$

where $[M_i]$ and $[M_0]$ are the mass matrix for in-plane and out-of-plane deformations. The second term is also expressed as:

$$\sum_i \frac{\partial}{\partial q} \left[\frac{1}{2} \iint \{\varepsilon\}^T \{\sigma\} dx dy \right] = [K_i]\{q_i\} + [K_0]\{q_0\} \tag{39}$$

where $[K_i]$ and $[K_0]$ are the stiffness matrices for in-plane and out-of-plane deformations.

The third term of Eq. (25) is obtained using Eqs. (36) and (37) as follows,

$$\sum_i \frac{\partial}{\partial q} \left[\sum_m \iint A \cdot J h dx dy \right] = \{F_m\} + [K_m]\{q\} \tag{40}$$

where $\{F_m\}$ and $[K_m]$ are the magnetic force vector and the magnetic stiffness matrix. The

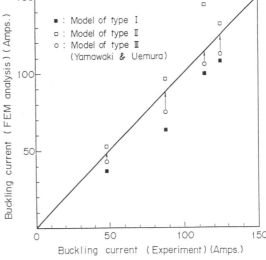

Fig. 4. Eight-coil superconducting full torus.

Fig. 5. Comparison of numerical results with experimental ones for buckling current.

magnetic force vector is usually a Lorentz force vector and appears when there is an unsymmetric array of coils.

Finally the system matrix of current carrying bodies is given by

$$[M]\{\ddot{q}\} + \{[K_e] - [K_m]\}\{q\} = \{F\} + \{F_m\}. \tag{41}$$

When the magnetic forces $\{F\}$ and $\{F_m\}$ are zero, Eq. (41) becomes an equation for an eigenvalue problem.

2) Experimental and Numerical Results

In order to verify the theory developed here we have conducted magnetoelastic instability experiments using superconducting toroidal, helical, and solenoidal coils.[5,6,8,9] An example of an eight-coil full torus is shown in Fig. 4. The comparison of numerical and experimental results is shown in Fig. 5. Types I, II, and III in the figure show the three modellings of composite superconducting coil structures. The results show good agreement and verify the present theory.

IV. MAGNETOMECHANICAL BEHAVIOR OF A FERROMAGNETIC MATERIAL

Magnetomechanical interaction problems of ferromagnetic material are classic ones. Some subjects, for example, the coupling problems of magnetic field and mechanical behavior, are still being investigated. Most experimental data on the behavior have been obtained with fields lower than the saturation field. In such an experiment the magnetomechanical deformation is proportional to the squared applied field Bo^2 and deformation shows simple linear behavior.

Recently ferromagnetic material has been considered for use in high field beyond the saturation field. Some ferromagnetic materials are candidates for the first wall of a tokamak fusion reactor because of their high swelling resistivity under heavy neutron irradiation. As the first wall is placed in high magnetic field, its magnetomechanical behavior in the field must be evaluated since if a high field beyond saturation is applied, strange mechanical behavior is observed.

In the past decades studies of magneto-solid mechanics have primarily focused on the magnetoelastic instability of a soft ferromagnetic plate.[2,10,11] The plate, when it is placed normal to an incident magnetic field, is known to buckle at a certain field intensity. This critical field must be taken into account in the ferritic first wall design of a tokamak fusion reactor. However, the magnetomechanical deformation problem is in common and more important.

There are two categories of ferromagnetic materials. One is soft magnetic material with constant relative permeability. The other is hard ferromagnetic material exhibiting hysteresis. When an applied field is small, we can treat mild steel and ferritic stainless steel as linear soft ferromagnetic materials. In the following, two methods of magnetostatic analysis are described. One is for a linear, soft, ferromagnetic material and the other is for a nonlinear, hard, ferromagnetic material.

IV-1. Numerical Analysis for a Linear Ferromagnetic Material

For magnetostatic analysis $\dot{D}$ in Eq. (4) becomes zero. Then a scalar magnetic potential exists such that

$$H = -\nabla \phi. \tag{42}$$

A substitution of Eqs. (42) and (7) into Eq. (2) yields

$$\nabla^2 \phi = 0. \tag{43}$$

The boundary conditions of a magnetic induction and a magnetic field can be written using the potential

$$\phi_1 = \phi_2 \tag{44}$$

$$\mu_r \frac{d\phi_1}{dn} = \frac{d\phi_2}{dn} \tag{45}$$

where subscripts 1 and 2 denote inside and outside a ferromagnetic material, and μ_r is relative permeability. The boundary condition at infinity is that a scalar potential is equal to a potential for an applied field

$$\phi_\infty = \psi(\infty). \tag{46}$$

Green's second theorem can be applied to derive the governing equations as follows:

$$\int_v (\zeta \nabla^2 \phi - \phi \nabla^2 \zeta) dV = \int_s \left(\zeta \frac{d\phi}{dn} - \phi \frac{d\zeta}{dn} \right) dS, \tag{47}$$

$$\zeta = \frac{1}{r}$$

where n is a normal to the surface, and r is the distance between p and q.

After lengthy calculation, the following integral equation is obtained as a governing equation[12]:

$$\phi_s(p) = \frac{2}{\mu_r + 1} \psi(p) - \frac{\mu_r - 1}{\mu_r + 1} \cdot \frac{1}{2\pi} \int_s \phi_s \frac{d}{dn}\left(\frac{1}{r}\right) dS. \qquad (48)$$

This equation can be solved numerically using a boundary element method. After potential distribution is obtained, potentials in a ferromagnetic material and air can be given by the following expression.

$$\phi_1(p) = \frac{\psi(p)}{\mu_r} - \frac{\mu_r - 1}{4\pi \mu_r} \int_s \phi_s(q) \frac{d}{dn}\left(\frac{1}{r}\right) dSq \qquad (49)$$

$$\phi_2(p) = \psi(p) - \frac{\mu_r - 1}{4\pi} \int_s \phi_s(q) \frac{d}{dn}\left(\frac{1}{r}\right) dSq. \qquad (50)$$

Equations (49) and (50) satisfy the boundary conditions expressed by Eqs. (44) and (45).

A numerical result is shown in Fig. 6. The figure shows the field concentration at edges of a ferromagnetic beam-plate. This concentration explains the discrepancy between experimental and theoretical results for the magnetoelastic buckling of a ferromagnetic beam-plate.

IV-2. Numerical Analysis for a Nonlinear Ferromagnetic Material

It is common to introduce a magnetic vector potential A for an analysis of magnetic field including current source. It is apparent that the magnetic vector potential can be applied as follows when the *B-H* relationship is nonlinear,

$$B = \nabla \times A. \qquad (51)$$

From Eqs. (4) and (7) we find,

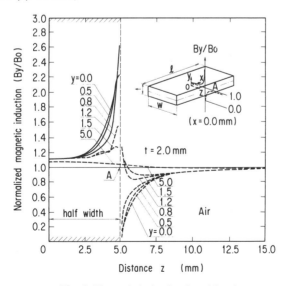

Fig. 6. Magnetic induction in midsection.

$$\nabla \times [\nu] B = J \tag{52}$$

where ν is a magnetic resitivity matrix. It goes without saying that the conservation law of B is satisfied automatically. From Eqs. (51) and (52), we find

$$\nabla \times ([\nu] \nabla \times A) = J. \tag{53}$$

When uniform field Bo is applied, the potential at infinity is given by

$$A = \frac{1}{2} Bo \times r. \tag{54}$$

The Newton-Raphson method can be conveniently applied to this kind of nonlinear analysis. For the purpose, the following functional corresponding to magnetic energy stored in the system is introduced:

$$F = \int \frac{1}{2} \nu B^2 \, dv. \tag{55}$$

The B-H relationship is expressed as

$$H = [\nu] B. \tag{56}$$

Once the B-H curve is obtained from an experiment, it is possible to express ν as a function of $|B|$.

$$\nu = f(|B|). \tag{57}$$

Since the finite element method is applied here, and thus the vector potential A is discretized with use of a nodal vector potential A_i, the functional F is an assemble of the nodal potential. If the nodal potential A_i gives a correct solution, it is required that differentiated value of $F(A)$ with A_i be zero,

$$\frac{\partial F(A)}{\partial A_i} = 0. \tag{58}$$

A Taylor expansion $F(A + \delta A)$ is given such that

$$\frac{\partial}{\partial A_i} F(A + \delta A) = \frac{\partial}{\partial A_i} F(A) + \sum_j \frac{\partial^2}{\partial A_i \partial A_j} F(A) \delta A_j. \tag{59}$$

Since $A + \delta A$ is a solution of Eq. (53), the left-hand side of Eq. (59) is zero, which yields

$$\sum_j \frac{\partial^2}{\partial A_i \partial A_j} F(A) \delta A_j = -\frac{\partial}{\partial A_j} F(A). \tag{60}$$

It is possible to determine δA from this equation to obtain a more accurate value as follows

$$A_i(k + 1) = A_i(k) + \delta A_i(k). \tag{61}$$

The ferromagnetic material receives magnetic force F and moment or torque C in external magnetic field Bo. They are expressed as follows[13]:

$$F = \int_v (M \cdot \nabla) Bo \, dv \tag{62}$$

$$C = \int_v [r \times (M \cdot \nabla)Bo + M \times Bo]dv. \tag{63}$$

M is induced magnetization of the material. If the applied field is uniform, the material experiences only the moment as follows,

$$C = \int_v M \times Bo \, dv. \tag{64}$$

As is known from Eq. (64), in the linear analysis the moment and deformation are proportional to a squared applied magnetic field Bo^2 and the material would break even for a small applied field due to high stress.

Numerical analysis was made for a rectangular plate of finite size, 75 mm in length, 20 mm in width, and 1 mm in thickness as shown in Fig. 7. Mesh division of space around the plate was made but is not shown here for simplicity. The number of elements is varied.

Figure 8 shows an initial part of the B-H curve of ferritic stainless steel (12Cr-1Mo) measured from its ring. An initial rise of the curve is approximated with a dotted line for simplicity in the range of 0 to 1.0 Tesla. H is approximated as follows as a function of B:

$$\mu_0 H = \begin{cases} C o B & (B \leq 1.0) \\ k_4 B^4 + k_3 B^3 + k_2 B^2 + k_1 B + k_0 & (1 < B \leq B_{sat}) \\ \mu_0 H_{sat} - B_{sat} + B & (B_{sat} \leq B) \end{cases} \tag{65}$$

where the saturated magnetic induction is 1.9 Tesla.

Figure 9 shows variations in magnetic torque (moment) induced for a beam-plate. The incident angle of the applied field to the plate was $\theta = 2°$. The magnetic torque is maximum at the magnetic induction of $Bo = 1.5$ Tesla. The first rise of the curve occurs due

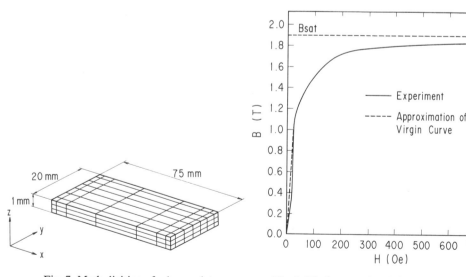

Fig. 7. Mesh division of a beam-plate.

Fig. 8. Virgin curve for 12Cr-1Mo B-H relation.

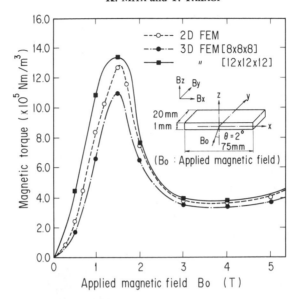

Fig. 9. Torque variation at shell center. $x=0$ mm; $y=10$ mm; $z=0$ mm.

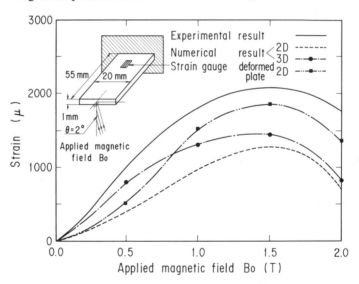

Fig. 10. Variation of strain of a beam-plate.

to the linear *B-H* relation for low magnetic field. It decreases until the magnetic induction becomes 3 Tesla. Then it becomes proportional to the magnetic induction. This strange behavior can be explained by the saturation effect. A tangential component of the magnetic induction in the beam-plate is magnified in the lower applied field because the permeability of ferromagnetic material is very large. In the higher applied field beyond the saturation field the tangential component becomes small because of the saturation effect. The results from three-dimensional analysis give higher torque than those from two-dimensional analysis because of consideration of the edge effect.

In order to verify the magnetic field analysis program developed here, an experiment was carried out. A beam-plate was set in a uniform magnetic field. The maximum field of 2.1 Tesla was generated by a normal magnet. The beam-plate was supported at one end (cantilever). Figure 10 shows variations of strain near a mechanical constraint and the free end of the plate with the applied field. Discrepancy between the experimental and numerical results from two-dimensional analysis was quite large. Numerical results from three-dimensional analysis were larger than those from two-dimensional analysis because it considers the edge effect. The discrepancy was still large, however, and could be attributed to the fact that the coupling effect of the magnetic field analysis and deformation analysis was not considered. The result from two-dimensional analysis considering deformation is also shown in the figure. This result agrees quite well with the experimental result. The result from three-dimensional analysis considering the deformation might agree better.

V. CONCLUSIONS

The basic theory for the study of magnetomechanical behavior of solids were briefly reviewed in this paper. Magnetomechanics problems for superconducting magnets and ferromagnetic materials were discussed.

For the design of electromagnetic devices, it is important to evaluate electromagnetic forces and stresses. Future work is needed for three-dimensional analysis and complicated structures which lead to highly optimized design of such devices.

Acknowledgments

We would like to acknolwedge the financial support of the Controlled Thermonuclear Reactor Committee (Chairman, Professor Y. Takahashi), Faculty of Engineering, University of Tokyo.

REFERENCES

1) Miya, K., An, S., and Ando, Y., Application of finite element method to electro-magnetomechanical dynamics of superconducting magnet coil and vacuum vessel. *Proc. 6th Symp. Engng. Problems of Fusion Research*, San Diego (1975), pp. 927–934.
2) Moon, F.C. and Pao, Y-H., Magnetoelastic buckling of a thin plate. *J. Appl. Mech.*, Vol. 35, No. 1 (1968), pp. 53–58.
3) Moon, F.C., *Magneto-Solid Mechanics*, John Wiley & Sons (1984).
4) Moon, F.C. and Swanson, C., Experiments on buckling and vibration of superconducting coils. *ASME J. Appl. Mech.*, Vol. 44 (1977), pp. 707–713.
5) Miya, K., Takagi, T., Uesaka, M., and Someya, K., Finite element analysis of magnetoelastic buckling of eight-coil superconducting full torus. *ASME J. Appl. Mech.*, Vol. 49 (1982), pp. 180–186.
6) Miya, K. and Uesaka, M., An application of finite element method to magnetomechanics of superconducting magnets for magnetic fusion reactors. *Nucl. Engng. and Design*, Vol. 72 (1982), pp. 275–296.
7) Wells, D.A., *Lagrangian Dynamics*, Schaum (McGraw Hill Inc.), Chap. 15.
8) Takaghi, T., Miya, K., Yamada, H., and Takagi, T., Theoretical and experimental study on the magnetomechanical behavior of superconducting helical coils for a fusion reactor. *Nucl. Engng. and Design/Fusion*, Vol. 1 (1984), pp. 61–71.
9) Miya, K., Takagi, T., and Takaghi, T., Finite element analysis of the magnetomechanical behavior of a toroidal coil with magnetic mirrors. *Nucl. Engng. and Design*, Vol. 74 (1982), pp. 339–346.
10) Popelar, C.H., Postbuckling analysis of a magnetoelastic beam. *ASME J. Appl. Mech.*, Vol. 39 (1972), pp. 207–211.

11) Miya, K., Takagi, T., and Ando, Y., Finite-element analysis of magnetoelastic buckling of ferromagnetic beam plate. *ASME J. Appl. Mech.*, Vol. **47** (1980), pp. 377–382.
12) Minato, A., Tone, T., and Miya, K., Three-dimensional analysis of magnetic field distortion of ferromagnetic beam-plates by the boundary element method. *Int. J. Num. Meth. Engng.*, Vol. **33** (1986), pp. 1201–1216.
13) Brown, W.F., Jr., *Magnetoelastic Interactions*, Springer Verlag (1966).

VII
WAVES AND VIBRATIONS

Stability of a Rotor Containing Viscous Fluid

Shin MORISHITA and Kiyotaka OKUZONO

Department of Naval Architecture and Ocean Engineering, Faculty of Engineering,
Yokohama National University, Yokohama

> An experimental investigation of the instability of a rotor partially or fully filled with viscous fluid has been performed. The rotor is assumed to be a rigid body and be supported by elastic plate springs, as the rotor is capable of moving in one direction. Whirl amplitudes and frequencies are measured, under various rotating speeds, fluid fill ratios, or fluid viscosity. These data are put into a micro-computer for analysis by FFT procedure. The experimental results are compared with the theoretical predictions by previous authors. It is found that an unstable whirl motion can appear even for a completely fully filled rotor, and that there must be a phenomenon different from what occurs in a partially filled rotor. Conical mode is also observed in addition to parallel mode, which has a very wide unstable range of rotating speeds.

I. INTRODUCTION

A circular cylinder mounted on a flexible shaft and partially filled with liquid undergoes unstable whirling motion under certain operating conditions. This self-excited vibration may be experienced in various mechanical components with circular cylindrical shells: turbo machinery, electric generators, or centrifugal separators.[1] It may be caused by either essentially placed liquid or trapped fluid such as lubricants or steam condensation, and even by powder. Once this happens, it is difficult to control this instability, and then the machinery components may be damaged.

Several authors have investigated this problem theoretically and experimentally.[2-10] J. A. Wolf considered an inviscid, incompressible fluid partially filling a cylinder, which was assumed to undergo a uniform circular whirling motion.[3] He showed that there is a range of rotating speeds which results in unstable motion. S. Saito investigated instability of a rotor partially filled with viscous, incompressible liquid, numerically and experimentally.[6,7] He applied the boundary layer theory to this problem, and get a stability diagram. K. Kaneko also showed analytical and experimental results.[8] However, there are still considerable difference between theoretical prediction and experimental results, and even among theoretical predictions themselves.

Therefore, it is important to know, first of all, under what conditions this phenomenon takes place. The authors conducted an experiment with a model of a rotor, and derived a stability diagram. It is compared with the theoretical predictions of previous authors.

II. APPARATUS AND INSTRUMENTATION

The experimental apparatus is shown in Fig. 1 and Fig. 2.[11] The rotor is supported by two ball bearings which are mounted on plate springs suspended from a frame. The rotor

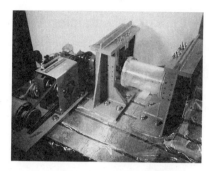

Fig. 1. General view of experimental apparatus.

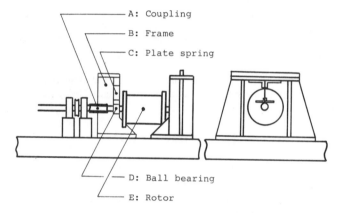

Fig. 2. Rotor and shaft construction.

is assumed to be a rigid body and plate springs represent the flexibility of an actual rotor; and its supports, though they are free to move only in one direction, are regarded as a one-degree-of-freedom system. Acrylic resin is used for the rotor, inner diameter 120 mm, length 200 mm and thickness 15 mm, so that a centrifugal free surface wave may be observed. The apparatus is supported through steel shafts attached at both sides of the rotor.

The shaft is driven by an AC motor through a flexible coupling so as not to prevent the whirling motion. The rotating speed is controlled by the inverter and the transmission consecutively, from 0 rpm to about 3700 rpm. Speed is measured by a counter composed of a photo transistor and light placed on each side of a disk with six holes. For each revolution, six pulses are recorded by the counter, so that when the counter is opened for a 10-second period, it displays the rotational speed in rpm. Plate springs are made of phosphor bronze, and their thicknesses are selected in accordance with the critical speed.

Whirl frequency and amplitude are determined by signals obtained from strain gauges attached to plate springs through a strain amplifier. The strain amplifier, which has a frequency response from 0 to 60 kHz, is calibrated in place.

The analog voltage outputs from the strain amplifier are fed to a micro-computer through an A/D converter. Each analog data bit is transformed into 12 bits of digital data,

Table 1. The characteristics of the oil.

Paraffin base oil		
Specific gravity		0.884
Coef. of visc. (cSt)	20°C	290
	40°C	97

and the minimum time required for the transformation is about 25 micro-seconds per channel.

From the data displayed instantaneously as a function of time, whirl amplitudes are measured. The micro-computer also performs a discrete Fourier transform to convert a finite segment of discrete time data into a discrete frequency spectrum, and then whirl frequencies are determined.

External damping originated from plate springs, and flexible coupling is 1.69×10^{-2} in logarithmic damping.

Water and oil are used as liquid input to the rotor, depending on their viscosity. The characteristics of the oil are given in Table 1.

III. PROCEDURE

The experiment is performed in variable depths of liquid in the rotor, and both the whirl frequency and the amplitude are measured. The depth, H, is chosen as follows in the case of water:

$$H = 1, 2, 4, 10, 20, 27, 39, 46, 50, 55, 60 \text{ mm}.$$

The dimensionless depth by radius of rotor is varied from 0.017 to 1.00 (full-fill condition).

When oil is used, we choose

$$H = 1, 2, 4, 10, 14, 24, 42, 59 \text{ mm}.$$

"Depth," as used here, means the thickness of the liquid when the center of the rotor and that of liquid are in concentrated condition.

As the occurrence of some hysteresis is anticipated, measurements are performed under the condition of both increasing and decreasing revolution.

IV. RESULTS AND DISCUSSIONS

As rotating speed is increased, the liquid in the rotor begins to rise along the inside wall, and then stick on the wall concentrically. As self-excited whirl motion can appear under such a coherent condition, the measurements are begun under the condition of waking revolution high enough for the liquid to stick to the wall, and then decreased. The liquid, once stuck to the wall, will not peel off easily. Figure 3 shows the upper limit of speed for "peel-off," and the lower limit for "stick-up."

The amplitude of whirl motion without liquid in the rotor is shown in Fig. 4. The first critical whirl motion is observed at 523 rpm($=\omega_p$), and the second at 883 rpm($=\omega_c$); these motions correspond to parallel mode and conical mode, respectively. Hereafter, di-

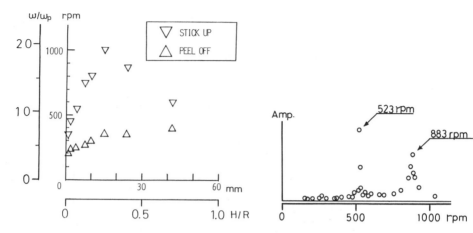

Fig. 3. Rotating speed at "peel off" or "stick up."

Fig. 4. Whirl amplitude whitout liquid.

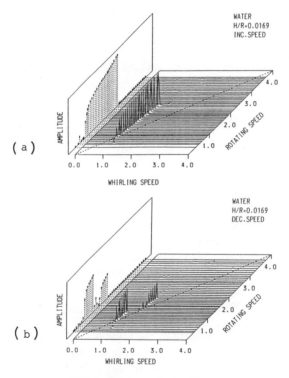

Fig. 5. Spectrum of whirl frequency.

mensionless parameters with reference to the rotating speed and the frequency are determined by the value of ω_p.

When the rotor is filled partially or fully with liquid, there appears self-excited whirl motion. The whirl amplitude and whirl frequency are shown in Figs. 5 to 8, as a function of rotating speed. The dashed line in each figure represents the condition in which whirl frequency is equal to rotating speed—that is, synchronous whirl condition. The whirl amplitudes are projected on the vertical plane standing on the left.

Figure 5 shows the results under the condition of water depth 1mm. When the dimensionless rotating speed is about 0.9, whirl motion increases rapidly. As its peak stands on the dashed line, it is regarded as synchronous whirl motion.

As rotating speed is increased further, the peak of the spectrum separates from the dashed line and whirl frequency remains at the first critical whirl speed. This is regarded as the occurrence of self-excited vibration. This condition is preserved until the rotating speed exceeds a certain value. As shown in Fig. 5(a), unstable whirl motion appears in the range of 0.5 to 2.3 of rotating speed, and then proceeds to a stable condition up to the capacity of this instrument, 3700 rpm.

In turn, when the rotating speed is decreased from the upper limit, the whirl amplitude begins to grow at about 1.93 of rotating speed. The peak of the spectrum stays at the second

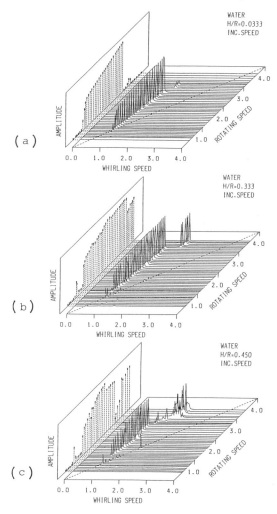

Fig. 6. Spectrum of whirl frequency.

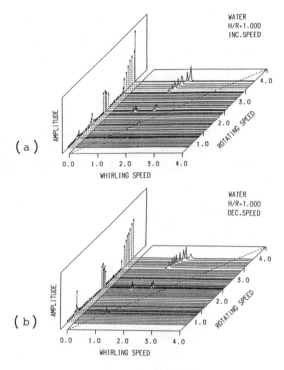

Fig. 7. Spectrum of whirl frequency.

critical whirl frequency, and the conical mode appears. This is also regarded as an unstable condition. When the rotating speed reaches about 1.59, the rotor is put into a stable condition again, and soom meets the second critical whirl frequency. If the rotating speed is decreased further, at about 1.28 the rotor falls into an unstable condition under which parallel mode is observed. When the rotating speed reaches 0.9, whirl motion presents ordinal first critical motion. The procedure mentioned above is shown in Fig. 5(b). Figure 6 shows the response at water depths of 2, 20, and 27mm. Though the phenomenon is essentially similar to those shown in Fig. 5, one peculiar condition is shown in Fig. 6(c): whirl frequency varies violently under self-excited conditions.

The results for full-fill condition are shown in Fig. 7. Previous authors, except for S.H. Crandall,[10] could not predict the self-excited whirl motion under this condition. Unstable state appears clearly, as shown in the figure. However, comparing this result with other figures, one can easily find the difference in the response. There seem to be two types of spectrum peak, a remarkable one and a small one; what is more interesting, they appear sometimes discontinuously and sometimes simultaneously. Whether these two types are essentially the same or not is the problem to be solved.

As shown in Figs. 6(b), (c) and Fig. 7, superharmonic resonance is observed when the rotating speed is lower than the first critical speed. This resonance tends to appear when there is a large quantity of liquid in the rotor.

Figure 8 shows the results in the case of oil. Under unstable conditions the spectrum

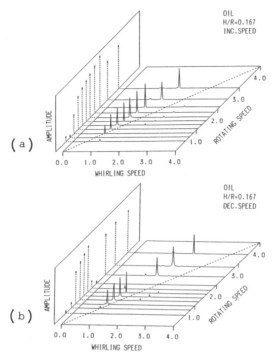

Fig. 8. Spectrum of whirl frequency.

Fig. 9. Centrifugal wave on the free surface (Parallel mode).

Fig. 10. Centrifugal wave on the free surface (Conical mode).

peaks do not stay at the first level attained at critical speed, but have a tendency to shift higher as the rotating speed is increased.

Centrifugal waves appearing on the free surface of the liquid are visualized in Figs. 9 and 10. The operating condition is 24 mm depth with water. Figure 9 shows the parallel mode and Fig. 10 the conical mode. In Fig. 9, the center of the cylindrical free surface

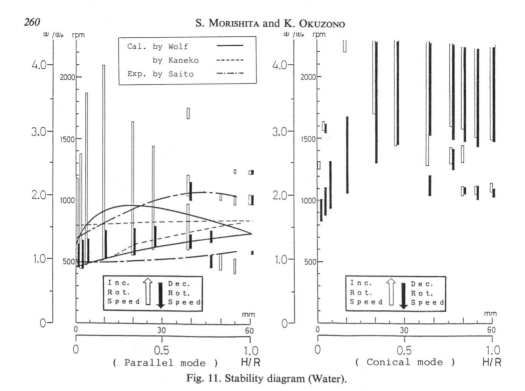

Fig. 11. Stability diagram (Water).

is observed to get out of the center position of the rotor. Under conical mode, the axis of the cylindrical free surface inclines to that of the rotor, as seen in Fig. 10. When the free surface is seen to be twisted, there must be a phase difference between whirl motion and free surface wave.

On the basis of the results shown in Figs. 5 to 8, it can be seen that the stability range of rotating speed is limited, as shown in Fig. 11 (water) and Fig. 12 (oil). In these figures, our experimental results are compared with theoretical predictions by J. A. Wolf and K. Kaneko, as well as with the previous experimental results of S. Saito. There are three pairs of curves in the figures corresponding to each prediction. For each pair, the regions lying between them indicate the rotating speed corresponding to the unstable state. For the parallel mode for water (Fig. 11 (d)), the experimental results show that the unstable ranges are much wider for increasing rotating speed than for decreasing speed, and that theoretical predictions show qualitative agreement with those for decreasing rotating speed. However, under the condition of greater depth, they do not agree with the experimental results at all, because the unstable region obtained by the experiment is divided into several parts and there is an unstable range for the fully filled condition. In this respect, the theoretical predictions of previous authors may not be sufficient, and there must still be room for study. As for the conical mode, shown in Fig. 11(b), it is impossible to predict the experimental results; they have very wide unstable ranges of rotating speed, and there also exist unstable discrete regions for the fully filled condition.

Figure 12 shows the results for oil. The situation is essentially the same as in the case of water, except that the hysteresis is much smaller for oil. From Figs. 11 and 12, it can

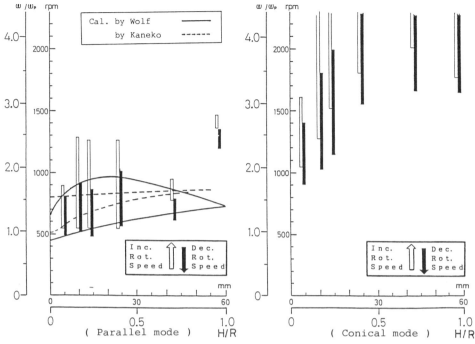

Fig. 12. Stability diagram (Oil).

be said that the lowest limit of the unstable discrete regions for parallel mode is the extension of unstable range under shallow conditions.

V. CONCLUSIONS

An experimental investigation of the asynchronous instability of a rotor partially and fully filled with liquid has been performed. The conclusions obtained are as follows:

1) There exist discrete unstable regions, especially at great depths, and unstable whirl motion appears even under fully filled condition.

2) Not only parallel mode but also conical mode is observed. The unstable region for conical mode is far wider than that for parallel mode.

It seems that there still is room for improving the theories proposed so far, by taking account of the occurrence of asynchronous instability without free surface as well as the effect of variation of depth in the axial direction.

REFERENCES

1) Ehrich, F.F. and Childs, D., Self-excited vibration in high-performance turbomachinery. *Mech. Eng.*, **106**, No. 5 (1984), pp. 66–79.
2) Ehrich, F.F., The influence of trapped fluid on high speed rotor vibration. *Trans. ASME, J. Eng. Ind.*, **89**, No. 4 (1967), pp. 806–812.

3) Wolf, J.A., Whirl dynamics of a rotor partially filled with liquid. *Trans. ASME, J. Applied Mech.* **35**, No. 4 (1968), pp. 676–682.
4) Daich, I.M. and Kazhdan, L.S., Vibration of a rotating rigid body with a cavity partly filled with an arbitrary viscous liquid. *Prikl. Mekh.*, **9**, No. 8 (1973), pp. 96–100.
5) Hendricks, S.L. and Morton, J.B., Stability of a rotor partially filled with a viscous incompressible fluid. *Trans. ASME, J. App. Mech.*, **46**, No. 4 (1979), pp. 913–918.
6) Saito, S., Someya, T., and Kobayashi, M., Self-excited vibration of a rotating hollow shaft partially filled with liquid (4th report). *Trans. JSME*, **48**, No. 427, C (1982), pp. 321–327.
7) Saito, S., Self-excited vibration of a rotating hollow shaft partially filled with liquid (5th report). *Trans. JSME*, **48**, No. 429, C (1982), pp. 656–661.
8) Kaneko, N. and Hayama, S., Self-excited vibration of a rotor partially filled with liquid. *Trans. JSME*, **51**, No. 464, C (1985), pp. 765–772.
9) Shimogo, T. and Yoshida, K., Vibration of rotor containing liquid. *Proc. Int. Conf. on ROTORDYNAMICS*, IFToMM (1986), pp. 453–458.
10) Crandall, S.H. and Mroszczyk, J., Whirling of a flexible cylinder filled with liquid. *Proc. Int. Conf. on ROTORDYNAMICS*, IFToMM (1986), pp. 449–452.
11) Morishita, S., Okuzono, K., and Fukuyama, K., Self-excited vibration of a rotor filled with liquid. *Trans. JSME*, **54**, No. 505, C (1988), pp. 2016–2023.

Natural Vibration Analysis of a Five-Span Continuous Rigid-Frame Bridge with V-Shaped Legs

Toshiro HAYASHIKAWA and Noboru WATANABE

Department of Civil Engineering, Hokkaido University, Sapporo

Analytical models of multispan continuous rigid-frame bridges with V-shaped legs are studied to enable accurate dynamic and earthquake response analyses. Three different analytical matrix methods for determining dynamic characteristics of in-plane vibrating rigid-frame bridges are presented. One method is the exact method based on the general solutions of the differential equations of motion for both axial and flexural vibrations, called the continuous mass method. The other two matrix methods are the lumped and consistent mass methods, based on the approximate finite element approach. The mathematical relationship between the exact and approximate methods is discussed, and the accuracy of the eigenvalues obtained by these methods is investigated. A numerical example based on an actual rigid-frame bridge with V-shaped legs is given to illustrate the applicability of the lumped, consistent, and continuous mass methods, and the computed results are also given in tabular form.

I. INTRODUCTION

Natural vibration analysis is performed of a five-span continuous rigid-frame steel bridge with V-shaped legs standing on two reinforced concrete piers as shown in Fig. 1. The boundary conditions are a fixed support at A1, pin supports at P1 and P2, and a movable support at A2. This rigid-frame bridge has a longitudinal incline of 2.600% from A1 to A2, and is an asymmetric bridge structure. The computation of natural frequencies and mode shapes presents a significant problem in the dynamic response analysis of bridge structures subject to seismic loads and is dependent on both the lower and the higher modes. Therefore, in designing such bridge structures, it is essential that the natural frequencies and modes be determined accurately.

From an analytical standpoint as well as an idealization of structures, it is convenient to divide the system of coordinates into two different basic types as shown in Fig. 2. The first type is the distributed coordinate system (or the distributed-parameter system)[1,2] which is applied to structures whose properties are continuously distributed in space, and to problems in which the forces are distributed. In this case, the basic relation between forces and displacements for a beam segment subjected to axial and flexural vibrations is obtained using general solutions of differential equations of motion. The exact equations lead to the dynamic flexural-stiffness matrix,[1] which is a function of the natural circular frequency of vibration. This approach is called the eigen stiffness matrix method,[3] or the continuous mass method.[4]

The second type of mass system to be distinguished is the discrete coordinate system (or the lumped-parameter system)[2,5] which defines forces and displacements at a set of discrete points in terms of components having specified directions. The analytical procedure

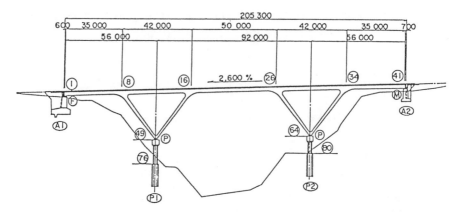

Fig. 1. General view of bridge.

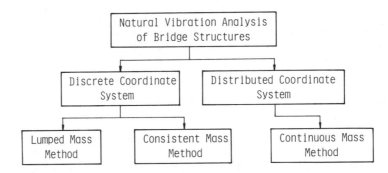

Fig. 2. Schematical description of natural vibration analysis of bridge structures.

for this type can greatly be simplified as an eigenvalue problem because the inertia forces are developed only at these points. In beam structures, the lumped mass matrix[1,6,7] is derived as a diagonal matrix by applying the mass and mass moment of half the beam to each nodal point. Moreover, the mass influence coefficients are evaluated using the cubic Hermitian polynomials as the interpolation functions (the displacement functions). The result is called the consistent mass matrix,[1,7] and it contains many off-diagonal terms due to the effect of mass coupling.

In this study, a procedure of natural vibration analysis is elucidated using the exact method[5] based on general solutions of differential equations of motion for longitudinal and lateral vibrations of a beam. The results are numerically and mathematically compared with the approximate method based on the finite element approach.[1,5] The purpose of this study is to determine a sufficient number of eigenvalues of multispan continuous rigid-frame bridges, and to facilitate accurate vibration analysis from lower modes to higher modes. Also, the numerical results are given in tabular form to investigate the applicability of the lumped, consistent, and continuous mass methods.

II. NATURAL VIBRATION ANALYSIS

It is assumed in the natural vibration analysis of rigid-frame bridge structures that the material is homogenous and isotropic, and that the element members of rigid-frame bridges have uniform cross sections and straight bars throughout each element span length. The behavior of beam segments is described by the Bernoulli-Euler beam theory, and the damping is neglected.

II-1. Distributed Coordinate System

The uniform beam segment of length L, bending stiffness EI, mass of the beam per unit length m, and cross-sectional area A is shown in Fig. 3. The axial force X, shear force Y, bending moment M, and the corresponding displacements u, v, θ at the ends of the beam segment are indicated in Fig. 3, in which all end forces and displacements shown are positive.

In general, a linear vibration of a plane rigid-frame member can be considered as composed of two independent vibrations: vibrations due to axial displacements and bending deformations in the vertical plane. It has already been stated that the use of the general solution of the differential equation of motion leads to the dynamic stiffness matrix.[1,3,5] Therefore, the terms of the dynamic stiffness K_{ae} and K_{fe} under the axial and flexural vibrations in vertical plane are established,[6,7] respectively,

$$F_a = K_{ae} U_a \qquad (1)$$

$$F_f = K_{fe} U_f \qquad (2)$$

in which

$$F_a = \{X_1, X_2\}^T, \quad F_f = \{Y_1, M_1, Y_2, M_2\}^T \qquad (3)$$

$$U_a = \{u_1, u_2\}^T, \quad U_f = \{v_1, \theta_1, v_2, \theta_2\}^T \qquad (4)$$

and

$$K_{ae} = EA\alpha \begin{bmatrix} \cot \alpha L & -\operatorname{cosec} \alpha L \\ -\operatorname{cosec} \alpha L & \cot \alpha L \end{bmatrix} \qquad (5)$$

$$K_{fe} = \frac{EI\beta}{1 - cC} \begin{bmatrix} \beta^2(sC + cS) & \beta s S & -\beta^2(s + S) & \beta(C - c) \\ & sC - cS & \beta(c - C) & S - s \\ & & \beta^2(sC + cS) & -\beta s S \\ \text{Symmetric} & & & sC - cS \end{bmatrix} \qquad (6)$$

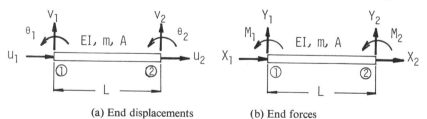

(a) End displacements   (b) End forces

Fig. 3. Positive end-displacements and forces of beam segment.

in which

$$\alpha = \sqrt{\frac{m\omega^2}{EA}}, \quad \beta = 4\sqrt{\frac{m\omega^2}{EI}}$$

$$c = \cos\beta L, \quad s = \sin\beta L, \quad C = \cosh\beta L, \quad S = \sinh\beta L \tag{7}$$

and ω is the natural circular frequency. Each of these square matrices K_{ae} and K_{fe} is designated as an eigen stiffness matrix,[3,6] because each includes the eigenvalue (natural circular frequency, ω) of the bridge structure. In the case of the distributed coordinate system, the stiffness and mass properties are combined to form the eigen stiffness matrix. The frequency equation of rigid-frame bridges for the continuum structure system can be expressed by means of the principle of superposition

$$\det |K(\omega)| = 0. \tag{8}$$

This equation is a transcendental equation of trigonometric and hyperbolic functions which contains the natural circular frequency ω. The solutions of Eq. (8) can be applied using the Regula-Falsi method[8] and by using a high-speed digital computer.

II-2. Discrete Coordinate System

In the discrete coordinate system the dynamic stiffness matrix for the axial or flexural action is expressed as a superposition of elastic and inertial forces forming, respectively, the stiffness and mass matrices. These matrices are derived from the assumed static displacement of a beam segment as shown in Fig. 3. The elastic relationship between end forces and displacements is given, respectively, by the following stiffness matrices

$$F_a = K_{as}U_a, \quad F_f = K_{fs}U_f \tag{9a·b}$$

and

$$K_{as} = \frac{EA}{L}\begin{bmatrix} 1 & -1 \\ -1 & 1 \end{bmatrix}, \quad K_{fs} = \frac{EI}{L^3}\begin{bmatrix} 12 & 6L & -12 & 6L \\ & 4L^2 & -6L & 2L^2 \\ & & 12 & -6L \\ \text{Symmetric} & & & 4L^2 \end{bmatrix}. \tag{10a·b}$$

By assigning the distributed mass of the beam of point masses, the lumped mass matrices for the axial and flexural vibrations are obtained as follows:

$$M_{al} = \frac{mL}{2}\begin{bmatrix} 1 & 0 \\ 0 & 1 \end{bmatrix}, \quad M_{fl} = \frac{mL}{24}\begin{bmatrix} 12 & 0 & 0 & 0 \\ & L^2 & 0 & 0 \\ & & 12 & 0 \\ \text{Symmetric} & & & L^2 \end{bmatrix}. \tag{11a·b}$$

These diagonal mass matrices M_{al} and M_{fl} are the simplest idealization of the inertial forces of a beam. The consistent mass matrices M_{ac} and M_{fc} are obtained by applying the same displacement functions which are used for formulating the static stiffness matrices K_{as} and K_{fs}.

$$M_{ac} = \frac{mL}{6}\begin{bmatrix} 2 & 1 \\ 1 & 2 \end{bmatrix}, \quad M_{fc} = \frac{mL}{420}\begin{bmatrix} 156 & 22L & 54 & -13L \\ & 4L^2 & 13L & -3L^2 \\ & & 156 & -22L \\ \text{Symmetric} & & & 4L^2 \end{bmatrix}. \quad (12\text{a·b})$$

The assemblage of the system mass matrix from the element mass matrices of Eqs. (11) and (12) can be carried out in exactly the same manner as that of the system stiffness matrix. Therefore, the frequency equation of the discrete coordinate system (the lumped and consistent mass matrix methods) may be given as follows:

$$\det |K - \omega^2 M| = 0 \qquad (13)$$

in which K and M are the system stiffness and mass matrices, respectively. The formulation of Eq. (13) is an important mathematical problem known as an eigenvalue problem. There are many numerical methods[9,10] dealing with eigenvalue and eigenvector problems. The computation of the eigenvalues in this study is carried out through a Householder-Bisection-Inverse Iteration Solution subroutine (DEIGAB and DEIGRS), a double-precision version of which is available from the mathematical subprogram library at the Hokkaido University Computing Center (HITAC M-682 and S-810 system).

III. NUMERICAL RESULTS

A numerical example is presented to demonstrate the applicability of the lumped, consistent, and continuous mass methods and to delineate some characteristics of the dynamic behavior of rigid-frame bridges with V-shaped legs. The numerical computations using data from the Shibechari Bridge located in Hokkaido, Japan, provide the basis for this example. The structural geometry of the bridge is a five-span continuous rigid-frame bridge with span lengths of 35.0 m, 42.0 m, 50.0 m, 42.0 m, and 35.0 m as shown in Fig. 1. The structural properties necessary for natural vibration analysis are given as follows: the material modulus of elasticity $E = 2.1 \times 10^7$ t/m² (206,000 MN/m²); the cross-sectional moment of inertia $I = 0.018 - 0.568$ m⁴; the weight of the beam per unit length $m = 1.267 - 5.700$ t/m (12,500 − 55,900 N/m); and the cross-sectional area $A = 0.100 - 2.500$ m².

The first 10 mode shapes resulting from the lumped and consistent mass methods for the five-span continuous rigid-frame bridge with V-shaped legs are shown in Fig. 4(a) and (b), respectively. As this bridge is an asymmetric structure, it shows singular mode configurations. By comparing the modal shapes of lumped and consistent mass methods for the rigid-frame bridge with V-shaped legs, it is found that there is a general similarity between the two methods. However, there is a considerable difference between the values of the natural periods computed by the lumped and consistent mass methods.

The natural circular frequencies computed by the lumped, consistent, and continuous mass methods, corresponding to the first 20 modes of the numerical example, are presented in Table 1. The computed values of the natural circular frequencies obtained by the continuous mass method are the exact solutions as far as the Bernoulli-Euler assumptions are concerned. On the other hand, the computed values obtained using the lumped and consistent mass methods are approximate solutions. In general, the values of the natural cir-

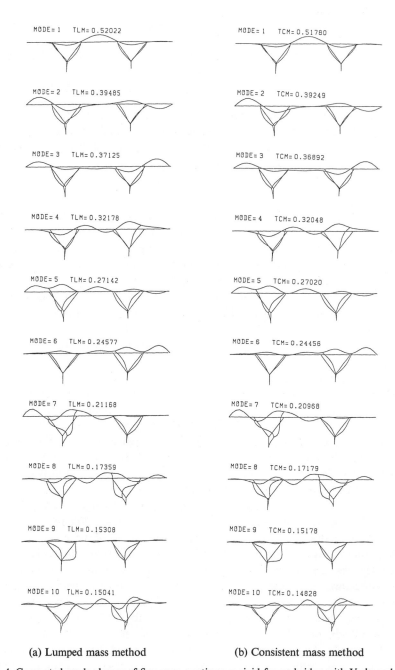

(a) Lumped mass method   (b) Consistent mass method

Fig. 4. Computed mode shapes of five-span continuous rigid-frame bridge with V-shaped legs.

Table 1. Natural circular frequencies computed by three different mass methods.

Mode order	Lumped mass method	Consistent mass method	Continuous mass method
1	12.0779	12.1343	12.1280
2	15.9130	16.0086	16.0000
3	16.9244	17.0312	17.0220
4	19.5263	19.6058	19.5950
5	23.1490	23.2537	23.2406
6	25.5647	25.6921	25.6773
7	29.6828	29.9651	29.9478
8	36.1952	36.5749	36.5529
9	41.0448	41.3972	41.3728
10	41.7743	42.3727	42.3452
11	45.6868	46.1153	46.0867
12	46.1368	46.5411	46.5132
13	48.4642	48.9259	48.8953
14	61.8550	63.0179	62.9664
15	63.0435	64.5331	64.4789
16	68.3102	69.7503	69.6821
17	69.2541	70.4185	70.3531
18	80.5942	82.7604	82.6762
19	88.1759	90.7033	90.6014
20	91.8246	95.2146	95.1020

cular frequencies obtained using the lumped mass method are small in comparison with those of the exact solutions, while the values of the natural circular frequencies from the consistent mass method are relatively large.

Figure 5 indicates the relationship between the natural circular frequency ratio ω/ω^* and the order of natural modes for the five-span continuous rigid-frame bridge with V-shaped legs. Here ω^* is the exact solution obtained by the continuous mass method, and ω is the approximate solution obtained by using the lumped and consistent mass methods. It is seen that by means of both the lumped and consistent mass methods the natural circular frequencies gradually approach the exact solutions as the number "N" of beam segments increases. In general, the natural circular frequencies obtained by the lumped and consistent mass methods are the lower and upper bounds to the exact solutions, respectively. It may also be pointed out that for the same number of beam segments, the consistent mass method provides better accuracy than the lumped mass method.

The mathematical relationship between the lumped, consistent, and continuous mass methods is established in this study. It is shown that the terms of the power series expansion of the coefficients in the eigen stiffness matrix of Eqs. (5) and (6) are precisely the stiffness and mass matrices in common use in the finite element method. The dynamic coefficients obtained in the first row and first column of Eqs. (5) and (6) are derived from a Taylor series expansion as follows:

$$k_{ae11} = EA\alpha \cot \alpha L = \frac{EA}{L} - \frac{mL}{3} \cdot \omega^2 - \frac{m^2 L^3}{45 EA} \cdot \omega^4 \cdots \quad (14)$$

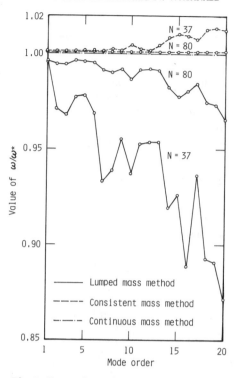

Fig. 5. Comparison of natural circular frequencies.

$$k_{f e11} = \frac{EI\,\beta^3(\sin \beta L \cosh \beta L + \cos \beta L \sinh \beta L)}{1 - \cos \beta L \cosh \beta L}$$

$$= \frac{12EI}{L^3} - \frac{13mL}{35} \cdot \omega^2 - \frac{59m^2L^5}{161700EI} \cdot \omega^4 \cdots \quad (15)$$

It should be recognized that the first two terms on the right-hand sides of Eqs. (14) and (15) are, respectively, the elements in the first row and first column of Eqs. (10) and (12). In general, the power series expansion of the eigen stiffness matrices of Eqs. (5) and (6) may be written in matrix notation as

$$K_{ae} = K_a - M_{a1}\,\omega^2 - M_{a2}\,\omega^4 \cdots \quad (16)$$

$$K_{fe} = K_f - M_{f1}\,\omega^2 - M_{f2}\,\omega^4 \cdots. \quad (17)$$

The stiffness matrix K_a and the first order mass matrix M_{a1} in the case of longitudinal vibrations agree precisely with the matrices K_{as} and M_{ac} shown by Eqs. (10a) and (12a), respectively. The matrices K_f and M_{f1} for lateral vibrations agree precisely with K_{fs} and M_{fc} of Eqs. (10b) and (12b), respectively. It is concluded, therefore, that the consistent mass method applies in the special case of the continuous mass method obtained by neglecting terms of higher than second order. Moreover, it can be estimated that the lumped mass method is a truncated result of the consistent mass method which is obtained by omitting the mass coupling. The relative relationship between the exact and approximate methods is demonstrated clearly from the aforementioned mathematical consideration, and it is also easily comprehensible from the computed results shown in Fig. 5.

IV. CONCLUSIONS

A natural vibration analysis of the five-span continuous rigid-frame bridges with V-shaped legs is presented by utilizing the lumped, consistent, and continuous mass methods. The numerical results are shown in Table 1 to investigate the applicability of the three different mass methods. The values of the natural circular frequencies obtained by the consistent mass method are the upper bounds to the exact solutions. The conventional lumped mass method yields eigenvalues that are considerably lower than the exact solutions in comparison with the consistent mass method. In general, the eigenvalues of rigid-frame bridges with V-shaped legs can be calculated more accurately by the consistent mass method than by the lumped mass method for the same number of beam segments.

The mathematical relationship between the exact method based on the differential equations of motion for axial and flexural vibrations and the approximate method based on a finite element approach is established in this study. The stiffness and mass matrices of the consistent mass method are represented by the first two terms of a power series expansion of the eigen stiffness matrices of Eqs. (5) and (6). Consequently, the consistent mass method is the special case of the continuous mass method obtained by neglecting terms of higher than the second order mass matrix. Moreover, the lumped mass method is a truncated result of the consistent mass method which is obtained by omitting the mass coupling. Finally, retrofitting procedures of existing bridges may lead to significant change in dynamic properties and consequently in seismic performance of the bridge structures.

Acknowledgments

The numerical computations involved in the present study were carried out using the digital computer HITAC M-682 and S-810 system installed at the Hokkaido University Computing Center.

REFERENCES

1) Clough, R.W. and Penzien, J., *Dynamics of Structures* (McGraw-Hill Book Co., 1975).
2) Jacobsen, L.S. and Ayre, R.S., *Engineering Vibrations* (McGraw-Hill Book Co., 1958).
3) Hayashikawa, T. and Watanabe, N., Dynamic behavior of continuous beams with moving loads. *J. Engng. Mech. Div., ASCE*, Vol. **107**, No. 1 (1981), pp. 229–246.
4) Ovunc, B.A., Dynamics of frameworks by continuous mass method. *J. Computers Structures*, Vol. **4** (1974), pp. 1061–1089.
5) Hurty, W.C. and Rubinstein, M.F., *Dynamics of Structures* (Prentice-Hall, Inc., Englewood Cliffs, N.J., 1964).
6) Hayashikawa, T. and Watanabe, N., Free vibration analysis of continuous beams. *J. Engng. Mech. Div., ASCE*, Vol. **111**, No. 5 (1985), pp. 639–652.
7) Paz, M., *Structural Dynamics, Theory and Computation* (Van Nostrand Reinhold Company, 1980).
8) Wendroff, B., *Theoretical Numerical Analysis* (Academic Press, 1966).
9) Crandall, S.H., *Engineering Analysis* (McGraw-Hill Book Co., 1956).
10) Ralston, A. and Wilt, H.S., *Mathematical Methods for Digital Computers* (John Wiley and Sons, 1967).

Dynamic Response Analysis of Specimen-Load Cell Systems in an Impact Loading Based upon Wave Propagation Theory

Masaaki ITABASHI* and Kozo KAWATA[2]*

* *Graduate School of Science University of Tokyo, Noda, Chiba,* [2]* *Department of Materials Science and Technology, Faculty of Industrial Science and Technology, Science University of Tokyo, Noda, Chiba*

It is known that when a dynamic load, e.g., with a tensile velocity faster than about 5m/s, is applied to the instrumentation with a load cell which works appropriately for quasistatic load measurement, the load cell output is superposed by disturbing high-frequency waves, causing difficulty in obtaining the real wave form of the phenomenon to be studied. This important fact has not yet been investigated systematically. The authors analyze the phenomenon from the standpoint of the one-dimensional wave propagation theory, and evaluate the amplitude and the frequency of the high-frequency wave for various combinations of specimen and load cell. From these results, effects of such conditions as tensile velocity and specimen dimensions are clarified.

I. INTRODUCTION

The instrumentation with a load cell for quasistatic load measurement, such as in ordinary material testing, has been established, with the merits of accuracy and easy maintenance. The main part of the load cell is a column or a cantilever of which elastic deformation proportional to the quasistatic load is transduced and output by the cemented strain gages. It is well known from experience that when a dynamic load with a tensile velocity larger than about 5 m/s, for example, is applied to such instrumentation, the output is quite different from the real load-time relation due to superposed disturbing waves. This fact shows the necessity of dynamic response analysis of load-measuring systems.

From this viewpoint, Kawata et al.[1] conducted FEM calculations on dynamic response of various load-measuring systems assuming test materials with their strain rate-sensitive constitutive equations. They concluded that a specimen-short load cell system did not give an exact load-time relation, since the dynamic load cell output is disturbed by the superposed high-frequency wave for the whole time duration. It must be stated that the disturbed load cell signal obstructs determination of exact dynamic mechanical properties of the material or the structure which is tested at high rates of strain.

The magnitudes of the amplitude and the frequency of the high-frequency wave may be changed by varying the combination of the specimen and the load cell. The evaluation of these values is important to determine the allowable range of the specimen-load cell system as a dynamic load-time relation acquisition system, although they have not yet been investigated based upon wave propagation theory as far as we know. Thus, in the present study the authors evaluate the high-frequency wave generated in the specimen-load cell system, and superposed to an essential load cell signal, with emphasis on its amplitude and frequency, based upon one-dimensional elastic wave propagation theory. For systematically varied combinations of specimen and load cell of column type, this effect is clearly shown, and errors caused by the high-frequency oscillations in the load cell output are discussed.

II. ANALYZED MODEL

Consider only two elements, i.e., the specimen and the load cell for the system to be analyzed. Their shapes are cylinders and they are inserted between rigid walls, as shown in the lower part of Figs. 1–3. Their materials are the same and perfectly elastic. The assumed loading condition is that since time $t=0$ the rigid wall connected with the specimen is pulled dynamically to the left-hand side in Figs. 1–3 with a constant velocity V for the whole time duration.

First, the size of the load cell is determined, considering typical practical conditions. The capacity, the maximum allowable stress, Young's modulus, the axial length, and the diameter of the elastic cylindrical column are taken as 49.0 kN (5tf), 100 MPa, 200 GPa, 50 mm, and 25 mm, respectively. The location of an infinitesimal strain gage adhering to the surface of it is taken at the center of the axial direction. Next, the specimen diameter

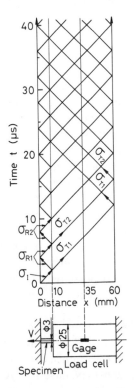

Fig. 1. The lagrangian diagram for $d=3$ mm and $l=10$ mm.

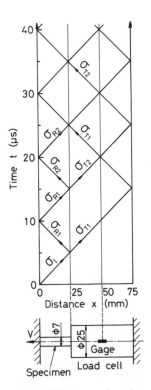

Fig. 2. The lagrangian diagram for $d=7$ mm and $l=25$ mm.

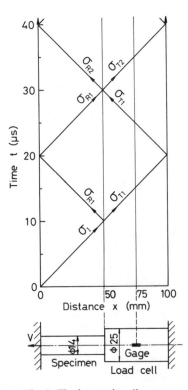

Fig. 3. The lagrangian diagram for $d=14$ mm and $l=50$ mm.

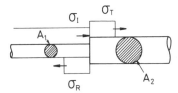

Fig. 4. The definition of a "shoulder" and reflected and transmitted waves generated there.

d is varied from 1 mm to 14 mm at an interval of 1 mm, and the specimen length l is taken as 10, 25, and 50 mm. In Figs. 1–3 corresponding to each specimen length, respectively, three lagrangian diagrams are shown.

III. ANALYSIS

In Fig. 4, when incident wave σ_I reaches a "shoulder," discontinuity of cross-sectional area, σ_I is divided into the reflected wave σ_R and the transmitted wave σ_T. According to the conditions to be satisfied at the shoulder with respect to the force and the particle velocity for both sides, the dividing ratios are given as follows:

$$\text{reflected wave} \quad \sigma_R = \frac{1-N}{1+N} \sigma_I \qquad (1)$$

$$\text{transmitted wave} \quad \sigma_T = \frac{2N}{1+N} \sigma_I \qquad (2)$$

where N is the area ratio of the area of the incident side A_1 to the transmitted side A_2, as the same material is taken for both sides in this case for simplicity.

When a material of density ρ and longitudinal wave velocity c is elongated dynamically at the particle velocity V', a tensile stress wave $\sigma = \rho c V'$ propagates to the opposite direction of the elongation. In this case, $V' = V$. The magnitude of incident stress wave σ_I in Figs. 1–3 is $\rho c V$ where ρ and c are assumed to be 8000 kg/m³ and 5000 m/s, respectively.

When σ_I in Figs. 1–3 reaches the interface between the specimen and the load cell, it is divided into σ_{R1} and σ_{T1}, as mentioned above. First, σ_{R1} propagates to the left rigid wall, and at this rigid wall, it reflected toward the shoulder again with the same sign and the same amplitude. When the σ_{R1} reaches the shoulder, σ_{R1} is divided into σ_{R2} and σ_{T2} in the ratios given by Eqs. (1) and (2), replacing σ_I in these by σ_{R1}. On the other hand, σ_{T1} travels to the rigid wall of the right-hand side and is reflected at the wall without any change. Thus, σ_{T1} reaches the shoulder from the opposite direction. Replacing N by $1/N$ and σ_I by σ_{T1}, Eqs. (1) and (2) give the dividing ratio of the reflected and the transmitted waves.

These procedures are repeated and each stress in the lagrangian diagrams is estimated. The stress at a distance x given in the system is equal to the summation of each stress wave passing this point at the given time t. Codes to conduct these procedures were developed by the authors and the stress in the specimen and the load cell output are compared. The former is represented by the stress at the center in the axial direction of the specimen. The latter is obtained being converted from the strain gage signal with the area ratio of the load cell to the specimen.

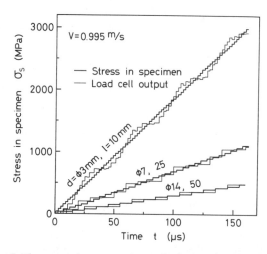

Fig. 5. The comparisons stress in specimen σ_S with load cell output.

IV. NUMERICAL RESULTS ON THE OUTPUT WAVE FORM

Figure 5 shows the comparisons of the stress in the specimen σ_S with the corresponding load cell output. All load cell outputs do not trace σ_S precisely and behave like sewing σ_S in a certain pitch without attenuation of the amplitude. This sewing action is the source of the high-frequency wave on the load cell signal. The time delay of the load cell output to σ_S is observed at the initial portion. The delay has its origin in the distance between the center of the specimen and the position of the strain gage on the load cell. The elastic stress wave takes a time equal to the delay time to pass through the distance. This phe-

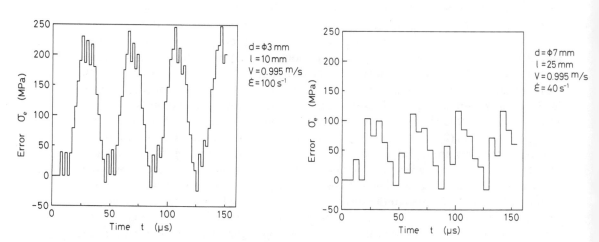

Fig. 6. Error σ_e vs. t relation for $d=3$ mm and $l=10$ mm.

Fig. 7. Error σ_e vs. t relation for $d=7$ mm and $l=25$ mm.

nomenon does not have an important role in this investigation and is cancelled by shifting the load cell output with respect to the abscissa to have the same rising point as σ_S.

V. DISCUSSION

The error of the load cell output σ_e is picked up by taking the difference between the shifted load cell output and the original loading profile. This treatment is shown in Figs. 6–8 and can evaluate the amplitude and the frequency of σ_e. σ_e is in the overestimating side for almost all cases and has a nonnegligible order of amplitude. The rigid wall at the load cell end contributes this phenomenon, as the reflector of propagated waves, and produces such an inconsiderable effect in quasistatic tension of specimens.[2] The one bar method[1] consisting of the specimen and the semi-infinite output bar, i.e., a very long load cell in practical terms, can measure precise dynamic stress-strain curves up to fracture at the strain rate of $1 \times 10^3 s^{-1}$ without disturbance.

The maximum amplitude of the error $\sigma_{emax} - \sigma_{emin}$ is defined as the difference of the maximum and the minimum values of σ_e in the range of $\sigma_S < 1000$ MPa. Figure 9 indicates that $\sigma_{emax} - \sigma_{emin}$ becomes small when thick d and short l are selected. σ_I is proportional to V and σ_S is a linear function of σ_I, so a relationship of the strain rate $\dot{\varepsilon}$ and $\sigma_{emax} - \sigma_{emin}$ is linear. In Fig. 9, to the extent of the above-mentioned consideration, it is recognized that these linear relationships hold with slight deviations.

Further, in the range of $t < 150$ μs, the frequency of σ_e is evaluated, counting several periods of the oscillations and tabulated in Table 1. The frequencies are independent of

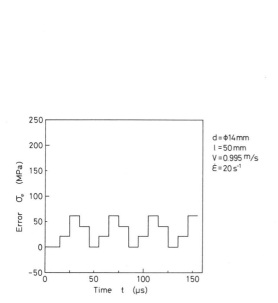

Fig. 8. Error σ_e vs. t relation for $d = 14$ mm and $l = 50$ mm.

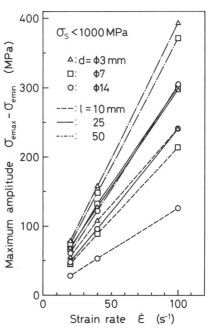

Fig. 9. The differences of maximum amplitude $\sigma_{emax} - \sigma_{emin}$ vs. $\dot{\varepsilon}$ relations for various specimen dimensions.

Table 1. The frequencies of oscillations of σ_e for various conditions.
($t < 150$ μs) (Unit: kHz)

d (mm)	$\dot{\varepsilon}$ (s^{-1})	l (mm) 10	25	50
$\phi 3$	20	25.0	25.0	25.0
	40	25.0	25.0	25.0
	100	25.0	25.0	25.0
$\phi 7$	20	28.8	25.0	25.0
	40	28.8	25.0	25.0
	100	28.8	25.0	25.0
$\phi 14$	20	34.1	30.0	25.0
	40	34.1	30.0	25.0
	100	34.1	30.0	25.0

Notes: d = specimen diameter
l = specimen length
$\dot{\varepsilon}$ = strain rate

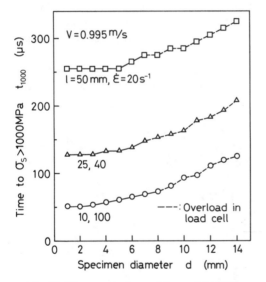

Fig. 10. Time required to satisfy $\sigma_S > 1000$ MPa vs. d relations at $V = 0.995$ m/s.

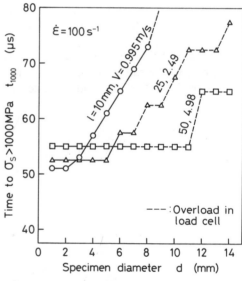

Fig. 11. Time required to satisfy $\sigma_S > 1000$ MPa vs. d relations at $\dot{\varepsilon} = 100$ s^{-1}.

$\dot{\varepsilon}$, because the elastic wave velocity is constant. Thicker or shorter specimen size gives higher frequency, except for the case of a thin diameter such as 3 mm.

From Fig. 9 and Table 1, the magnitude of $\dot{\varepsilon}$ varies $\sigma_{e\max} - \sigma_{e\min}$ proportionally, but not the frequency of σ_e. The behavior of σ_e depends on specimen size. Smaller amplitude and higher frequency of σ_e appear when thicker d or shorter l is selected. In specimen size selection, however, the load cell capacity should also be considered. Figures 10 and 11 show the time necessary to satisfy $\sigma_S > 1000$ MPa versus d. In these figures the symbols connected by broken lines indicate that the load cell is loaded beyond the allowable load. In these cases, 7 mm is the limit for the specimen diameter and it should be thinner than this value.

VI. CONCLUSIONS

Based upon the one-dimensional elastic wave propagation theory, the authors analyzed the dynamic response of specimen-load cell systems for high-velocity tension by a constant long stress wave. The following facts are derived. High-frequency oscillation is superposed on the essential result, and this superposed oscillation is based upon the waves reflected at the rigid wall of the load cell side. The load cell output generally indicates a tendency to overestimate, compared with the actual stress of the specimen. For analyzed examples, the magnitude of the overestimation is often beyond 200–300 MPa and it is difficult to remove this error. The amplitude of the oscillations of the load cell signal decreases to some extent by selecting a lower loading velocity, a thicker specimen diameter, or a shorter specimen length. The frequency of the signal oscillations, typically 30 kHz, increases by adopting a thicker specimen diameter or a shorter specimen length. These theoretical results are valuable also in understanding the behavior of similar systems, for example, by Clark and Duwez[3] who observed experimentally the oscillations of several tens of kHz in a specimen-load cell system with a specimen length of 8 inches (203 mm).

The one-dimensional solutions give results close to reality so long as the length of the bar is much greater than its diameter. When a more detailed consideration is needed, it is desirable to introduce the transverse inertia, as Skalak[4] discussed in the impact of two cylindrical elastic bars.

REFERENCES

1) Kawata, K., Hashimoto, S., Kurokawa, K., and Kanayama, N., A new testing method for the characterization of materials in high-velocity tension. *Mechanical Properties at High Rates of Strain 1979*, Harding, J., ed., Inst. of Phys., Bristol and London (1979), pp. 71–80.
2) Itabashi, M., High velocity deformation of several structural steels. *Master Thesis, Science University of Tokyo* (1986).
3) Clark, D.S. and Duwez, P.E., Discussion of the forces acting in tension impact tests of materials. *J. Appl. Mech.*, Vol. 15, No. 9 (1948), pp. 243–247.
4) Skalak, R., Longitudinal impact of a semi-infinite circular elastic bar. *J. Appl. Mech.*, Vol. 24, No. 3 (1957), pp. 59–64.

Self-Excited Oscillation of a Closed-Engine-Governor Loop: Equilibrium Instability and Limit Cycle

Yoshihiko KAWAZOE

Department of Mechanical Engineering, Saitama Institute of Technology, Okabe, Saitama

Most of the past research work on self-excited oscillation called low-speed hunting of a closed-engine-governor loop has been devoted to discriminating the divergence of a small disturbance given at an equilibrium state, resulting in a hunting estimation different from that of the actual system. In addition, it is impossible to predict whether the limit cycle will occur or not. The present work linearly estimates the instability of the equilibrium state of a pneumatically governed diesel engine, and compares it with the limit cycle given by numerical simulation with respect to the frequency and increment of amplitude as a step to explaining analytically the mechanism of limit cycle evolution. The results show that the linear theory considering a phase lag of the governing pressure responding to the engine speed fluctuation surely predicts the instability of an equilibrium state at higher idling speeds, but not the stability at extreme lower speeds, and gives the frequency almost correctly; but the calculated value of the logarithmic increment of amplitude at the largest hunting state is twice that of the nonlinear simulation.

I. INTRODUCTION

On a governed engine the idling speed cannot remain constant, and is followed by a low-frequency noise of its own.[1-10] This fluctuation of engine speed is called low-speed hunting. This phenomenon is a problem of control of a closed loop composed of the engine crankshaft system, governor, fuel injection pump, and combustion torque as shown in Fig. 1, and is known to be a self-excited oscillation. Nevertheless, most of the past research work on engine speed hunting has been devoted to discriminating the divergence of a small disturbance given at an equilibrium state, resulting in a hunting estimation different from that of the acutual system. In addition, it is impossible to predict whether the limit cycle will occur or not.

The cause of the low-speed hunting of pneumatically governed diesel engines has recently been revealed by the present author[9,11-14] to be the phase lag of suction pressure for displacing the fuel control rack. Further, computer simulation yields a transient process during which a small oscillation develops into a sustained oscillation with a large amplitude. However, for a better understanding of the hunting it is desirable to obtain the amplitude and frequency of the limit cycle analytically.

In the present report a linear approximation is provided considering the phase lag of governing pressure responding to the engine speed fluctuation, and is compared with a numerical, nonlinear simulation concerning the frequency and increment of amplitude as a step to explaining analytically the mechanism of limit cycle evolution.

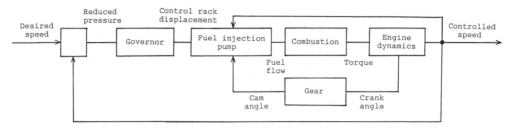

Fig. 1. Block diagram of a closed-engine-governor loop.

II. ENGINE GOVERNING SYSTEM AND LOW-SPEED HUNTING PHENOMENA

A Bosh-type individual fuel injection pump with its control rack locked delivers an increased quantity of fuel in each injection with an increased engine speed except at high-speed running; thus engine torque also increases with increasing engine speed, so that the equilibrium state of the idling speed is statically unstable. To get rid of this, a governor is provided.

As shown in Fig. 2, a pneumatic governor controls fuel delivery by displacing the fuel control rack with the reduced pressure taken at a narrow passage called a subventuri beside a throttle valve. Reduced pressure due to increased engine speed displaces the rack in the direction of decreasing fuel delivery through a diaphragm combined with a spring. The engine speed depends on the throttle valve opening. Figure 2 also shows an example of recorded hunting behavior of a pneumatically governed four-cylinder, four-stroke engine of the swirl-chamber type with a total stroke volume amounting to 1986 cm³ (idling at 800 rpm), where P_v, P_d, and X are subventuri pressure, diaphragm chamber pressure, and fuel control rack displacement, respectively. The engine hunts with a frequency of about 2 Hz. The control rack moves to and fro with a large amplitude, while the motion com-

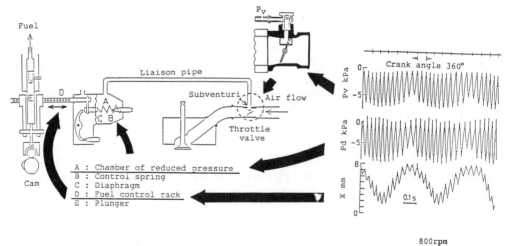

Fig. 2. Fuel control mechanism of a pneumatically governed engine and an example of recorded hunting behavior.

prises a component of the hunting frequency and a component of a higher frequency caused by the suction stroke of each piston.

III. THEORY

III-1. Nonlinear Simulation[14]

Let N_e rpm denote the engine speed. Then the rate of engine speed increase during idling is given by the expression

$$J_e \frac{dN_e}{dt} = T_e \qquad (1)$$

where I_e is the moment of inertia of the crankshaft system, $J_e = 2\pi I_e/60$, and T_e the accelerating torque. T_e is assumed to be constant throughout each 180° crank angle for a four-cylinder engine, and becomes a function of fuel rack displacement and engine speed. The quantity of fuel in each injection is assumed to be determined by the engine speed N_e and rack displacement X at the moment when the corresponding piston reaches its top dead center. The subroutine program gives a map of the measured torque as a function of the rack displacement and engine speed, and the torque T_{ei-1} corresponding to the values of X_{i-1} and N_{ei-1} is found by interpolation as

$$T_{ei-1} = T_e(X_{i-1}, N_{ei-1}). \qquad (2)$$

The response of rack displacement X to the governing pressure P can be written as

$$m_e \frac{d^2X}{dt^2} + C_e \frac{dX}{dt} + k(X + L_o - L) = A_d P \qquad (3)$$

where m_e is the equivalent mass of moving parts of the pneumatic governor system, C_e the equivalent viscous damping coefficient, k the stiffness of the rack spring, L_o the length of rack spring at $X=0$, L the free length of rack spring, and A_d the effective diaphragm area.[12] At a steady running condition without hunting, the mean value of the governing pressure shows a linear relation to the mean engine speed at a given throttle opening. In transients, however, the mean governing pressure P_π for each 180° crank angle at each throttle opening requires the following equation, so as to satisfy a lot of experimental data, where N_e, α, and T_p are the engine speed, the sensitivity of governing pressure to the steady-state engine speed at each throttle opening, and the time constant of the first order, respectively.

$$T_p \frac{dP_\pi}{dt} + P_\pi = -\alpha N_e. \qquad (4)$$

Since the higher frequency component caused by the suction stroke of each piston somewhat affects the governing of fuel delivery, the wave form of reduced pressure between the instant of the i-1th injection t_{i-1} and that of t_i is given as

$$P = P_{\pi i-1}(1 - \cos \omega_{mi-1} t) \qquad (5)$$

where ω_{mi-1} is the frequency caused by the suction stroke depending on N_e, and the origin of t is put on the instant when the piston leaves its top dead center.

Figure 3 shows the calculated transient behavior on a phase plane plotting rack dis-

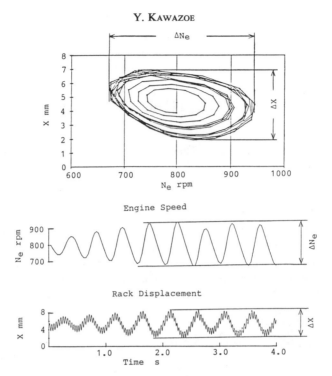

Fig. 3. Calculated limit cycle in a phase plane (800 rpm).

placement at the instant of injection versus engine speed, indicating development of a limit cycle of the closed-engine-governor loop responding to a small step increment of the throttle opening at an initial engine speed of 800 rpm.

III-2. Linear Approximation

The following equation is a linearized differential equation of a closed-engine-governor loop derived from the equation of motion of the engine, the relation of developed torque to the fuel delivered per stroke, the equation of motion of the fuel control rack, and the characteristics of pressure reduction at the subventuri

$$A_4 \frac{d^4 n}{dt^4} + A_3 \frac{d^3 n}{dt^3} + A_2 \frac{d^2 n}{dt^2} + A_1 \frac{dn}{dt} + A_0 n = 0 \qquad (6)$$

where

$$A_4 = T_p m_e J_e$$

$$A_3 = T_p \left[- m_e \left(\frac{\partial T_e}{\partial N_e} \right)_0 + C_e J_e \right] + m_e J_e$$

$$A_2 = T_p \left[- C_e \left(\frac{\partial T_e}{\partial N_e} \right)_0 + k J_e \right] - m_e \left(\frac{\partial T_e}{\partial N_e} \right)_0 + C_e J_e$$

$$A_1 = - T_p k \left(\frac{\partial T_e}{\partial N_e} \right)_0 + \left[- C_e \left(\frac{\partial T_e}{\partial N_e} \right)_0 + k J_e \right]$$

$$A_0 = - k \left(\frac{\partial T_e}{\partial N_e} \right)_0 - A_d \left(\frac{\partial P}{\partial N_e} \right)_0 \left(\frac{\partial T_e}{\partial X} \right)_0 \qquad (7)$$

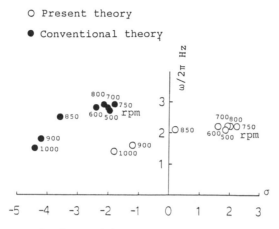

Fig. 4. Representative characteristic root $s=\sigma+j\omega$ of linear approximate system.

In Eq. (6) a component of the higher frequency caused by the suction stroke is neglected, and the fuel delivery process is assumed to be continuous. Phase lag of the subventuri pressure is considered as a time constant T_p of the first order, where as it has not been considered in past research work.

A characteristic equation derived from Eq. (6) has two real roots and a pair of conjugate complex roots. These two real roots have large negative values; accordingly stability depends on the complex roots. In Fig. 4 are shown representative characteristic roots $s=\sigma+j\omega$ plotted on a complex plane relative to the equilibrium engine speed as a parameter. When the value of σ is positive, it gives an increment of small oscillation. From Fig. 4, the dynamic linear system could be unstable when the engine speed lies below 850 rpm. The frequency of oscillation in the linearly unstable region is 2.1–2.2 Hz in good agreement with the experimental values. On the other hand, the unstable region does not appear and the value of frequency is higher than the experimental value in the conventional linear theory without considering the lag of the subventuri pressure.

IV. RESULTS OF COMPARISON BETWEEN LINEAR THEORY AND NONLINEAR SIMULATION

Calculated results with a nonlinear model are shown in Fig. 5, which indicates that the calculated engine speed fluctuation, i.e., the amplitude of the limit cycle is large over the range 650–820 rpm of mean engine speed with the peak at 750 rpm in good agreement with the experimental results. Figure 5 also shows the calculated behavior of the nonlinear model responding to a small disturbance of 10 rpm at the engine speeds of 500, 800, 850, and 1000 rpm, where P, N_e, T_e, and X are governing pressure, engine speed, shaft torque, and fuel rack displacement, respectively. The calculated hunting frequency is about 2 Hz, in good agreement with the experimental value. The magnitude of logarithmic increment of amplitude of the closed loop is consistent with the amplitude of limit cycle.

Figure 6 shows the results of linear approximation, and indicates the calculated behavior of the linearized engine-governor loop responding to a small disturbance of 10 rpm at

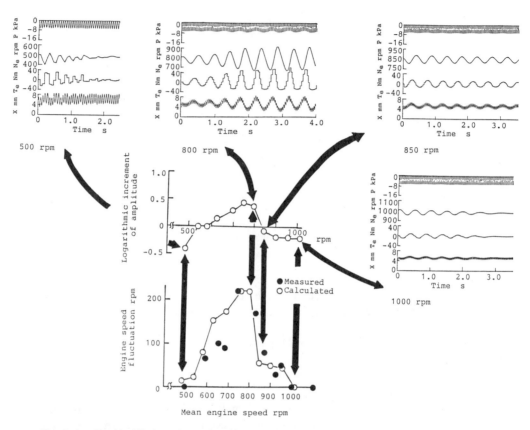

Fig. 5. Amplitude of limit cycle and logarithmic increment of amplitude of nonlinear simulation.

the equilibrium engine speeds of 500, 800, and 850 rpm and the values of logarithmic increment of amplitude calculated with the linear model, where circles represent the linear simulation and deltas stand for the results from the representative characteristic root at each engine speed. Comparison between the result of the linear model and that of the non-linear model shows that the logarithmic increment of the linear system is twice that of the nonlinear model at the engine speed where large hunting occurs.

V. CONCLUSIONS

The magnitude of logarithmic increment of amplitude of the closed loop with the nonlinear model is consistent with the amplitude of the limit cycle. The linear approximation considering a phase lag of the governing pressure responding to engine speed fluctuation predicts the instability of an equilibrium state at higher idling speeds but not stability at extreme lower speeds and gives the hunting frequency almost correctly but the calculated value of logarithmic increment of amplitude at the largest hunting state is twice that of the nonlinear simulation.

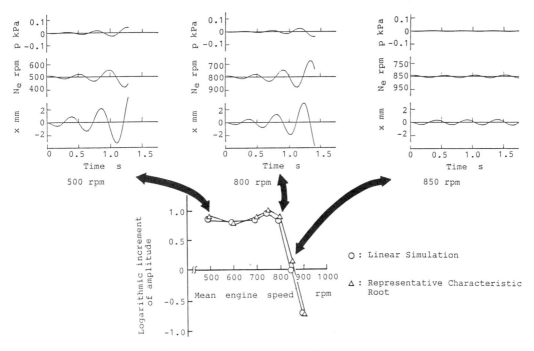

Fig. 6. Logarithmic increment of amplitude of linear approximate system.

Acknowledgments

The author is grateful to Professor Emeritus K. Tsuda, Professor H. Sakai and Mr. Y. Ohtake of the Faculty of Engineering, University of Tokyo, for their suggestions and encouragement. This study was supported in part by Grant-in-Aid for Scientific Research No. 61550195, 1986 from the Ministry of Education, Science and Culture, Japan.

REFERENCES

1) Webb, C.R., The pneumatic governor. *Auto. Engng.*, Vol. **47**, No. 4 (1957), pp. 146–151.
2) Takahashi, T., Governing of internal combustion engines (Part 1). *J. Jpn. Soc. Mech. Engrs.*, Vol. **62**, No. 483 (1959), pp. 565–582 (in Japanese).
3) Takahashi, T., Governing of internal combustion engines (Part 2). *J. Jpn. Soc. Mech. Engrs.*, Vol. **62**, No. 484 (1959), pp. 755–768 (in Japanese).
4) Takahashi, T., in: *Handbook of Internal Combustion Engines*, Hatta, K. and Asanuma, T. (eds.), Asakura Publishing (1960), pp. 236–245 (in Japanese).
5) Kaneko, Y., Speed control characteristics of the high speed diesel engine. *Shin Mitsubishi Juko Giho*, Vol. **5**, No. 3 (1963), pp. 247–253 (in Japanese).
6) Fujihira, U., Stability problems in the fuel injection system for diesel engines. *Internal Combustion Engine*, Vol. **4**, No. 37 (1965), pp. 23–29 (in Japanese).
7) Ishimaru, T., Idling hunting of diesel engine mounted on road vehicles. *J. Soc. Auto. Engrs. Jpn.*, Vol. **19**, No. 11 (1965), pp. 845–850 (in Japanese).
8) Welbourn, D.B. et al., Governing of compression-ignition oil engines. *Proc. Inst. Mech. Engng.*, Vol. **173**, No. 22 (1959), pp. 575–604.
9) Kawazoe, Y., Low speed hunting of the pneumatically governed compression ignition engine (1st Re-

port, various experiments for identifying the cause). *Bull. JSME*, Vol. **28**, No. 243 (1985), pp. 2022–2027.
10) Kamata, M. et al., Idle speed fluctuations of diesel engines with mechanical governors. *Trans. Soc. Auto. Engrs. Jpn.*, No. 32 (1986), pp. 53–60 (in Japanese).
11) Kawazoe, Y., Low speed hunting of the pneumatically governed compression ignition engine (2nd report, stability criticism of the closed engine-governor loop based on measured frequency responses of the two open systems composing the loop). *Bull. JSME*, Vol. **28**, No. 244 (1985), pp. 2365–2369.
12) Kawazoe, Y., Dynamics and equivalent damping of the fuel control system of the pneumatically governed compression ignition engine. *Bull. JSME*, Vol. **29**, No. 248 (1986), pp. 501–507.
13) Kawazoe, Y., Low speed hunting of the pneumatically governed compression ignition engine (3rd report, computer simulation of the low speed hunting). *Bull. JSME*, Vol. **29**, No. 253 (1986), pp. 2211–2217.
14) Kawazoe, Y., Low speed hunting of the pneumatically governed compression ignition engine (4th report, effect of simulated venturi diameter on phase lag of the fuel control rack response). *Bull. JSME*, Vol. **29**, No. 250 (1986), pp. 1233–1238.

Free Vibration Analysis of a Distributed Flexural Vibrational System by the Transfer Influence Coefficient Method

Takahiro Kondou,* Atsuo Sueoka,[2]* Deok Hong Moon,[3]* Hideyuki Tamura,[3]* and Toshimi Kawamura*

*Department of Electronic and Mechanical Engineering, Fukuoka Institute of Technology, Fukuoka, [2]*Department of Mechanical Engineering, Kyushu University, Fukuoka, [3]*Department of Mechanical Engineering, Power Division, Kyushu University, Fukuoka.*

The transfer matrix method for a distributed flexural vibrational system posesses such defects as cancelling attributable to the sum and difference of the hyperbolic and trigonometric functions appearing in the elements of the transfer matrices, and occurrence of numerical instability due to numerical imbalance among elements and multiplication of the matrices. The authors apply the concept of the transfer influence coefficient method to a distributed flexural vibrational system, and formulate an algorithm for the analysis of free vibration with high speed and with high accuracy which succeeds in overcoming all these defects of the transfer matrix method. The validity of the present algorithm is demonstrated by a relatively simple computing forward critical speeds of a rotating shaft, and is also compared with that of the transfer matrix method on a personal computer.

I. INTRODUCTION

Our previous reports[1,2] presented a newly developed algorithm of free vibrational analysis for a discrete parameter system, in which its validity was clarified in comparison with the transfer matrix method (TMM).[3] This algorithm was called the transfer influence coefficient method (TICM), and was based on the concept of the recurrent transmission of the dynamic influence coefficients at every station. The TICM is suitable for personal computer use, and is easily applicable to a distributed flexural vibrational system as demonstrated below.

It is well known that the TMM has defects in analyzing a distributed flexural vibrational system such as cancelling attributable to the sum and difference of the hyperbolic and trigonometric functions appearing in the elements of the transfer matrices, and occurrence of numerical instability due to numerical imbalance among the elements and multiplication of the matrices. These defects are especially apparent in computing higher order natural frequencies and when the intermediate elastic supports of structures are very stiff.

In this report, the authors apply the TICM to the distributed flexural vibrational system, and formulate an algorithm for free vibrational analysis with high speed and high accuracy, successfully improving all the defects of the TMM mentioned above. The validity of the present algorithm is demonstrated by a relatively simple example computing the critical speeds of a rotating shaft, and is also compared with that of the TMM on a personal computer.

II. TRANSFER INFLUENCE COEFFICIENT METHOD

An algorithm is formulated for computing the dynamic influence coefficients, the natural frequencies, and the characteristic modes of straight-line structures constructed from uniformly distributed beams and concentrated disks, where the structure is partitioned into n elements in such a way that every discontinuous point concerning the shearing force or the bending moment becomes a node. The left and right ends of the system are defined as nodes 0 and n, respectively, and continuity of the lateral and angular displacements is assumed at every node. The concept for applying the TICM to the flexural distributed system is illustrated in Fig. 1. The system ① shown in Fig. 1(a) represents the already established section from node 0 to node $j-1$. Another uniformly distributed beam ② of length l_j, flexural rigidity $(EI)_j$, and linear density μ_j as in Fig. 1(b) is connected to the right end of this established system ①, and, in addition, a concentrated disk with lumped mass m_j and lumped inertia moment J_j supported by a linear spring k_j and a rotational spring K_j is connected rigidly to the right end of the beam as shown in Fig. 1(c). This is the fundamental procedure for obtaining the dynamic influence coefficient of the whole system by connecting successively the distributed beams and the concentrated disks supported by the springs. The algorithm is formulated for the Euler-Bernoulli beam in the text, and the results to be corrected for the Timoshenko beam are summarized in the Appendix.

II-1. Dynamic Influence Coefficient

The relationships between lateral and angular displacements $d_j = {}^t(y,\theta)_j$ and shearing forces and bending moments $\bar{f}_j = {}^t(\bar{F}, \bar{N})_j$, $f_j = {}^t(F, N)_j$ at node j shown in Fig. 1(b) and (c) are, respectively, given by

$$d_j = U_j \bar{f}_j, \qquad d_j = U_j f_j. \tag{1}$$

The dynamic influence coefficient matrices $\bar{U}_j$ and U_j are both symmetric, and are indicated as

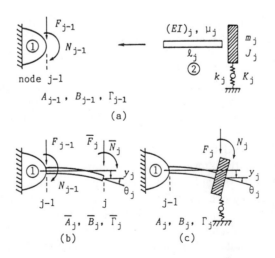

Fig. 1. Convention in series connection of element j.

$$\bar{U}_j = \begin{bmatrix} \bar{A} & \bar{\Gamma} \\ \bar{\Gamma} & \bar{B} \end{bmatrix}_j, \quad U_j = \begin{bmatrix} A & \Gamma \\ \Gamma & B \end{bmatrix}_j$$

where $(\bar{A}, \bar{B}, \bar{\Gamma})_j$ and $(A, B, \Gamma)_j$ correspond to the dynamic influence coefficients on the left- and the right-hand sides at node j, respectively. Here, the notations with bars represent the physical quantities just before the disk at node j and the same notations without bars represent those just after the disk at node j.

II-2. Rule in Series Connection of Dynamic Infulence Coefficients

The rule for connecting the uniformly distributed beam to the right-hand side of system ① shown in Fig. 1(a) can analytically be derived from the fundamental equation of the beam. In this report, however, considering that a number of transfer matrices are obtained for many kinds of structure elements,[3] a method of transformation from the TMM to the TICM is presented in order to make good use of the transfer matrices.

The relationship between the state variable $\bar{z}_{j-1}$ on the right-hand side of node $j-1$ of the uniformly distributed beam and the state variable $\bar{z}_j$ on the left-hand side of node j is expressed by virtue of the transfer matrix T_j as

$$\bar{z}_j = T_j z_{j-1}. \tag{2}$$

In order to convert relation (2) to that of the TICM, $\bar{z}_j(z_{j-1})$ is separated into the displacement vector $d_j(d_{j-1})$ and the force vector $\bar{f}_j(f_{j-1})$, and the transfer matrix T_j is rearranged by using the 2×2 submatrices P_j, Q_j, R_j, and S_j corresponding to this separation. Then, Eq. (2) becomes

$$\begin{bmatrix} d \\ \bar{f} \end{bmatrix}_j = \begin{bmatrix} P & Q \\ R & S \end{bmatrix}_j \begin{bmatrix} d \\ f \end{bmatrix}_{j-1} \tag{2'}$$

where

$$\lambda_j^4 = \mu_j \omega^2 / e_j, \quad a_j = l_j^2 / e_j, \quad e_j = (EI)_j \quad \beta_j = \lambda_j l_j$$

$$P_j = \begin{bmatrix} c_0 & l c_1 \\ \beta^4 c_3/l & c_0 \end{bmatrix}_j, \quad Q_j = \begin{bmatrix} -a l c_3 & a c_2 \\ -a c_2 & a c_1/l \end{bmatrix}_j,$$

$$R_j = \begin{bmatrix} -\beta^4 c_1/a l & -\beta^4 c_2/a \\ \beta^4 c_2/a & \beta^4 l c_3/a \end{bmatrix}_j,$$

$$S_j = \begin{bmatrix} c_0 & -\beta^4 c_3/l \\ -l c_1 & c_0 \end{bmatrix}_j,$$

$c_{0j} = (\cosh \beta_j + \cos \beta_j)/2, \qquad c_{1j} = (\sinh \beta_j + \sin \beta_j)/2\beta_j,$

$c_{2j} = (\cosh \beta_j - \cos \beta_j)/2\beta_j^2, \qquad c_{3j} = (\sinh \beta_j - \sin \beta_j)/2\beta_j^3,$

and ω is a circular frequency to be obtained. Substituting Eq. (1) into Eq. (2)′ yields

$$\bar{U}_j = (P_j U_{j-1} + Q_j)(R_j U_{j-1} + S_j)^{-1}. \tag{3}$$

If $\bar{U}_j$ in Eq. (3) is computed by using the submatrices P_j, Q_j, R_j, and S_j obtained from the transfer matrix, the numerical instability due to cancelling occurs often in obtaining the natural frequencies of higher order of very long structures, because the cancelling is attributable to the sum and difference of the hyperbolic and trigonometric functions with the same argument β_j appearing in $c_{0j} \sim c_{3j}$ in the submatrices, and the argument β_j becomes large in this case. Therefore, we calculate Eq. (3) explicitly, and rewrite it by using only the bounded continuous functions. By this modification, numerical instability is avoidable, and hence the dynamic influence coefficients are computed with high accuracy. The modified results are arranged in the recurrent formulae as follows:

$$\hat{G}_j \bar{u}_j = H_j u_{j-1} \tag{4}$$

where

$$\bar{u}_j = {}^t(\bar{A}, \bar{B}, \bar{\Gamma}, 1, \bar{C})_j, \qquad \bar{C}_j = \bar{A}_j \bar{B}_j - \bar{\Gamma}_j^2,$$

$$u_j = {}^t(A, B, 2\Gamma, C, 1)_j, \qquad C_j = A_j B_j - \Gamma_j^2,$$

and $H_j = [h_{km}]_j$ is the 5×5 symmetrical matrix whose elements are given as

$$h_{11} = v_2/\lambda^2, \qquad h_{12} = v_1, \qquad h_{13} = w_1/\lambda,$$
$$h_{14} = -e\lambda w_2, \qquad h_{15} = w_2/\lambda^3 e, \qquad h_{22} = -\lambda^2 v_2,$$
$$h_{23} = -\lambda w_2, \qquad h_{24} = -e\lambda^3 w_1, \qquad h_{25} = w_1/\lambda e,$$
$$h_{33} = v_1/2, \qquad h_{34} = -e\lambda^2 v_2/2, \qquad h_{35} = v_2/2\lambda^2 e,$$
$$h_{44} = e^2 \lambda^4 w_4/2, \qquad h_{45} = w_3/2, \qquad h_{55} = w_4/2e^2\lambda^4,$$
$$v_1 = \cos\beta, \qquad v_2 = v_3 v_6, \qquad v_3 = \sin\beta,$$
$$v_4 = v_1 v_6, \qquad v_5 = \text{sech}\,\beta, \qquad v_6 = \tanh\beta,$$
$$w_{1,2} = (v_3 \pm v_4)/2, \qquad w_{3,4} = v_5 \pm v_1.$$

The subscript j is omitted in the above expressions (as in the Appendix). The functions cos, sin, tanh, and sech used in the above expressions take all limited values between -1 and 1. Equation (4) represents the field transmission rule of the dynamic influence coefficients from the right-hand side of node $j-1$ to the left-hand side of node j, and the factor $\hat{G}_j$ is computed first in Eq. (4). When the disk or the linear and rotational springs are attached to the left end of the system, the dynamic influence coefficient matrix at node 0 is

$$U_0 = \begin{bmatrix} 1/S & 0 \\ 0 & 1/M \end{bmatrix}_0 \tag{5}$$

where $S_j = k_j - m_j \omega^2$ and $M_j = K_j - J_j \omega^2$. Since numerical inconvenience occurs in Eq. (5) when $S_0 = 0$ and/or $M_0 = 0$, Eq. (4) for $i = 1$ must be modified using Eq. (5) as

$$\hat{G}_1 \bar{u}_1 = H_1 u_0 \tag{6}$$

where we define $u_0 \equiv {}^t(S, M, 0, 1, SM)_0$. All the boundary conditions on the left end of the system can be treated systematically by this modification, even for the case of $S_0 = 0$ and/or $M_0 = 0$. When the values of the spring constants k_0 and K_0 are infinitely large, we only need to substitute adequately large finite values into the corresponding spring constants in the numerical computation, because the dynamic influence coefficients are obtained as limited values for the case when the numerators and denominators are both infinitely large.

Next, the point transmission rule between the left- and right-hand sides of node j is obtained from the balance of the shearing forces and the moments as follows:

$$A_j = (\bar{A} + \bar{C}M)_j/\bar{G}_j, \qquad B_j = (\bar{B} + \bar{C}S)_j/\bar{G}_j,$$
$$\varGamma_j = \bar{\varGamma}_j/\bar{G}_j, \qquad C_j = \bar{C}_j/\bar{G}_j \qquad (7)$$

where

$$\bar{G}_j = M_j\bar{C}_jS_j + \bar{A}_jS_j + \bar{B}_jM_j + 1. \qquad (7)'$$

Balancing the shearing forces and the moments at node j yields

$$\bar{f}_j = f_j - \text{diag}(S,M)_j d_j. \qquad (8)$$

As described above, the transmission rule from the right-hand side of node j to that of node $j+1$ is expressed separately by two steps of Eqs. (4) and (7). Since Eq. (6) is more general than Eq. (4) at the first step from the left end of the system to the left-hand side of node 1, the present algorithm is programmed using Eq. (6). When there is neither a disk nor a spring at node j, we obtain $A_j = \bar{A}_j$, $B_j = \bar{B}_j$, $\varGamma_j = \bar{\varGamma}_j$, $\bar{G}_j = 1$, since $S_j = 0$ and $M_j = 0$ in Eq. (7).

III. FREQUENCY EQUATION

III-1. Summary of Frequency Equation

The frequency equation corresponding to each boundary condition at the right end of the system can be expressed separately. However, we can also deal with all the boundary conditions at the right end systematically by the equation[1]

$$\bar{H}_n \equiv \bar{G}_n/(M_nS_n) = \bar{C}_n + \bar{A}_n/M_n + \bar{B}_n/S_n + 1/M_nS_n = 0 \qquad (9)$$

except for the free and simply supported ends without a disk, where we need only to set $(k_n, K_n) = (0,0)$ for the free end, $(k_n, K_n) \to (\infty, 0)$ for the simply supported end, and $(k_n, K_n) \to (\infty, \infty)$ for the fixed end. In the numerical computation, an adequately large finite value is practically used for the infinitely large spring constant, so that the natural frequencies are saturated within the prescribed effective digits. In the present conditions, $M_n \neq 0$ and $S_n \neq 0$ even when $k_n = 0$ and $K_n = 0$, respectively, since $m_n \neq 0$ and $J_n \neq 0$. When the right end of the system is the free and simply supported end without a disk, both M_n and S_n, and only M_n is zero, respectively. Therefore, the frequency equation cannot be unified by Eq. (9), and it must be used individually for each boundary condition. That is, we obtain the frequency equation $1/\bar{C}_n = 0$ for the free end, and from Eq. (4) this is equivalent to the equation

$$\hat{G}_n = 0. \qquad (10)$$

For the simply supported end, we obtain

$$\bar{A}_n = 0. \qquad (11)$$

In the present stage, three kinds of frequency equations of Eqs. (9), (10), and (11) must be properly used depending on the three conditions described above. However, these equations are finally unified by combination with the method of elimination of false roots treated in the following section.

III-2. Mechanism of Occurrence of False Roots and their Elimination

When the bisection method is used to obtain the solution of the frequency equation for the discrete parameter system, false roots are generated. In our previous report,[1] the mechanism of their occurrence was discussed and a simple, effective method for their elimination was presented. Although false roots also exist in the distributed flexural system, as described below, they can be eliminated by the same procedure as for the discrete parameter system.

First, we consider the case where a disk or linear and rotational springs exist at every node (that is, $S_j \neq 0$ and $M_j \neq 0$) and where the frequency equation is given by Eq. (9). Since the components $\bar{A}_n$, $\bar{B}_n$, and $\bar{C}_n$ of the function $\bar{G}_n$, which comprises a main part of the function $\bar{H}_n$, have the common denominator $\hat{G}_n$ from Eq. (4), the zeros of $\hat{G}_n$ correspond to the poles of $\bar{G}_n$ (and hence the poles of $\bar{H}_n$). As the roots of $\hat{G}_n = 0$ are ordinarily single roots, the sign of $\bar{G}_n$ (and hence the sign of $\bar{H}_n$) changes before and after the poles. Here, we call such poles asymmetric poles, and a pole before and after which the sign is not changed is a symmetric pole. Then, the zeros of M_n and S_n also correspond to the asymmetric poles of $\bar{H}_n$. Therefore, not only the zeros but also the asymmetric poles of $\bar{H}_n$ are obtained as roots by the bisection method unless special improvements are used. The former and the latter correspond to real roots (that is, natural circular frequencies) and so-called false roots, respectively. In order to eliminate such false roots, we transform all asymmetric poles into symmetric ones by multiplying the signum functions of $\hat{G}_n$, M_n, and S_n to $\bar{H}_n$. By the multiplication, the asymmetric poles of $\bar{H}_n$ caused by the zeros of $\hat{G}_n$, M_n, and S_n are all transformed into symmetric ones. However, if $\hat{G}_n$ has asymmetric poles, these become the new false roots in a new function. In fact, since the components A_{n-1}, B_{n-1}, Γ_{n-1}, and C_{n-1} of $\hat{G}_n$ have the common denominator of $\bar{G}_{n-1}$, the zeros of $\bar{G}_{n-1}$ become the asymmetric poles of $\hat{G}_n$ in the same manner as described above. These asymmetric poles can be transformed into symmetric ones by multiplying additionally the signum function of $\bar{G}_{n-1}$ to a new function. Such corresponding relationships between the zeros and the asymmetric poles are established between any $\hat{G}_j$ and $\bar{G}_j$, and between any $\bar{G}_{j-1}$ and $\hat{G}_j$. Since there exist no asymmetric poles in $\hat{G}_1$ [cf. Eq. (6)], we can obtain the function whose sign changes only at the zeros of $\bar{H}_n$ by multiplying the signum functions of $\hat{G}_1$, $\bar{G}_1$, ..., $\hat{G}_{n-1}$, $\bar{G}_{n-1}$, M_n, and S_n to $\bar{H}_n$. We define the resulting function obtained in such a way as $\bar{H}'_n$. As the bisection method needs only the sign of the function, we use the following signum function of $\bar{H}'_n$ instead of $\bar{H}_n$:

$$\operatorname{sgn}(\bar{H}'_n) = \prod_{j=1}^{n-1} \{\operatorname{sgn}(\hat{G}_j) \cdot \operatorname{sgn}(\bar{G}_j)\}$$

$$\times \operatorname{sgn}(\hat{G}_n) \cdot \operatorname{sgn}(\bar{H}_n) \cdot \operatorname{sgn}(M_n) \cdot \operatorname{sgn}(S_n)$$

$$= \prod_{j=1}^{n} \{\operatorname{sgn}(\hat{G}_j) \cdot \operatorname{sgn}(\bar{G}_j)\}. \qquad (12)$$

When no disk and no springs exist at node j, the zeros of $\hat{G}_j$ correspond directly to the asymmetric poles of $\hat{G}_{j+1}$ not through $\bar{G}_j$. Then, the discussion described above need not be exchanged, because $\bar{G}_j = 1$. When the frequency equations are given by Eqs. (10) and

(11), the sign of Eq. (12) changes only at the zeros of $\hat{G}_n$ and $\bar{A}_n$, respectively, because $\bar{G}_n = 1$ for the former and $\bar{G}_n$ corresponds to the case of $M_n = 0$ and $S_n \to \infty$ for the latter. Thus applying the bisection method to Eq. (12), it is found to be the frequency equation applicable to every boundary condition by adjusting the spring constants adequately, and by which the false roots are all eliminated. By this procedure, we can compute the natural frequencies all at once over a wide range of frequencies as for the TMM.

IV. INTERMEDIATE STIFF ELASTIC SUPPORTS

As described in a previous report[1] for the discrete parameter system, the TICM is applicable readily and uniformly to distributed flexural systems not only with very soft intermediate supports but also with rigid ones, by merely changing adequately the corresponding linear spring constants. This is because the point transmission rules at node j supported by the linear spring constant k_j are the same for both systems in form. The results of the intermediate rigid support obtained by putting $k_j \to \infty$ (that is, $S_j \to \infty$) in Eq. (7) are as follows:

$$A_j \to 0, \quad B_j \to \bar{C}_j / (\bar{A}_j + \bar{C}_j M_j), \quad \Gamma_j \to 0. \tag{13}$$

Then, the reaction force defined as $R_j = F_j - \bar{F}_j$ becomes

$$R_j \to F_j + \bar{\Gamma}_j N_j / (\bar{A}_j + \bar{C}_j M_j). \tag{14}$$

In conclusion, the dynamic influence coefficients and the reaction force at the node of the intermediate rigid support can be obtained by setting their values directly according to Eqs. (13) and (14), or by numerically substituting an adequately large value into the corresponding linear spring constant.

V. COMPUTATION OF CHARACTERISTIC MODES

First, we consider the case in which the disk or the linear spring is connected with the right end of the system, and its right-hand side is free, that is, the case of $\bar{F}_n \neq 0$, $F_n = 0$, and $N_n = 0$. Then, the following expressions are obtained from Eqs. (1) and (8) under the assumption of $\bar{F}_n = -1$:

$$\left. \begin{array}{l} y_n = 1/S_n, \quad \theta_n = -\bar{\Gamma}_n / (1 + \bar{B}_n M_n) \\ \bar{F}_n = -1, \quad \bar{N}_n = -M_n \theta_n. \end{array} \right\} \tag{15}$$

Such a setting conforms rationally to the fixed and the simply supported end, because $y_n \to 0$ and $\theta_n \to 0$ as $k_n(S_n) \to \infty$ and $K_n(M_n) \to \infty$, respectively. The characteristic modes for all boundary conditions can also be obtained by controlling the values of the spring constants k_n and K_n in Eq. (15). If the state variables d_n, f_n, and $\bar{f}_n$ at node n are given, then those from node $n-1$ to node 0 are computed from Eqs. (1) and (8) and the equation

$$f_{j-1} = (R_j U_{j-1} + S_j)^{-1} \bar{f}_j \equiv D_j \bar{f}_j \tag{16}$$

where the elements D_{km} defined as $D_j = [D_{km}]_j / \hat{G}_j$ are represented by

$$D_{11} = e\,\lambda\,w_8\,B + e\,\lambda^2\,w_6\,\Gamma + w_5,$$
$$D_{12} = \lambda(e\,\lambda\,w_6\,B + e\,\lambda^2\,w_7\,\Gamma + w_8).$$
$$D_{21} = -e\,\lambda^2\,w_6\,A - e\,\lambda\,w_8\,\Gamma + w_7/\lambda,$$
$$D_{22} = -e\,\lambda^3\,w_7 A - e\,\lambda^2\,w_6\,\Gamma + w_5,$$
$$w_{5,6} = (1 \pm v_1 v_5)/2, \qquad w_{7,8} = (v_6 \pm v_3 v_5).$$

The subscript j of D_j means the substitution of $j-1$ into the subscripts of A, B, and Γ, and the substitution of j into those of other notations in the above equations. Once a natural frequency has been computed, the corresponding characteristic mode can be computed from Eqs. (1), (8), (15), and (16) by repartitioning the structure more closely using the same rule as described in Section 2 and by obtaining the influence coefficients at each node from Eqs. (4) and (6).

Next, when the right end is free without a disk and the springs, we obtain $f_n = \bar{f}_n = 0$, and the following equation holds good from Eq. (10) when the natural frequencies are computed:

$$\hat{G}_n = \mathrm{sech}\beta_n \cdot \det(R_n U_{n-1} + S_n) = 0. \tag{17}$$

Since $\mathrm{sech}\beta_n$ never vanishes, Eq. (17) is equivalent to $\det(R_n U_{n-1} + S_n) = 0$. Therefore, the ratio of F_{n-1} to N_{n-1} which belong to f_{n-1} can be obtained as D_{12n}/D_{22n} from Eq. (16) set as $j=n$, and then d_{n-1} is computed from Eq. (1). The displacement vector d_n is obtained by

$$d_n = (P_n U_n + Q_n) f_{n-1}. \tag{18}$$

The other procedures are the same as described above.

VI. RESULTS OF NUMERICAL COMPUTATION AND DISCUSSION

In order to compare the computational accuracy and the computing time of the TICM with those of the TMM, the forward critical speeds for a uniform beam under certain conditions are computed as a fundamental problem. The beam is a solid steel shaft 960 mm in length and 20 mm in diameter. For the calculation, a 16-bit personal computer (NEC PC9801 VX21 with numeric data processor, FORTRAN77) is used with single and double precisions. We search for the critical speeds using the bisection method. The criterion of the convergence of the forward critical speeds was set as $\varepsilon \leq 10^{-7}$ and $\varepsilon \leq 10^{-9}$ for single and double precisions, respectively, where ε is the absolute value of relative error.

VI-1. Control of Boundary Conditions by Virtue of Spring Constants

As an example, we calculated the forward critical speeds for various boundary conditions at both ends of the rotating shaft mentioned above. We call this problem example 1. The variables of single precision were used for the TICM and those of double precision for the TMM. The rotating shaft was regarded as a Euler-Bernoulli beam. For this example, the shaft was not partitioned (that is, $n=1$), and the two lowest critical speeds were searched for with an initial frequency interval of 10Hz from an initial value of 10Hz. In the TICM, the method of elimination of false roots was adapted in programming by using Eq. (12) as the frequency equation, and, in addition, the boundary conditions were con-

Table 1. Control of boundary conditions by spring constants (Euler-Bernoulli beam).

Boundary condition	TMM (D)	TICM (S)	Poles in TICM
Free-free	98.900792	98.9008	(−)
	272.62384	272.624	(−)
Hinged-hinged	43.628447	43.6285	Hinged-free
	174.51379	174.514	
Fixed-fixed	98.900792	98.9008	Fixed-free
	272.62384	272.624	
Fixed-free	15.542496	15.5425	(−)
	97.403159	97.4032	(−)
Free-fixed	15.542496	15.5425	Free-free
	97.403159	97.4032	
Hinged-free	68.155961	68.1560	(−)
	220.86897	220.869	(−)

trolled by the spring constants. Here, we set k_j[N/m]$=0$ and K_j[N·m/rad]$=0$ for the free end, $k_j=10^{20}$, $K_j=10^{20}$ for the fixed end, and $k_j=10^{20}$ and $K_j=0$ for the simply supported (hinged) end, where $j=0$ or n. In the TMM, the critical speeds were computed by applying the frequency equation corresponding to each boundary condition. In Table 1, we see the numerical computation results, where S and D in parentheses show those with single and double precisions, respectively. The boundary conditions under which the false roots occur are also indicated in Table 1. Such false roots are given as the critical speeds for the boundary condition shown in the fourth column, and can easily be eliminated by the method described in Section 3. The mark "(−)" denotes that no false root exists, because the rotating shaft does not have a disk (that is, $\bar{G}_1=1$) and it is not partitioned. The critical speeds from the TICM and the TMM coincide with each other. Especially for the boundary condition "fixed-fixed", $\bar{C}_1$ obtained from the fifth equation of Eqs. (4) and (6) is more accurate than that from $\bar{A}_1\bar{B}_1 - \bar{\Gamma}_1^2$. Although the critical speeds of higher order for the boundary conditions, "fixed-fixed" and "free-fixed" become entirely coincident with those for the boundary conditions "fixed-free" and "free-free" within the prescribed effective digit, respectively, the method of elimination of false roots enables us to obtain them separately. In such a way, we can execute without problem the unification of the frequency equation by virtue of the control of the spring constants and the elimination of false roots.

VI-2. Computational Examples of Intermediate Stiff Elastic Supports

As a computational example of the critical speeds for a multispan shaft, we compute the forward critical speeds for the case when the unifom rotating shaft mentioned above is partitioned in the ratio of 6:1:3 and four nodes involving the right and left ends are elastically supported only with common linear spring constant k. We call this problem example 2. For this problem, the shaft is regarded as a Timoshenko beam (shape factor $\kappa_j=0.886$[1]), and the gyro effect is also taken into account.[3]

The computational results for $k=10^8$[N/m] are shown in Table 2. In the table, the TIMR shows the method by which the dynamic influence coefficients are obtained not from Eqs. (4) and (6) but from Eq. (3) directly using the elements of the transfer matrix. The values in the table denote the critical speeds (Hz) which can be obtained with acceptable accuracy

Table 2. Critical speeds (Hz) of example 2 at $k = 10^8$ [N/m] (Timoshenko beam).

No.	TICM (D)	TICM (S)	TIMR (D)	TIMR (S)	TMM (D)	TMM (S)
1	169.97287	169.973	169.97287	169.973	169.97287	169.974
2	550.61933	550.619	550.61933	550.619	550.61933	–
3	616.79230	616.792	616.79230	616.792	616.79230	–
4	1142.2093	1142.21	1142.2093	1142.22	1142.2093	1142.2
5	1918.5053	1918.51	1918.5053	*	1918.5053	–
6	1941.6352	1941.64	1941.6352	*	1941.6352	–
7	2861.4355	2861.44	2861.4355	*	2861.4355	–
8	3235.3950	3235.40	3235.3950	**	3235.3951	–
9	3790.3485	3790.35	3790.3485	**	–	–
10	4082.1098	4082.11	4082.1098	**	–	–
11	5188.9575	5188.96	5188.9575	**	–	–
12	5917.1595	5917.16	5917.1595	**	–	–
13	6609.3809	6609.38	6609.3810	**	–	–
14	7679.6598	7679.66	7679.6599	**	–	–
15	8486.2060	8486.21	*	**	–	–
16	9508.7657	9508.77	*	***	–	–
17	10559.846	10559.9	*	**	–	–
18	11762.779	11762.8	*	**	–	–
19	12904.983	12905.0	*	**	–	–
20	14165.322	14165.3	*	***	–	–

without partitioning each span more closely. For critical speeds of higher order, the errors of the corresponding numerical roots computed by the TIMR and the TMM are increased into the prescribed effective digits, and the correct roots (defined later) do not numerically become obtainable any longer. Then, we try to obtain them accurately by partitioning each span more closely. The mark "*" denotes that the correct root can numerically be computed only by such a partition, and its number represents the number of partitions. That is, the left- and the right-hand spans are partitioned in the ratio of 2:1, so that the length of the beam elements of both spans are equal to each other, and the number of partitions of the right-hand span is represented by that of the mark "*". The intermediate span is not partitioned. For example, the mark "***" means that the total number of partitions is equal to 10. The correct root can usually be obtained in the TIMR if each span is partitioned more closely than the number of partitions indicated in the table, but it could accidentally be obtained only for the number of partitions shown in the table in the TMM. The mark "—" shows that the correct roots can never be obtained, however closely each span is partitioned. Since the results of these three methods using quadruple precision ($\varepsilon \leq 10^{-9}$) on a large-sized computer (HITAC M260-H) coincided entirely with one another up to the 20th critical speed, these were considered as the correct roots. The results are summarized as follows:

(1) The TICM maintains sufficiently high accuracy up to the critical speeds of higher order, and it is not necessary to partition each span, even if single precision is used in the computation.

(2) The TIMR (D) and (S) can improve the computational accuracy by partitioning the rotating shaft adequately to shorten the length of each element and by decreasing the value of the common argument in hyperbolic and trigonometric functions.

Table 3. Critical speeds (Hz) of example 2 at $k=\infty$[N/m] (Timoshenko beam).

No.	TICM (D)	TICM (S)	TIMR (D)	TIMR (S)	TMM (D)	TMM (S)
1	173.18691	173.187	173.18691	173.187	173.18691	173.187
2	561.85782	561.858	561.85782	561.859	561.85782	561.858
3	655.91185	655.912	655.91185	655.912	655.91185	655.912
4	1187.8812	1187.88	1187.8812	*	1187.8812	1187.90
5	2013.1581	2013.16	2013.1581	*	2013.1581	–
6	2162.5466	2162.55	2162.5466	**	2162.5466	2162.55
7	3088.8822	3088.88	3088.8822	**	3088.8822	–
8	4249.9751	4249.98	4249.9751	**	4249.9751	–
9	4504.1696	4504.17	4504.1696	**	4504.1696	–
10	5671.2394	5671.24	5671.2394	***	5671.2394	–
11	6416.0325	6416.03	6416.0326	**	6416.0325	–
12	7619.6750	7619.68	*	**	7619.6749	–
13	8064.1229	8064.12	*	**	8064.1229	–
14	9498.1963	9498.20	*	***	*	–
15	11433.734	11433.7	*	**	–	–
16	11708.318	11708.3	*	**	–	–
17	13594.198	13594.2	*	***	–	–
18	15706.246	15706.3	*	***	–	–
19	16035.270	16035.3	*	**	–	–
20	17826.676	17826.7	*	**	–	–

(3) The TMM (D) and (S) cannot be expected to improve the computational accuracy by partitioning the rotating shaft.

(4) In general, the TICM has the highest accuracy, and the TIMR and the TMM become more inaccurate in that order.

Table 3 indicates results similar to those in Table 2 for the case of the intermediate rigid support, that is, $k=\infty$. For the TICM and the TIMR, $k=10^{20}$[N/m] was then selected in the numerical computation. Since it immediately causes numerical instability[1] in the TMM to substitute such a large value into k, another algorithm of TMM (elimination method[3]) was applied which can treat the intermediate rigid support. Therefore, the program of the TMM used for obtaining the results in Table 3 is different from that in Table 2. In this case, the results of these three methods by quadruple precision coincided entirely with one another and with those of the TICM (D) up to the 20th critical speed. The results are summarized as follows:

(5) The TMM (elimination method) is comparable with the TIMR in accuracy. Once a root becomes unobtainable, we cannot expect improvement of the accuracy by partitioning the rotating shaft for the TMM.

(6) The TICM maintains sufficiently high accuracy even for intermediate rigid supports, and requires only the substitution of a large finite value into k so that the roots are numerically saturated, and hence the TICM does not require any special algorithms.

(7) The effect of the variables used (single, double, and quadruple precisions) on computational accuracy is remarkable, especially for the TMM and the TIMR.

On the basis of the results described above, we now consider why the difference in the accuracy of these four methods (including the TMM (elimination method)) occurs.

First, the difference between the TMM and the TMM (elimination method) is discussed.

The transfer matrix is multiplied by the point matrix at each node in the former, but the point matrix is not used in the latter, since the state variables are adequately exchanged at each node with the intermediate rigid support.[3] If the spring constants are relatively large, cancelling or overflow due to multiplication by the point matrix occurs in the former, because the spring constants are linearly included in the point matrix.[3] In the latter, however, the spring constants do not appear entirely. This is the reason why the difference of the accuracy between them indicated in summaries (4) and (5) occurs. In the TICM and the TIMR, such a decrease in computational accuracy does not occur, because the dynamic influence coefficients obtained from Eq. (7) are numerically normalized even for large values of spring constants.

Next, the difference in accuracy among the TIMR and the TMM (standard and elimination methods) indicated in summaries (2), (3), and (5) is considered. This is attributable to a difference in the field transmission rules. That is, the dynamic influence coefficients of the TIMR obtained from Eq. (3) are normalized by $\det(R_j U_{j-1} + S_j)$ at each node. If the beam is partitioned more closely, the accuracy of the elements of submatrices P, Q, R, and S in Eq. (3) is improved by decreasing the value of the common argument in the hyperbolic and trigonometric functions. Consequently, the accuracy of successive transmission in the TIMR can be improved by partitioning the beam even for critical speeds of higher order. On the other hand, the transfer matrix of the beam before partition (that is, the field matrix of such a beam) is theoretically the same as the resulting matrix obtained by multiplying successively the field matrices of the closely partitioned beams. Therefore, once cancelling, overflow, or numerical instability occur due to successive multiplication or numerical imbalance among elements of the transfer matrices, the accuracy of the successive transmission in the TMM cannot be improved, even if the elements of the transfer (field) matrix become more accurate by partitioning each shaft. In the TICM, the field transmission rules (Eqs. (4) and (6)) are given by using only finite valued functions, so that the elements of the matrix H_j in Eqs. (4) and (6) can be accurately computed without partition, and the transfer influence coefficients can be obtained with high accuracy in the process of the successive transmission through normalizing them by $\hat{G}_j$ at each node.

In order to make the characteristics described above more clear, the relationships between the forward critical speeds (kHz) and k[N/m] are shown in Fig. 2. In this problem, the shaft is not partitioned. The solid lines denote the correct solution obtained by the TICM with single precision. The small white circles and the small black circles in Fig. 2 show that the correct solutions can be obtained by the TMM with single and double precisions, respectively. The correct solution denoted by the small black circles cannot be obtained by TMM with single precision any longer. The crosses show that correct solutions cannot be obtained even if double precision is used. That is, in the region involving small black circles and crosses, numerical instability[1] occurs when single precision is used, and many solutions are obtained near a true critical speed, depending on the initial incremental width of the frequency in the bisection method. The small circles on the line $k = \infty$ show the results obtained from the TMM (elimination method). From Fig. 2, the region with small black circles and crosses expands in a statistical manner in the direction not only of increasing k, but also of higher frequency, which indicates a defect of the TMM.

The method of elimination of false roots was also executed without trouble in example 2. This method is effective even for intermediate elastic and rigid supports, and is applicable not only to the TICM but also to the TIMR.

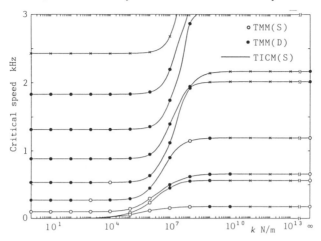

Fig. 2. Relationship between critical speeds and linear spring constant k for example 2 (Timoshenko beam).

In this report, the characteristics of four methods were compared with one another on the basis of a simple problem. The results obtained show that the TICM possesses the highest reliability, especially when the vibrational analysis for a complex and unknown structure is executed using a personal computer.

VI-3. *Computational Example of Characteristic Modes*

Figure 3 shows the characteristic modes of orders 1st to 5th for the Timoshenko beam of example 2 at $k=10^9[N/m]$ obtained by the TICM with single precision. The arrows show the positions of the elastic supports. The dynamic influence coefficients and the characteristic modes are both computed by partitioning the shaft into 50 elements with same length after the corresponding critical speed is computed without partition.

VI-4. *CPU Time*

Next, CPU times in example 2 are compared between the TICM, TIMR, and TMM (standard and elimination methods). Since the eight lowest critical speeds can at least be obtained for both $k=10^8[N/m]$ and $k=\infty[N/m]$ by every method with double presicion when the shaft is not partitioned (see Tables 2 and 3), the CPU times required to compute the eight lowest critical speeds and the corresponding modes all at once are summarized in Table 4. If the method of elimination of the false roots is not applied for the TICM and TIMR, their CPU times cannot substantially be compared with those of the TMM. Table

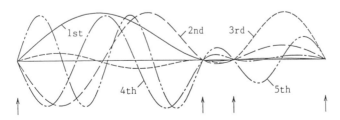

Fig. 3. Characteristic modes of example 2 (Timoshenko beam).

Table 4. Comparison of CPU time (s) of example 2.

Beam	k_j	ε	TICM (S)	TICM (D)	TIMR (D)	TMM (D)
Euler-Bernoulli	10^8	10^{-7}	6.7	7.5	7.6	13.6
		10^{-9}	–	8.3	8.3	14.9
	∞	10^{-7}	8.0	9.0	9.1	15.1
		10^{-9}	–	9.7	9.8	16.3
Timoshenko	10^8	10^{-7}	8.6	9.8	9.1	15.2
		10^{-9}	–	10.8	10.0	16.6
	∞	10^{-7}	10.1	11.7	10.9	16.7
		10^{-9}	–	12.7	11.8	18.1

4 shows the results of the CPU times (second) for both the Euler-Bernoulli and the Timoshenko beams. The figures denoted in the table are the averaged values of 10 measurements. The critical speeds were sought from an initial frequency of 10 Hz with an initial step of 10 Hz. The results are summarized as follows:

(1) In the Euler-Bernoulli beam, the CPU time of the TICM is the same as that of the TIMR, and the computing speed of the TICM is about 1.8 times as fast as that of the TMM.

(2) In the Timoshenko beam, the TIMR becomes slightly faster than the TICM. This is due to the same reason that the expressions for obtaining the elements of H_j for the TICM (see Appendix) are more complicated than those for the TIMR, that is, the elements of the transfer matrix.[3] The computing speed of the TICM is about 1.5 times as fast as those of the TMM.

On the other hand, in the computation using BASIC language without the numeric data processor, it was confirmed that the ratios of the computing speed of the TICM to that of the TMM increased to about 4.0 for the Euler-Berunoulli beam and to about 2.6 for the Timoshenko beam.

The results described above show that the TICM is a superior method of structural analysis also from the point of view of computing speed.

VII. CONCLUSIONS

The authors applied the concept of the transfer influence coefficient method described in a previous report[1] to the distributed flexural vibrational system, and they formulated an algorithm for the free vibrational analysis. The present method was compared with the transfer matrix method for computational accuracy and speed on a personal computer. As a result, it became clear that the present method of vibrational analysis has high speed and accuracy which can effectively cope with all defects of the transfer matrix method. such as occurrence of cancelling attributable to the sum and difference of the hyperbolic and trigonometric functions.

REFERENCES

1) Sueoka, A., Kondou, T., Yokomichi, I., Ayabe, T., and Tamura, H., A method of vibrational analysis using a personal computer (a suggested transfer influence coefficient method). *Trans. Jpn. Soc. Mech. Engrs.*, Vol. **52**, No. 484 C (1986), pp. 3090–3099 (in Japanese).

2) Sueoka, A., Tamura, H., Ayabe, T., and Kondou, T., A method of high speed structural analysis using a personal computer (comparison with the transfer matrix method). *Bull. Jpn. Soc. Mech. Engrs.*, Vol. **28**, No. 239 (1985), pp. 924–930.
3) Pestel, E.C. and Leckie, F.A., *Matrix Methods in Elastomechanics*, McGraw-Hill Book Co. (1963).

APPENDIX

In the Timoshenko beam with circular section, the elements of the symmetrical matrix $H_j=[h_{km}]_j$ in Eqs. (4) and (6) become

$$h_{11} = \Lambda^2 \, l^2 \, \lambda_1^2 \, \lambda_2^2 \, x_3/b, \qquad h_{12} = \Lambda^2 \, x_1,$$

$$h_{13} = \Lambda \, l (\lambda_2^2 \, x_2 + \lambda_1^2 \, x_4),$$

$$h_{14} = \Lambda \, l \{(\lambda_2^2 - \tau) \, \lambda_1^2 \, x_4 - (\tau + \lambda_1^2) \, \lambda_2^2 \, x_2\}/a,$$

$$h_{15} = a \, \Lambda \, l \, \{(\lambda_2^2 - \tau) \, \lambda_2^2 \, x_2 - (\tau + \lambda_1^2) \, \lambda_1^2 \, x_4\}/b,$$

$$h_{22} = -\Lambda^2 \, bx_3/l^2, \qquad h_{23} = \Lambda b(x_4 - x_2)/l,$$

$$h_{24} = \Lambda \, b\{(\tau - \lambda_2^2) \, x_2 - (\tau + \lambda_1^2) \, x_4\}/al,$$

$$h_{25} = a \, \Lambda \, \{(\tau + \lambda_1^2) \, x_2 + (\lambda_2^2 - \tau) \, x_4\}/l,$$

$$h_{33} = \Lambda^2 \, x_5/2 - 2bx_6 + b(\sigma + \tau) \, x_3,$$

$$h_{34} = b[(\tau - \sigma) \, x_6 - \{(\sigma + \tau)\tau + 2\lambda_1^2 \, \lambda_2^2\} \, x_3]/a,$$

$$h_{35} = a[(\tau - \sigma) \, x_6 + \{(\sigma + \tau)\sigma + 2\lambda_1^2 \, \lambda_2^2\} \, x_3],$$

$$h_{44} = b[2bx_6 + \{\Lambda^2 \, \tau - b(\sigma + \tau) \, x_3\}/a^2,$$

$$h_{45} = 2bx_6 + \Lambda^2 \, x_1 - b(\sigma + \tau) \, x_3,$$

$$h_{55} = a^2[2bx_6 + \{\Lambda^2 \, \sigma - b(\sigma + \tau) \, x_3\}/b,$$

where

$$\Lambda = \lambda_1^2 + \lambda_2^2, \qquad \sigma = \mu \, l^2 \, \omega^2/\kappa \, G' A'$$

$$b = \beta^4, \qquad \tau = -ha \, \mu \, d^2 \, \omega^2/16,$$

$$\lambda_{2,1}^2 = [b + (\sigma - \tau)^2/4]^{1/2} \pm (\sigma + \tau)/2,$$

$$h = \begin{cases} -1: \text{flexural vibration} \\ 1: \text{forward critical speed} \\ -3: \text{backward critical speed} \end{cases}$$

$$x_1 = \cos\lambda_2, \qquad x_2 = \sin\lambda_2/\lambda_2, \qquad x_3 = x_2 x_7$$

$$x_4 = x_1 x_7, \qquad x_5 = \text{sech}\lambda_1, \qquad x_6 = x_5 - x_1$$

$$x_7 = \tanh\lambda_1/\lambda_1,$$

and d and $\kappa G'A'$ denote the diameter and the shearing stiffness of the shaft, respectively. The elements of the matrix $\boldsymbol{D}_j = [D_{km}]_j/\hat{G}_j$ in Eq. (16) are given as

$$D_{11} = bc_2 \Gamma/a + c_0 - \tau c_2 + l\{(b + \tau^2) c_3 - \tau c_1\}B/a,$$

$$D_{12} = b(c_1 - \sigma c_3) \Gamma/a l + bc_2 B/a + b c_3/l,$$

$$D_{21} = -b c_2 A/a + l\{c_1 - (\sigma + \tau) c_3\}$$
$$\quad - l\{(b + \tau^2) c_3 - \tau c_1\} \Gamma/a,$$

$$D_{22} = - b(c_1 - \sigma c_3) A/al - bc_2 \Gamma/a + c_0 - \sigma c_2,$$

$$c_0 = \Lambda(\lambda_2^2 + \lambda_1^2 x_8), \qquad c_1 = \Lambda(\lambda_2^2 x_7 + \lambda_1^2 x_9),$$

$$c_2 = \Lambda(1 - x_8), \qquad c_3 = \Lambda(x_7 - x_9),$$

$$x_8 = x_1 x_5, \qquad x_9 = x_2 x_5.$$

Characteristics of Acoustic Radiation from Plate Girder Railway Bridge

Toshiyuki SUGIYAMA, Yasuharu FUKASAWA, and Yukihito GOMI*

*Department of Civil Engineering, Yamanashi University, Yamanashi, * Yamanashi Prefectural Office, Yamanashi*

The accuracy of radiation efficiency of a rectangular panel is discussed based on the results of free and forced vibration experiments. A rectangular panel has one opposite clamped and one opposite simply supported edge. Sensitiveness of radiation efficiency to the boundary conditions of a rectangular panel is also discussed. Close agreement between experimental and theoretical values has been obtained. It can be concluded that radiation efficiency is not very sensitive to differences in boundary condition. In the latter part of this study, we show a method to calculate precisely the acoustic spectrum based on the measured acceleration spectrum of a panel. This discussion is made from the point of view of the clearness of a physical interpretation, and its accuracy is confirmed by measured field data on a plate girder railway bridge.

I. INTRODUCTION

In order to grasp the characteristics of vibration of a bridge and acoustic radiation from it, the following approach has generally been used. First we measure the acceleration of vibration of main girders and noise radiated by them. Next, we investigate the relation between these two quantities. In this approach, radiation efficiency (radiation resistance) is necessarily used.[1,2] Radiation efficiency is a dimensionless quantity that relates the vibration of a member (structure) with the noise radiated by it. For rectangular panels with particular boundary conditions, we can obtain radiation efficiency theoretically under the appropriate assumptions. The accuracy of this calculated value has no sufficiently beent verified experimentally, however. Furthermore, a method to apply this dimensionless quantity to measured field data has not yet been fully established.

One of the purposes of this study is to investigate the accuracy of the theoretical radiation efficiency of rectangular panels experimentally. Free and forced vibration experiments are conducted. In the former experiment, vibration is caused by hitting a panel with a metal ball. A rectangular panel with one opposite clamped and one opposite simply supported edge is used. In order to estimate the sensitiveness of radiation efficiency to the boundary conditions of a panel, two rectangular panels are used: one is a panel with one opposite clamped and one opposite freely supported edge; the other is a fully clamped panel. The values of radiation efficiency of these two panels cannot be obtained theoretically.

Another purpose of this study is to discuss the way to calculate precisely the acoustic spectrum by use of the measured acceleration spectrum of a panel. This discussion is made from the point of view of the clearness of a physical interpretation. We attempt to confirm the accuracy of the method proposed here based on the measured field data on a plate girder railway bridges.

II. ACCELERATION SPECTRUM AND ACOUSTIC SPECTRUM

When a panel vibrates in a single mode, the sound radiated by it is composed of only one frequency component. However, the sound radiated by a rectangular panel generally includes various frequency components. Therefore when the measured field data filtered in third octave bands are used, the following assumption seems acceptable. The acoustic spectrum corresponding to its third octave band central frequency is a sort of total of each acoustic spectrum whose frequency exists between two cut-off frequencies. Considering this, the expression defined by Eq. (1) is reasonable to relate the acceleration spectrum of vibration with the acoustic spectrum.

$$SPL = AL + 10 \log_{10}(\sum_{i=1}^{s} b_i \cdot K_i/f_i^2) + 22.3 + 10 \log_{10} \psi \qquad (1)$$

where SPL is the acoustic spectrum and AL is the acceleration spectrum. Both of SPL and AL are filtered in third octave bands. The constant b_i, included in the second term on the right-hand side of Eq. (1), relates each acceleration spectrum of vibration mode with measured AL and is given by

$$b_i = d_i^2/d_o$$

in which $d_i = A_{mn}/(R^2 \cdot m^2 + n^2)^2$, $d_o = \sum_{i=1}^{s} d_i$, $A_{mn} = 16/(\pi^2 \cdot mn)$, and $R = 1.25$. Parameters m and n represent the (m,n)-th vibration mode of a rectangular panel. s is the number of vibration modes existing between the two cut-off frequencies under consideration. In other words, b_i is the ratio of acceleration spectrum of the i-th vibration mode to the measured acceleration spectrum. Constant A_{mn} is the amplitude of the (m,n)-th deflection mode under the condition that a uniformly distributed unit load acts on the whole panel. K_i and f_i are radiation efficiency and natural frequency corresponding to i-th vibration mode, respectively. The fourth term on the right-hand side is the correction factor which depends on the distance between the panel and the microphone of the noise meter.

The following expression is adopted as it has been used formerly.[3]

$$SPL = AL + 10 \log_{10}(K_m/F_m^2) + 22.3 + 10 \log_{10} \psi \qquad (2)$$

where F_m is the third octave band center frequency and K_m is average radiation efficiency. A few studies have shown that Eq. (2) gives a close approximation of the experimental value. However, the actual calculation of radiation efficiency is rather complicated because the values of K_m depend on the following two relations. One is the relation between wave number in the panel and that in the acoustic field, and the other is the relation between the critical frequency of a panel and each frequency of vibration mode of a panel. It is assumed for obtaining K_m that all relative velocities corresponding to each vibration mode take the same value. This assumption is not necessarily reasonable from a physical standpoint because some modes are generally dominant during vibration. On the other hand, using the method proposed here, we can calculate the values of the acoustic spectrum using only Eq. (1) for various vibration modes. In other words, there is no complication in the calculation process; moreover, physical interpretation of this method is very clear.

III. FORCED AND FREE VIBRATION EXPERIMENTS

The panel used in the experiments was a steel board 500 mm long, 400 mm wide, and 2 mm thick. This panel is supported by two steel columns and installed on the vibration table as shown in Fig. 1. A schematic diagram of the experimental equipment is given in Fig. 2. Measurements were conducted as follows. In the forced vibration experiment, the data obtained through a noise meter and vibration level meter were directly filtered by a third octave band analyzer. In the free vibration experiment, data were filtered by the third octave band analyzer after being recorded in a data recorder. The microphone of the noise meter was set up at the position where the perpendicular distance from the panel center was 10 cm. In the free vibration experiment, first we raised a metal ball 2 cm vertically along the orbit of a simple pendulum and then hit the panel with it. The diameter and weight of the ball were 50 mm and 575 gW, respectively. A block diagram of the

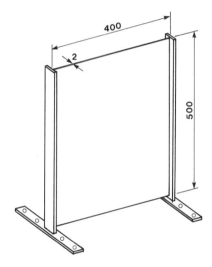

Fig. 1. Rectangular panel used in experiments.

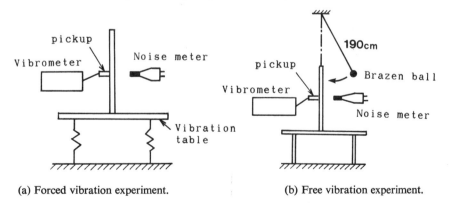

(a) Forced vibration experiment. (b) Free vibration experiment.

Fig. 2. Diagram of experimental equipment.

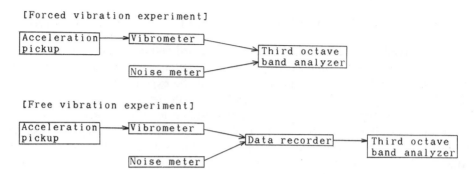

Fig. 3. Block diagram for measurement.

measurement system is presented in Fig. 3. Since the acceleration spectrum AL and the acoustic one SPL are obtained from experiments, we can calculate the experimental value of radiation efficiency by rearranging Eq. (1) as

$$K = f^2 \cdot 10^{\{0.1 \times (SPL - AL - 30.21)\}}. \tag{3}$$

IV. EXPERIMENTAL RESULTS AND DISCUSSION

Figure 4 shows the radiation efficiency of a panel with one opposite clamped and one opposite simply supported edge. These values were obtained from the forced vibration experiment. In this figure the abscissa is the third octave band center frequency and the ordinate is radiation efficiency (common logarithms of radiation efficiency multiplied by 10). The theoretical values of radiation efficiency, which are calculated discretely for each vibration mode, are shown with broken lines for clarity. From Fig. 4, it is found that the

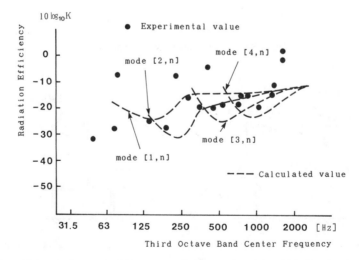

Fig. 4. Radiation efficiency of a panel with one opposite clamped and one opposite simply supported edge in the forced vibration experiment.

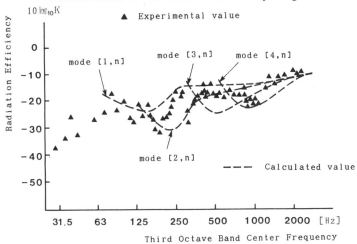

Fig. 5. Radiation efficiency of a panel with one opposite clamped and one opposite simply supported edge in the free vibration experiment.

agreement between theoretical and experimental values at frequencies from 100 Hz to 2000 Hz is indeed remarkable. However, there exists a considerable discrepancy at the low-frequency range.

The values of radiation efficiency obtained from the free vibration experiment are shown in Fig. 5. The boundary condition of the panel is the same as shown in Fig. 4. It is found by comparing Fig. 5 with Fig. 4 that the sound radiated by a panel in the case of free vibration includes more frequency components than that of forced vibration. Regarding the value of radiation efficiency, the same facts as obtained from Fig. 4 are recognized. The following is considered to be the reason for the discrepancy at the low-frequency re-

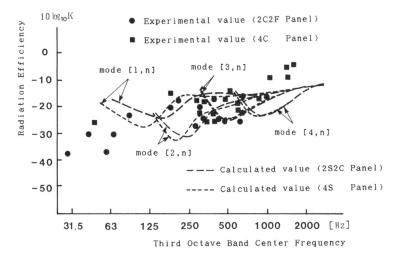

Fig. 6. Radiation efficiency of a panel with one opposite clamped and one opposite freely supported edge and of a panel with fully clamped edges in the forced vibration experiment.
C: Clamped edge; F: freely supported edge; S: simply supported edge.

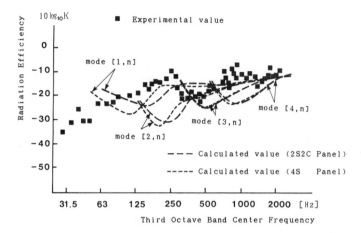

Fig. 7. Radiation efficiency of a panel with fully clamped edges in the free vibration experiment.
C: clamped edge; S: simply supported edge.

gion. The boundary condition of the panel used in these experiments is not identical to that assumed in the theoretical approach because of the effect of initial deformation due to welding. The characteristics of low-frequency vibration modes of a panel are affected significantly by this discrepancy. It is, however, very difficult to determine which mode corresponds to the frequency under consideration in our experiments. Further discussion is needed on this point.

Figures 6, 7, and 8 illustrate the experimental values of radiation efficiency for a panel with one opposite clamped and one opposite freely supported edge and for a fully clamped panel. Theoretical radiation efficiency for these two panels cannot be calculated. The broken lines and the dotted lines in these figures correspond to the following cases, respectively:

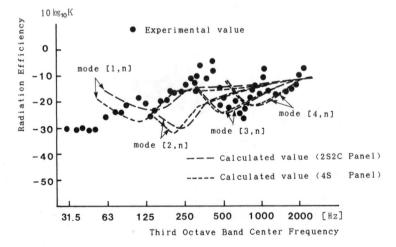

Fig. 8. Radiation efficiency of a panel with one opposite clamped and one opposite freely supported edge in the free vibration experiment.
C: clamped edge; S: simply supported edge.

broken lines, radiation efficiency of a panel with one opposite clamped and one opposite simply supported edge; and dotted lines, radiation efficiency of a panel with four simply supported edges.

We can obtain radiation efficiency of the latter two panels theoretically. In Figs. 6, 7, and 8, there are no significant differences among these four panels regarding the values of radiation efficiency. In other words, radiation efficiency is not sensitive to the difference in boundary condition.

V. FIELD MEASUREMENT OF ACOUSTIC AND ACCELERATION SPECTRUMS

Figure 9 gives a schematic diagram of the *Arakawa* railway bridge, a plate girder railway bridge constructed on the *JR-Chuo* line. The dimensions of the bridge and each panel are given in this figure. Figure 10 shows a block diagram of the field measurement. The microphone of the noise meter was set up at a position 10 cm perpendicular from the center of a web panel divided by two vertical stiffeners. An acceleration pickup was attached at the center of the web. All data were filtered in third octave bands.

VI. COMPARISON OF MEASURED FIELD RESULTS WITH CALCULATED RESULTS

When calculating the acoustic spectrum using Eq. (1), boundary condition of a panel assumes that a panel has four simply supported edges or one opposite clamped and one opposite simply supported edge. For the latter panel the sides divided by vertical stiffeners are assumed to be the simply supported edges. It should be noted that Eq. (2) is an approximate expression applicable only to a simply supported panel.

Figures 11, 12, and 13 show the measured field acoustic spectrum and the calculated

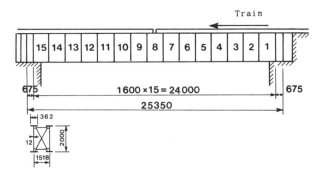

Fig. 9. Schematic diagram of *Arakawa* railway bridge.

Fig. 10. Block diagram for field measurement.

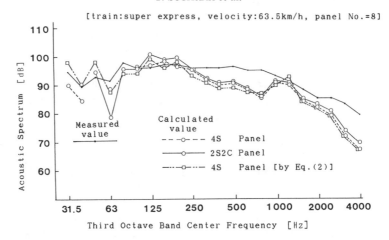

Fig. 11. Comparison of calculated acoustic spectrum with measured acoustic spectrum.
C: clamped edge; S: simply supported edge.

ones from measured field acceleration spectrum using Eqs. (1) and (2). The measurement conditions (i.e., kind of train passing over the bridge, train velocity, and the position of measured panel) are written in each figure. From these figures, the following facts are recognized:

i) The acoustic spectrum calculated using Eq. (1) is approximately the same as that using Eq. (2) and these two spectra conform well with the measured field one.

ii) Calculated acoustic spectra are not very sensitive to the boundary condition of a panel.

iii) Acoustic characteristics of the web are not affected very much by differences in the kind of train passing over the bridge or the velocity of train.

iv) Acceleration spectrum measured at the center of web panel is usable when we cal-

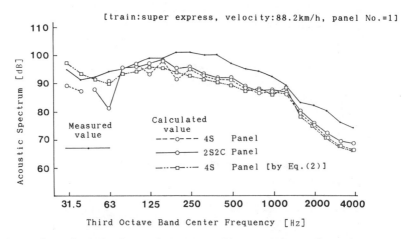

Fig. 12. Comparison of calculated acoustic spectrum with measured acoustic spectrum.
C: clamped edge; S: simply supported edge.

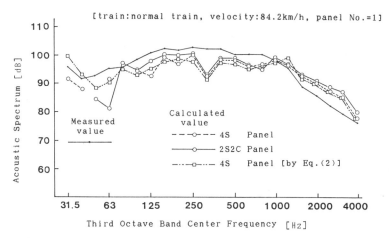

Fig. 13. Comparison of calculated acoustic spectrum with measured acoustic spectrum.
C: clamped edge; S: simply supported edge.

culate the acoustic spectrum using Eq. (1), although this value does not necessarily represent the mean accelration spectrum of the total panel.

From these facts, we can conclude that the acoustic spectrum can be calculated from the measured acceleration spectrum using Eq. (1) with the same accuracy as Eq. (2). We also conclude that the calculated acoustic spectrum is not very sensitive to the boundary condition of a panel.

It is also found that in most cases the measured field acoustic spectra are larger than the calculated ones. The reason is considered to be as follows. In addition to the sound radiated by the web panel, the former includes sound caused by the contact of train wheels with rails and includes the impact sound radiating from joints in the rails.

VII. CONCLUSIONS

The accuracy of theoretically calculated radiation efficiency of a rectangular panel has been tested through free and forced vibration experiments, and a method to calculate precisely the acoustic spectrum based on the measured acceleration spectrum of a panel has been proposed. Its accuracy has been confirmed by the measured field data on a plate girder railway bridge. The results obtained in this study are summarized as follows.

1) Theoretically calculated radiation efficiency of a rectangular panel agrees well with the experimental results.

2) Effects of the difference regarding boundary conditions of a panel on radiation efficiency are negligible.

3) Using the method proposed in this study, we can accurately calculate the acoustic spectrum radiated by a panel based on the measured acceleration spectrum of a panel.

REFERENCES

1) Maidanik, G., Response of ribbed panels to reverberant acoustic fields. *J. Acoust. Soc. Am.*, Vol. **34**, No. 6 (1962), pp. 809–826.

2) Wallace, C.E., Radiation resistance of a rectangular panel. *J. Acoust. Soc. Am.*, Vol. **51**, No. 3 (1972), pp. 946–952.
3) Saito, Y., Mizuguchi, F., Fukatsu, S., and Matsuura, T., Acoustic radiation from mechanically vibrated structural members. *Mitsubishi Heavy Industry Co. Ltd. Tech. Rep.*, Vol. **15**, No. 1 (1978), (in Japanese).

Chaos in Autonomous Elastic Systems

Kohji SUMINO, Hiroyuki SOGABE, and Yasuyuki WATANABE

Department of Mathematical Engineering, Faculty of Industrial Technology, Nihon University, Chiba

In this paper, chaotic phenomena in two kinds of autonomous elastic system are shown numerically from the results obtained by phase-space analysis. One is an infinitely long slightly curved elastic cylindrical panel with the upper surface exposed to a supersonic flow, and the other is the Reut model with two rods, two masses, and two springs.

I. CHAOTIC PHENOMENA IN CURVED PANELS

It was shown by Dowell[1,2] that chaotic flutter vibrations appear on the surface of flat elastic panels subjected to appropriate combined actions of in-plane compression and supersonic flow and such phenomena do not take place on the panels without in-plane compression.

The authors consider here infinitely long and slightly curved (in the streamwise direction) elastic cylindrical panels with the upper surfaces exposed to supersonic flow and all these have two simply supported (hinged) immovable edges as shown in Fig. 1. The governing partial differential equation for the above problem is given as follows:

$$Dw_{xxxx} - \left[Eh/2l(1 - \nu^2) \int_0^l (w_x^2 - 2wZ_{xx}) \, dx + P \right](w_{xx} + Z_{xx})$$
$$+ \rho V/M(w_t + Vw_x + VZ_x) + mw_{tt} = \Delta q, \qquad (1)$$

where notations have been expressed in a previous paper and H is the rise of an undeformed panel,

$$z(x) = H[1 - 4(x/l - 1/2)^2] \qquad (2)$$

indicates a shallow cylindrical surface,

$$Nx = Eh/2l(1 - \nu^2) \int_0^l (w_x^2 - 2wz_{xx}) \, dx \qquad (3)$$

is the inplane stress resultants, and boundary conditions are

$$w(0, t) = w(l, t) = 0,$$
$$w_{xx}(0,t) = w_{xx}(l,t) = 0. \qquad (4)$$

$$W = \sum_i a_i(t) \sin \frac{i\pi x}{l}. \qquad (5)$$

By applying Galerkin method with modal expansion (5), which satisfies the boundary conditions of Eq. (4), to Eq. (1), the following set of ordinary differential Eq. (6) is obtained

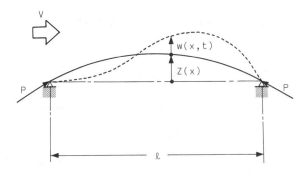

Fig. 1. Curved plate.

$$(j\pi)^4 Aj + Rx(j\pi)^2 Aj + \sqrt{\lambda\mu/M}\,\dot{A}j + \ddot{A}j$$

$$+ \sum_{i=1}^{n}\left[\frac{2ij\lambda}{j^2 - i^2}\{1-(-1)^{i+j}\} + \frac{24\,\Gamma^2}{ij\,\pi^2}\{1-(-1)^i\}\{1-(-1)^j\}\right]Ai$$

$$+ \sum_{i=1}^{n}\left[12\,\Gamma\,\frac{j^2\pi}{i}\{1-(-1)^i\}\,AjAi + 6\,\Gamma\,\frac{i^2\pi}{j}\{1-(-1)^j\}A_i^2\right]$$

$$+ 3\sum_{i=1}^{n}(ij)^2\,\pi^4\,A_jA_i^2 + \frac{2}{j\pi}(Rx\,\Gamma - q)\{1-(-1)^j\}$$

$$+ \frac{1}{j\pi}\lambda\,\Gamma\,\{1+(-1)^j\} = 0 \quad (j = 1 \sim n) \tag{6}$$

where nondimensional notations $\Gamma = 8H/h$, $A_i = a_i/h$, $W = w/h$, $Rx = Pl^2/D$, $\lambda = \rho V^2 l^3/MD$, $\mu = \rho l/m$, $q = \Delta q l^4/hD$, $x = l\xi$, $\tau = t\sqrt{D/ml^4}$, and $(\dot{\ }) = d(\ )/d\tau$ are used.

Then Eq. (6) in the case $n = 2$, $\mu = 0.01$ are solved by a numerical time-marching procedure and the four-dimensional phase-plane trajectories $(A_1,\dot{A}_1)$, $(A_2,\dot{A}_2)$, Poincare maps, and power spectrum are obtained. Accordingly, these are compared with Dowell's results on flat panels in the same way, where in computation the authors adopt the time step $\Delta\tau = 0.001$, initial value $(A_1,\dot{A}_1,A_2,\dot{A}_2) = (0.1,0,0,0)$, and $A_i = A_{i/\xi=0.75}$.

I-1. Results of Numerial Analysis

(i) The critical value (nondimensional parameter) λ_{cr} of flow velocity which leads to flutter motion in the panel with rising H (but static pressure difference $q = 0$) is shown in Table 1 when H is varied from 0 to $2\,h$.

(ii) The trajectories of flutter vibrations which appear in the panel with $H = 1.0\,h$ are all limit cycles, and hence no chaotic flutter takes place. However in the panel with slightly larger rising H (for example $1.05\,h$), strange attractors come out on the phase-plane $(A_1, \dot{A}_1)$ and $(A_2,\dot{A}_2)$ which means that chaotic flutter occurs (Fig. 2).

Table 1. Critical flow velocity.

H/h	0	1.0	1.5	2.0
λ	275	240	190	140

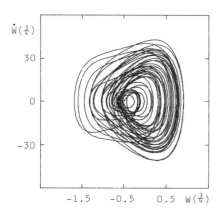

Fig. 2. Phase-plane trajectory for curved plate ($H/h = 1.05$, $\lambda = 250$, $Rx = 0$, $q = 0$).

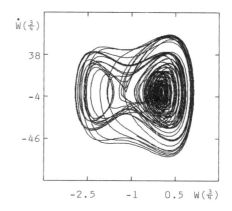

Fig. 3. Phase-plane trajectory for curved plate ($H/h = 1.5$, $\lambda = 190$, $Rx = 0$, $q = 0$).

Table 2. Flutter classification.

H/h	$\lambda = 150$	$\lambda = 160$	$\lambda = 190$	$\lambda = 200$	$\lambda = 220$	$\lambda = 250$	$\lambda = 270$
1.5	PA	PA	ST	ST	ST	LS	LS
2.0	L3	LI	LI	ST	ST	ST	ST

PA: point attractor;
ST: strange attractor;
LS: limit cycle, single loop;
L3: limit cycle, three loops;
LI: limit cycle, indistinguishable.

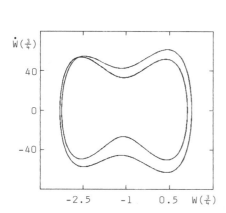

Fig. 4. Phase-plane trajectory for curved plate ($H/h = 1.5$, $\lambda = 250$, $Rx = 0$, $q = 0$).

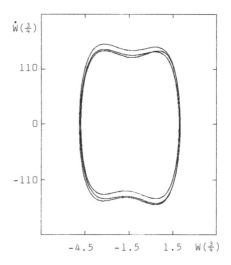

Fig. 5. Phase-plane trajectory for curved plate ($H/h = 2$, $\lambda = 150$, $Rx = 0$, $q = 0$).

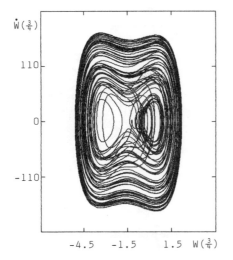

Fig. 6. Phase-plane trajectory for curved plate ($H/h = 2$, $\lambda = 250$, $Rx = 0$, $q = 0$).

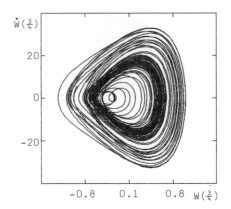

Fig. 7. Phase-plane trajectory for curved plate ($H/h = 1$, $\lambda = 250$, $Rx = -0.1\pi^2$, $q = 0$).

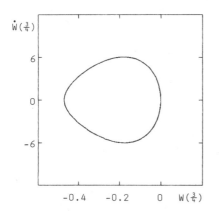

Fig. 8. Phase-plane trajectory for curved plate ($H/h = 1$, $\lambda = 250$, $Rx = -0.1\pi^2$, $q = 10$).

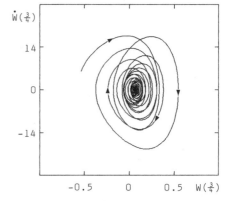

Fig. 9. Phase-plane trajectory for curved plate ($H/h = 1$, $\lambda = 250$, $Rx = -0.1\pi^2$, $q = 100$).

(iii) For the panels with $H = 1.5\,h$ and $2.0\,h$, Table 2 shows whether flutter vibration that occurred is a pure flutter (with limit cycle trajectory) or a chaotic flutter (with strange attractive trajectory) when λ values are varied (and $q = 0$). From results (ii) and (iii), it can be seen that chaotic flutter easily occurs on the surface of a curved panel exposed to a supersonic flow, but with no initial stress.

(iv) The static pressure difference has a stabilization effect. For example, in the panel with $H = h$, $\lambda = 250$ when nondimensional parameter $q\,(= q_L - q_U$ are the static pressure difference of the lower and upper surface) is varied, the trajectories are $q = 0$ (chaos) (Fig. 7), $q = 10$ (limit cycle flutter) (Fig. 8), and $q = 100$ (point-attractor, stable) (Fig. 9.).

Such behavior may be effective in stabilizing the outer skin of a flying object by increasing inner pressure.

Chaos in Autonomous Elastic Systems

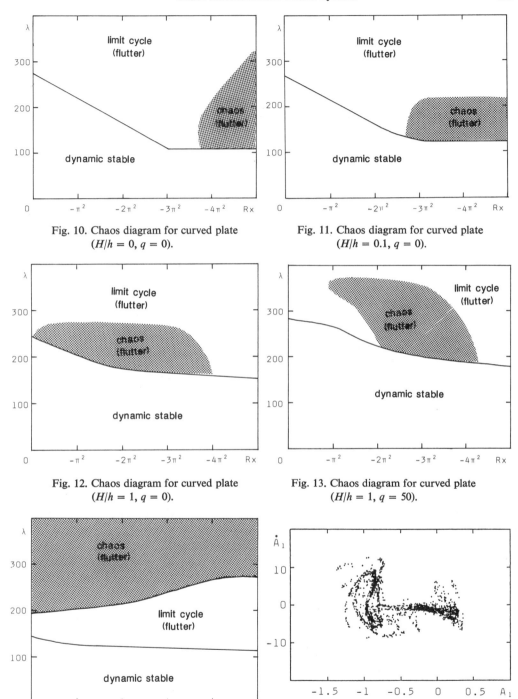

Fig. 10. Chaos diagram for curved plate ($H/h = 0$, $q = 0$).

Fig. 11. Chaos diagram for curved plate ($H/h = 0.1$, $q = 0$).

Fig. 12. Chaos diagram for curved plate ($H/h = 1$, $q = 0$).

Fig. 13. Chaos diagram for curved plate ($H/h = 1$, $q = 50$).

Fig. 14. Chaos diagram for curved plate ($H/h = 2$, $q = 0$).

Fig. 15. Poincaré map of chaotic motion of curved plate ($H/h = 1$, $\lambda = 250$, $Rx = -\pi^2$, $q = 0$).

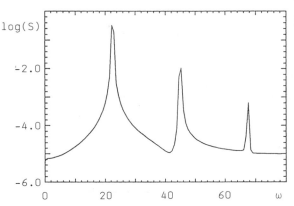

Fig. 16. Frequency spectrum of curved plate ($H/h = 1$, $\lambda = 250$, $Rx = 0$, $q = 0$).

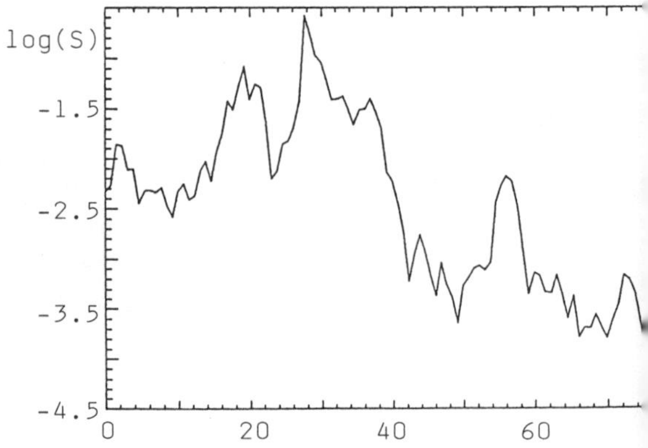

Fig. 17. Frequency spectrum of curved plate ($H/h = 1$, $\lambda = 250$, $Rx = -\pi^2$, $q = 0$).

(v) Although it may be impossible to give rigorously in-plane uniform prestress to a curved panel, the authors also consider curved panels prestressed in nearly such states (because of shallow rise) in the same way Dowell has done for flat panels. In this case the in-plane stress resultant becomes $N_x + N_x^0$, where $N_x^0 = D \cdot Rx/l^2$ is the in-plane uniform prestress resultant. Figures 10~14 show the classifications of whether vibrations of the panels are dynamically stable or unstable (pure flutter or chaotic flutter) on the $\lambda \sim Rx$ plane as H and q are varied. It is seen that the regions where chaotic vibrations appear transfer to the $R_x \sim$ decreasing mode as H increases, and to the λ-increasing mode as q increases on the $\lambda \sim Rx$ plane. Moreover it should be noted that in the panel with $H/h = 2$, the chaotic region spreads upward (of larger λ) compared to the nonchaotic (limit cycle) region on the ($\lambda \sim Rx$) plane by examining about 100 (λ, Rx) points (Fig. 14).

(vi) The Poincaré maps of a limit cycle (in $Rx = 0$) and a chaotic strange attractor (in $Rx = -\pi^2$) in the panel with $H/h = 1$, $\lambda = 250$, and $q = 0$, and the power spectra of the same states are shown in Figs. 15~17.

II. NONLINEAR REUT MECHANICAL MODEL

One of the authors reported on the chaotic vibrations occurring in the Ziegler-Hermann model in a previous paper. Here the authors consider a geometrically nonlinear Reut mechanical model as shown in Fig. 18 where the mechanism of this model and notations have been expressed in the previous paper. Lagrange equations of motion for this model are given as follows:

$$ml^2[(1 + \beta)\ddot{\varphi}_1 + \ddot{\varphi}_2 \cos(\varphi_2 - \varphi_1) - \dot{\varphi}_2^2 \sin(\varphi_2 - \varphi_1)]$$
$$+ \mu_1\dot{\varphi}_1 - \mu_2(\dot{\varphi}_2 - \dot{\varphi}_1) + e_L(2\varphi_1 - \varphi_2) + e_N(2\varphi_1\varphi_2 - \varphi_2^2) + F_{\varphi_1} = 0 \qquad (7)_1$$

$$ml^2[\ddot{\varphi}_2 + \ddot{\varphi}_1 \cos(\varphi_2 - \varphi_1) + \dot{\varphi}_1^2 \sin(\varphi_2 - \varphi_1)]$$
$$+ \mu_2(\dot{\varphi}_2 - \dot{\varphi}_1) + e_L(\varphi_2 - \varphi_1) + e_N(\varphi_2 - \varphi_1)^2 + F_{\varphi_2} = 0 \qquad (7)_2$$

$$F_{\varphi_1} = - Pl \sin(\varphi_1 - \alpha \varphi_2) \qquad (8)_1$$

$$F_{\varphi_2} = Pl[\cos(1 - \alpha)\varphi_2(\sin \varphi_1 \sec \varphi_2 + \tan \varphi_2) - \sin(1 - \alpha) \varphi_2] \qquad (8)_2$$

$$0 \leq |\alpha| \leq 1, \quad \theta_1 = \varphi_1, \quad \theta_2 = \varphi_2 - \varphi_1. \tag{9}$$

The above external force is only conservative at $\alpha = 1$ and is always nonconservative when $\alpha \neq 1$. The linearized differential Eqs. (7) and (8) were studied in detail by Herrmann and coworkers.[3,4] It is known in the original case $\alpha = 0$ (and for example $C_1 = 1$, $C_2 = 5$), where load parameter $K = pl/e_\nu$ is varied, A ($0 \leq K < 3.2$), B ($3.2 \leq K < 19.6$), and

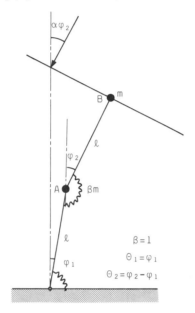

Fig. 18. Reut model.

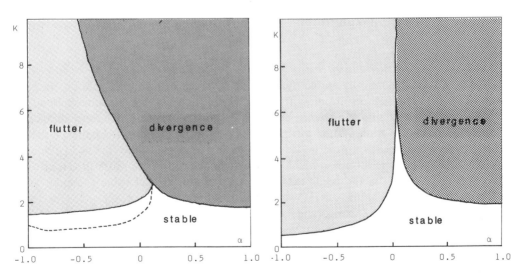

Fig. 19 (a). Stability diagram for Reut model. ($C_1 = C_2 = 0$, ... $C_1 = C_2 = 0.01$).

Fig. 19 (b). Stability diagram for Reut model. ($C_1 = C_2 = 0$).

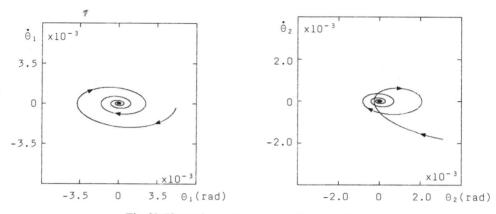

Fig. 20. Phase-plane trajectories for Reut model
($\alpha = 0$, $K = 1.5$, $C_1 = 1$, $C_2 = 5$).

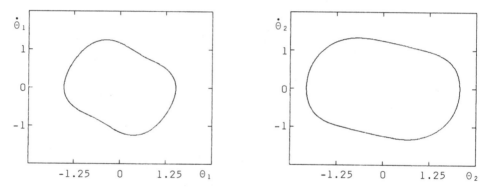

Fig. 21. Phase-plane trajectories for Reut model
($\alpha = 0$, $K = 7$, $C_1 = 1$, $C_2 = 5$).

C ($19.6 \leq K$) are dynamically stable, flutter, and divergence regions of motion, respectively (Fig. 19). The authors examine nonlinear Eqs. (7) and (8) by the same method as in[5]. Figure 20 shows a point attractor on the $(\theta_1\text{-}\dot\theta_1)$ and $(\theta_2\text{-}\dot\theta_2)$ phase planes which means a stable (decreasing vibration) state in the case of $(\alpha = 0, K = 1.5) \in A$. Figure 21 shows a limit cycle which means a pure flutter vibration in the case of $(\alpha = 0, K = 7) \in B$, and Fig. 22 shows a strange attractor which means a chaotic flutter vibration in the case of $(\alpha = 0, K = 21.9) \in C$. In the nonlinear analysis, the flutter critical value coincides very well with that in the linear analysis. However region of $(\alpha = 0, C_1 = 1, C_2 = 5, K > 19.6)$ always means a limit cycle or a strange attractor, and no divergence region exists. Moreover Fig. 23 and Fig. 24 display a divergence instability in the case of $K = 10$, $\alpha = 1$, $C_1 = 1$, $C_2 = 5$ and a limit cycle in the case of $K = 15$, $\alpha = 1$, $C_1 = 1$, $C_2 = 5$, respectively. But the process of transfer from limit cycle to strange attractor or the reverse cannot clearly be shown on a phase plane diagram due to violent changing (Figs. 25–27).

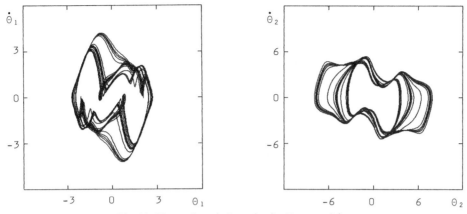

Fig. 22. Phase-plane trajectories for Reut model
($\alpha = 0$, $K = 21.9$, $C_1 = 1$, $C_2 = 5$).

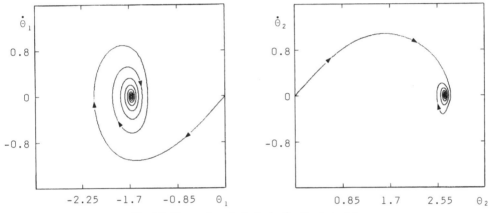

Fig. 23. Phase-plane trajectories for Reut model
($\alpha = 1$, $K = 10$, $C_1 = 1$, $C_2 = 5$).

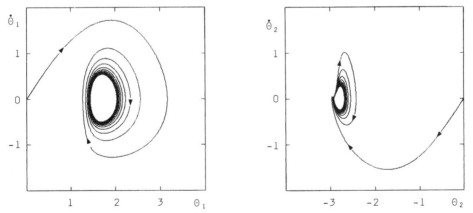

Fig. 24. Phase-plane trajectories for Reut model
($\alpha = 1$, $K = 15$, $C_1 = 1$, $C_2 = 5$).

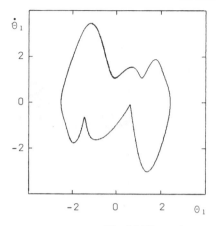 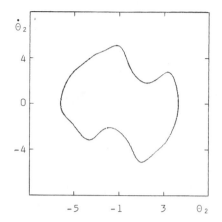

Fig. 25. Phase-plane trajectories for Reut model
($\alpha = 0$, $K = 21$, $C_1 = 1$, $C_2 = 5$).

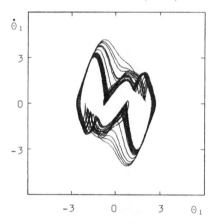 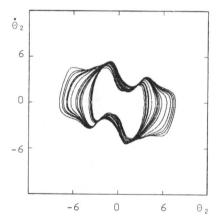

Fig. 26. Phase-plane trajectories for Reut model
($\alpha = 0$, $K = 22$, $C_1 = 1$, $C_2 = 5$).

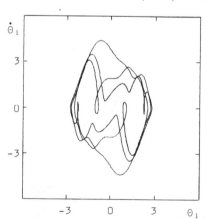 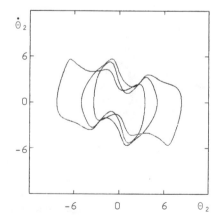

Fig. 27. Phase-plane trajectories for Reut model
($\alpha = 0$, $K = 23$, $C_1 = 1$, $C_2 = 5$).

III. CONCLUSIONS

It is known through numerical (two modes) analysis that chaotic motion easily occurs on the surface of an infinitely long curved panel with an appropriate small rise exposed to a supersonic flow but with no in-plane precompression, and the maximum amplitudes of such chaotic vibrations are about the same size as those of the adjacent limit cycle (pure flutter) vibrations. However, these results should be compared with those by four or six mode approximations and finite-length curved panels should be examined. And experimental verification of the Reut model may be expected in future.

Acknowledgment

This study was supported by a grant for scientific research from the Ministry of Education, Japan.

REFERENCES

1) Dowell, E.H., Flutter of a buckled plate as an example of chaotic motion of a deterministic autonomous system. *J. Sound Vib.*, Vol. 85 (1982), pp. 333–344.
2) Dowell, E.H., Observation and evolution of chaos for an autonomous system. *J. Appl. Mech.*, Vol. 51 (1984), pp. 664–673.
3) Herrmann, G. and Jong, I.C., On the destabilizing effect of damping in nonconservative elastic system. *J. Appl. Mech.*, Vol. 32 (1965), pp. 592–597.
4) Herrmann, G. and Jong, I.C., On nonconservative stability problems of elastic system with slight damping. *J. Appl. Mech.*, Vol. 33 (1966), pp. 125–133.
5) Sumino, K., Nonlinear phenomena and chaos in elastic system. *Theoret. Appl. Mech.*, Vol. 35 (1987), pp. 147–155.
6) Sugiyama, Y. et al., A theoretical and experimental study on effect of damping in nonconservative stability problems. *Mech. Materials* (1985).

VIII
ELASTICITY, BUCKLING, AND FATIGUE

Expression of Three-Dimensional Deviatoric Tensor through Complex Number ω

Masao SATAKE

Department of Civil Engineering, Tohoku University, Sendai

A new expression of a three-dimensional (3-D) deviatoric tensor is proposed by using complex number $\omega = 1/2(-1 + \sqrt{3}\,i)$. It is explained that, by virtue of this expression, similar expressions in 3-D for isotropic and anisotropic decompositions of a symmetric tensor are given as in 2-D. Such anisotropic decompositions introduced here are considered to be especially important for analysis in mechanics of anisotropic materials, such as geologic and composite materials.

I. INTRODUCTION

In this paper, the isotropic and anisotropic decompositions of a 2-D symmetric tensor are first explained. Isotropic decomposition is very commonly used in continuum mechanics. It is considered, however, that anisotropic decomposition may also play a very important role in the analysis of mechanics of materials with anisotropic properties. In order to extend the isotropic decomposition to 3-D and to give similar expressions as in 2-D, a new expression of a 3-D deviatoric tensor using the complex number ω is proposed. It is shown that the absolute value and the argument of this complex quantity are closely related to the invariants and Lode's parameter of the deviatoric tensor. Finally, the anisotropic decomposition in 3-D is derived by Cardano's formula.

II. ISOTROPIC AND ANISOTROPIC DECOMPOSITIONS OF 2-D SYMMETRIC TENSOR

The isotropic decomposition of a 2-D symmetric tensor is commonly used in continuum mechanics. In the following, we explain it by taking the stress tensor as an example.

A 2-D stress tensor $\boldsymbol{\sigma}$ is decomposed as

$$\boldsymbol{\sigma} = p\,\boldsymbol{I} + q\,\boldsymbol{J}, \tag{1}$$

where

$$\boldsymbol{I} = \begin{pmatrix} 1 & \\ & 1 \end{pmatrix}, \quad \boldsymbol{J} = \begin{pmatrix} 1 & \\ & -1 \end{pmatrix} \tag{2}$$

with respect to the principal axes of $\boldsymbol{\sigma}$. $\boldsymbol{I}$ is the unit tensor and $\boldsymbol{J}$ is a tensor called the unit deviatoric tensor, respectively.

$$\left. \begin{aligned} p &= \frac{1}{2}\boldsymbol{\sigma} \cdot\cdot\, \boldsymbol{I} = \frac{1}{2}(\sigma_1 + \sigma_2) \\ q &= \frac{1}{2}\boldsymbol{\sigma} \cdot\cdot\, \boldsymbol{J} = \frac{1}{2}(\sigma_1 - \sigma_2) \end{aligned} \right\} \tag{3}$$

where $\cdot\cdot$ denotes the double inner product ($\boldsymbol{\alpha} \cdot\cdot \boldsymbol{\beta} = \alpha_{ij}\beta_{ij}$ for $\boldsymbol{\alpha} = (\alpha_{ij})$ and $\boldsymbol{\beta} = (\beta_{ij})$) of two tensors, and σ_1 and σ_2 are two principal stresses of $\boldsymbol{\sigma}$, where $\sigma_1 \geq \sigma_2$ is assumed. p and q are called the mean stress and the deviatoric stress, respectively.

This isotropic decomposition is clearly based on the following orthogonality conditions:

$$\left.\begin{array}{l} \boldsymbol{I} \cdot\cdot \boldsymbol{I} = \boldsymbol{J} \cdot\cdot \boldsymbol{J} = 2 \\ \boldsymbol{I} \cdot\cdot \boldsymbol{J} = 0 \end{array}\right\}. \qquad (4)$$

Introducing a parameter φ defined by

$$\sin \varphi = \frac{q}{p} \qquad (5)$$

which is usually called the angle of internal friction in the mechanics of granular materials, we can write

$$\left.\begin{array}{l} \sigma_1 = p + q = p(1 + \sin \varphi) \\ \sigma_2 = p - q = p(1 - \sin \varphi) \end{array}\right\} \qquad (6)$$

$$\sqrt{\frac{\sigma_1}{\sigma_2}} = \sqrt{\frac{1 + \sin \varphi}{1 - \sin \varphi}} = \tan\left(\frac{\pi}{4} + \frac{\varphi}{4}\right). \qquad (7)$$

Thus it is seen that φ is a parameter that describes the anisotropy of $\boldsymbol{\sigma}$.

Table 1. Decomposition of 2-D work increment.

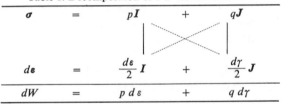

Table 1 shows the decomposition of a 2-D work increment $dW = \boldsymbol{\sigma} \cdot\cdot d\boldsymbol{\varepsilon}$ and the corresponding decompositions of stress tensor $\boldsymbol{\sigma}$ and strain increment tensor $d\boldsymbol{\varepsilon}$, where coaxiality is assumed. From the orthogonality conditions, it is seen that the work is only done along the solid lines and is not done along the dotted lines.

In a similar manner, we can introduce the anisotropic decomposition expressed as

$$\boldsymbol{\sigma} = p'\boldsymbol{L} + q'\boldsymbol{M} \qquad (8)$$

where

$$\boldsymbol{L} = \begin{pmatrix} \lambda & \\ & \lambda^{-1} \end{pmatrix}, \quad \boldsymbol{M} = \begin{pmatrix} \lambda^{-1} & \\ & -\lambda \end{pmatrix} \qquad (9)$$

with respect to the principal axes of $\boldsymbol{\sigma}$. $\boldsymbol{L}$ and $\boldsymbol{M}$ are called the anisotropic tensor and the accompanied tensor, respectively, and $\lambda (0 < \lambda \leq 1)$ is called the anisotropic parameter. Similar to Eq. (3), we have

$$\left.\begin{array}{l} p' = \frac{1}{l^2}\boldsymbol{\sigma} \cdot\cdot \boldsymbol{L} = \frac{1}{l^2}(\lambda \sigma_1 + \lambda^{-1}\sigma_2) \\ q' = \frac{1}{l^2}\boldsymbol{\sigma} \cdot\cdot \boldsymbol{M} = \frac{1}{l^2}(\lambda^{-1}\sigma_1 - \lambda \sigma_2) \end{array}\right\} \qquad (10)$$

where

$$l = \sqrt{\lambda^2 + \lambda^{-2}} \tag{11}$$

in which p' and q' are called the anisotropic mean stress (or weighted mean stress) and the anisotropic deviatoric stress, respectively.

In this case, the following orthogonality conditions are also to be noted:

$$\left.\begin{array}{l} \boldsymbol{L} \cdot\cdot \boldsymbol{L} = \boldsymbol{M} \cdot\cdot \boldsymbol{M} = l^2 \\ \boldsymbol{L} \cdot\cdot \boldsymbol{M} = 0 \end{array}\right\}. \tag{12}$$

It is easy to see that Eqs. (8), (9), (10), and (12) reduce to Eqs. (1), (2), (3), and (4), respectively, when $\lambda = 1$.

If we apply the same anisotropic decomposition to the strain increment $d\boldsymbol{\varepsilon}$, we have

$$d\boldsymbol{\varepsilon} = \tfrac{1}{l^2}(d\varepsilon' \boldsymbol{L} + d\gamma' \boldsymbol{M}) \tag{13}$$

where

$$\left.\begin{array}{l} d\varepsilon' = d\boldsymbol{\varepsilon} \cdot\cdot \boldsymbol{L} = \lambda\, d\varepsilon_1 + \lambda^{-1} d\varepsilon_2 \\ d\gamma' = d\boldsymbol{\varepsilon} \cdot\cdot \boldsymbol{M} = \lambda^{-1} d\varepsilon_1 - \lambda\, d\varepsilon_2 \end{array}\right\} \tag{14}$$

and $d\varepsilon_1$ and $d\varepsilon_2$ ($d\varepsilon_1 > d\varepsilon_2$) are the principal strain increments. From the orthogonality conditions (12) we have

$$dW = \boldsymbol{\sigma} \cdot\cdot d\boldsymbol{\varepsilon} = p'\, d\varepsilon' + q'\, d\gamma' \tag{15}$$

which is a relation similar to that explained in Table 1.

III. ISOTROPIC DECOMPOSITION IN 3-D

To extend the decompositions of a symmetric tensor explained in the previous chapter to the case of 3-D, we introduce the 3-D unit deviatoric tensor defined as

$$\boldsymbol{J} = \begin{pmatrix} 1 & & \\ & \omega & \\ & & \omega^2 \end{pmatrix} \tag{16}$$

where

$$\omega = \tfrac{1}{2}(-1 + \sqrt{3}\, i) \tag{17}$$

is a root of the equation $x^3 = 1$. Using $\boldsymbol{J}$ and the 3-D unit tensor $\boldsymbol{I}$, we obtain the isotropic decomposition of a 3-D stress tensor $\boldsymbol{\sigma}$ expressed as

$$\boldsymbol{\sigma} = p\boldsymbol{I} + q\bar{\boldsymbol{J}} + \bar{q}\boldsymbol{J} \tag{18}$$

where

$$\left.\begin{array}{l} p = \tfrac{1}{3}\boldsymbol{\sigma} \cdot\cdot \boldsymbol{I} = \tfrac{1}{3}(\sigma_1 + \sigma_2 + \sigma_3) \\[4pt] q = \tfrac{1}{3}\boldsymbol{\sigma} \cdot\cdot \boldsymbol{J} = \tfrac{1}{3}(\sigma_1 + \omega\sigma_2 + \omega^2 \sigma_3) \\[4pt] \bar{q} = \tfrac{1}{3}\boldsymbol{\sigma} \cdot\cdot \bar{\boldsymbol{J}} = \tfrac{1}{3}(\sigma_1 + \omega^2\sigma_2 + \omega\sigma_3) \end{array}\right\} \tag{19}$$

where σ_1, σ_2 and σ_3 are the principal stresses ($\sigma_1 \geq \sigma_2 \geq \sigma_3$ is assumed) and – denotes the conjugate complex number or a tensor which has the conjugate complex numbers as its elements. q is called the *complex deviatoric stress*.

Quite similarly as in Eq. (4), we have the following relations:

$$\left.\begin{array}{l} \boldsymbol{I} \cdot\cdot \boldsymbol{I} = \boldsymbol{J} \cdot\cdot \bar{\boldsymbol{J}} = 3 \\ \boldsymbol{I} \cdot\cdot \boldsymbol{J} = \boldsymbol{J} \cdot\cdot \boldsymbol{J} = 0 \end{array}\right\}. \tag{20}$$

As q is a complex number, we can write it as

$$q = |q| e^{i\alpha} \tag{21}$$

where $|q|$ and α are the absolute value and argument of q, respectively. We can easily derive the relations

$$\left.\begin{array}{l} |q| = \sqrt{\dfrac{S_2}{3}} \\[2mm] \cos 3\alpha = \dfrac{1}{2|q|^3} \sqrt{S_3} \end{array}\right\} \tag{22}$$

$$\alpha = \tan^{-1} \dfrac{\mu}{\sqrt{3}} + \dfrac{\pi}{6} \tag{23}$$

where μ is the Lode's parameter of $\boldsymbol{\sigma}$ and S_2 and S_3 are the invariants of the deviatoric stress tensor $\boldsymbol{\sigma}'$ expressed as

$$\left.\begin{array}{l} S_2 = -\sigma_1'\sigma_2' - \sigma_2'\sigma_3' - \sigma_3'\sigma_1' \\ S_3 = \sigma_1'\sigma_2'\sigma_3' \end{array}\right\} \tag{24}$$

where σ_1', σ_2', and σ_3' are the principal deviatoric stresses. We can also derive the inverse relations expressed as

$$\left.\begin{array}{l} \sigma_1' = 2|q| \cos \alpha \\[2mm] \sigma_2' = 2|q| \cos\left(\alpha - \dfrac{2}{3}\pi\right) \\[2mm] \sigma_3' = 2|q| \cos\left(\alpha + \dfrac{2}{3}\pi\right) \end{array}\right\}. \tag{25}$$

Figure 1 illustrates this relation in the Gaussian plane. By using the Lode's parameter μ, we can write

$$\left.\begin{array}{l} \sigma_1' = \dfrac{3 - \mu}{\sqrt{3 + \mu^2}} |q| \\[2mm] \sigma_2' = \dfrac{2\mu}{\sqrt{3 + \mu^2}} |q| \\[2mm] \sigma_3' = -\dfrac{3 + \mu}{\sqrt{3 + \mu^2}} |q| \end{array}\right\}. \tag{26}$$

Introducing a parameter φ defined by

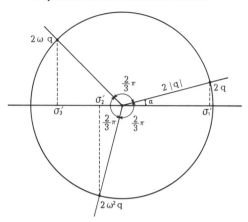

Fig. 1. Illustration of $2q$, $2\omega q$, $2\omega^2 q$ in the Gaussian plane, which gives the three principal deviatoric stresses.

$$\sin \varphi = \frac{|q|}{p} \qquad (27)$$

similarly as in Eq. (5), we can write

$$\left.\begin{array}{l} \sigma_1 = p\,(1 + 2\sin\varphi\cos\alpha) \\ \sigma_2 = p\left\{1 + 2\sin\varphi\cos\left(\alpha - \frac{2}{3}\pi\right)\right\} \\ \sigma_3 = p\left\{1 + 2\sin\varphi\cos\left(\alpha + \frac{2}{3}\pi\right)\right\} \end{array}\right\}. \qquad (28)$$

Heret three parameters p, φ, and α are employed to express the three principal stresses.

Table 2. Decomposition of 3-D work increment.

$\boldsymbol{\sigma}$	$=$	$p\boldsymbol{I}$	$+$	$(q\,\boldsymbol{J}$	$+$	$\bar{q}\,\boldsymbol{J})$
$d\boldsymbol{\varepsilon}$	$=$	$\frac{d\varepsilon}{3}\boldsymbol{I}$	$+$	$\frac{1}{3}(d\gamma\,\bar{\boldsymbol{J}}$	$+$	$\overline{d\gamma}\,\boldsymbol{J})$
dW	$=$	$p\,d\varepsilon$	$+$	$(\overline{q}\,d\gamma$	$+$	$q\,\overline{d\gamma})$

Table 2 shows the decomposition of a work increment dW in 3-D. Similarly as is explained in Table 1, the work is only done along the solid lines in this case too.

IV. ANISOTROPIC DECOMPOSITION IN 3-D

In the analysis for anisotropy in 3-D, we introduce an anisotropic tensor defined by

$$\boldsymbol{L} = \begin{pmatrix} \lambda_1 & & \\ & \lambda_2 & \\ & & \lambda_3 \end{pmatrix} \qquad (29)$$

where
$$\lambda_1 \lambda_2 \lambda_3 = 1, \quad 0 < \lambda_1 \leq \lambda_2 \leq \lambda_3. \tag{30}$$

Next we derive a complex tensor

$$M = \begin{pmatrix} \mu_1 & & \\ & \mu_2 & \\ & & \mu_3 \end{pmatrix} \tag{31}$$

having the following properties:
$$L \cdot\cdot M = \lambda_1 \mu_1 + \lambda_2 \mu_2 + \lambda_3 \mu_3 = 0 \tag{32}$$
$$\operatorname{tr} M^2 = M \cdot\cdot M = \mu_1^2 + \mu_2^2 + \mu_3^2 = 0 \tag{33}$$
$$|M| = \mu_1 \mu_2 \mu_3 = 1. \tag{34}$$

It is easily shown that M is expressed in the form (see Appendix)
$$M = L^{-1}(u\, J + v\, \bar{J}) \tag{35}$$

where u and v are given by Eq. (A6) or (A9).

Using L and M, we can give the following anisotropic decomposition of a 3-D stress tensor:
$$\boldsymbol{\sigma} = p' L + q' \overline{M} + \bar{q}'\, M \tag{36}$$

where
$$\left. \begin{array}{l} p' = \dfrac{1}{l^2} \boldsymbol{\sigma} \cdot\cdot L = \dfrac{1}{l^2}(\lambda_1 \sigma_1 + \lambda_2 \sigma_2 + \lambda_3 \sigma_3) \\[6pt] q' = \dfrac{1}{m^2} \boldsymbol{\sigma} \cdot\cdot M = \dfrac{1}{m^2}(\mu_1 \sigma_1 + \mu_2 \sigma_2 + \mu_3 \sigma_3) \end{array} \right\} \tag{37}$$

and
$$\left. \begin{array}{l} l^2 = \operatorname{tr} L^2 = \lambda_1^2 + \lambda_2^2 + \lambda_3^2 \\ m^2 = \operatorname{tr}(M \cdot \overline{M}) = \mu_1 \bar{\mu}_1 + \mu_2 \bar{\mu}_2 + \mu_3 \bar{\mu}_3 \end{array} \right\}. \tag{38}$$

If $L \to I$, it is seen that $M \to J$, $p' \to p$, and $q' \to q$, where p and q are given in Eq. (19). In this case, the orthonality conditions are written as

$$\left. \begin{array}{l} L \cdot\cdot L = l^2, \ M \cdot\cdot \overline{M} = m^2 \\ L \cdot\cdot M = M \cdot\cdot M = 0 \end{array} \right\}. \tag{39}$$

Assuming that $d\boldsymbol{\varepsilon}$ is coaxial with $\boldsymbol{\sigma}$ and giving it the similar anisotropic decomposition expressed as

$$d\boldsymbol{\varepsilon} = \frac{1}{l^2} d\varepsilon'\, L + \frac{1}{m^2}(d\gamma'\, \overline{M} + \overline{d\gamma'}\, M) \tag{40}$$

we have
$$\begin{aligned} dW &= \boldsymbol{\sigma} \cdot\cdot d\boldsymbol{\varepsilon} \\ &= p'\, d\varepsilon' + q'\, \overline{d\gamma'} + \overline{q'}\, d\gamma' \end{aligned} \tag{41}$$

V. APPLICATION OF ANISOTROPIC DECOMPOSITION

The isotropic decomposition of a symmetric tensor is very common in the analysis of continuum mechanics. However, when the material has some anisotropy, anisotropic decomposition will become useful for the analysis of the mechanical properties. In this section, we will give a brief explanation of an example in which anisotropic decomposition plays a very important role.

In granular materials, the strain increment is generally prescribed some constraint condition, i.e. the so-called stress-dilatancy relation. It is simply expressed as

$$\boldsymbol{L} \cdot\cdot d\boldsymbol{\varepsilon} = 0 \tag{42}$$

where $\boldsymbol{L}$ is a symmetric tensor depending on the stress and the stress path. In the plastic deformation of metals, it is considered that $\boldsymbol{L}$ in Eq. (42) reduces to the unit tensor $\boldsymbol{I}$, and consequently we have

$$\boldsymbol{I} \cdot\cdot d\boldsymbol{\varepsilon} = \operatorname{tr} d\boldsymbol{\varepsilon} = d\varepsilon = 0. \tag{43}$$

In the plastic analysis of granular materials, the modified associated flow rule proposed by Kanatani[2] is applied, which is expressed as

$$d\boldsymbol{\varepsilon} = \left(\frac{\partial}{\partial \boldsymbol{\sigma}}\right)^* \Phi \, d\Lambda \tag{44}$$

where Φ is the plastic potential and Λ is a scalar parameter. The notation $(\partial/\partial \sigma)^*$ denotes a differentiation, in which the constraining stress is kept constant. The constraining stress is defined as a stress that does work against the strain that vanishes under a given constraint condition.

Using the anisotropic tensor $\boldsymbol{L}$ defined in Eq. (42), we apply the anisotropic decomposition, such that

$$\left.\begin{aligned} \boldsymbol{\sigma} &= p'\boldsymbol{L} + (q'\overline{\boldsymbol{M}} + \overline{q}'\boldsymbol{M}) \\ d\boldsymbol{\varepsilon} &= \frac{1}{l^2}d\varepsilon' \, \boldsymbol{L} + \frac{1}{m^2}(d\gamma' \, \overline{\boldsymbol{M}} + \overline{d\gamma}' \, \boldsymbol{M}) \end{aligned}\right\} \tag{45}$$

where $\boldsymbol{M}$ is the accompanied (complex) tensor of $\boldsymbol{L}$ and the parentheses indicate the anisotropic deviatoric parts. From Eq. (42) and the definition of $d\varepsilon'$, we have

$$d\varepsilon' = d\boldsymbol{\varepsilon} \cdot\cdot \boldsymbol{L} = 0 \tag{46}$$

and

$$dW = q' \, \overline{d\gamma}' + \overline{q}' \, d\gamma'. \tag{47}$$

In this case, p' is considered as the constraining stress, because it does work against $d\varepsilon'$, which vanishes from the constraint condition of Eq. (46). Thus we have

$$\left(\frac{\partial}{\partial \boldsymbol{\sigma}}\right)^* = \frac{\partial p'}{\partial \boldsymbol{\sigma}} \frac{\partial}{\partial p'} + \frac{\partial q'}{\partial \boldsymbol{\sigma}} \frac{\partial}{\partial q'} + \frac{\partial \overline{q}'}{\partial \boldsymbol{\sigma}} \frac{\partial}{\partial \overline{q}'}$$

$$= \frac{\partial q'}{\partial \sigma} \frac{\partial}{\partial q'} + \frac{\partial \overline{q'}}{\partial \sigma} \frac{\partial}{\partial \overline{q'}} \qquad (48)$$

because p' is kept unchanged. As we have

$$\frac{\partial q'}{\partial \sigma} = \frac{1}{m^2} M \qquad (49)$$

from Eq. (37), Eq. (44) reduces to

$$d\varepsilon = \frac{1}{m^2}\left(\frac{\partial \Phi}{\partial q'} M + \frac{\partial \Phi}{\partial \overline{q'}} \overline{M}\right) d\Lambda \qquad (50)$$

It is easy to see from Eq. (39) that the strain increment given by Eq. (50) satisfies the constraint condition of Eq. (42). Further it is found that

$$d\gamma' = \frac{\partial \Phi}{\partial \overline{q'}} d\Lambda \qquad (51)$$

VI. CONCLUSIONS

The isotropic and anisotropic decompositions of a symmetric tensor are fundamentally important in the analysis of continuum mechanics. It is first explained that the 2-D anisotropic decomposition with the anisotropic parameter is introduced similar to isotropic decomposition. In 3-D, it is recognized that the use of the complex number ω makes similar isotropic and anisotropic decompositions as in 2-D possible. As is seen in Section V, such decompositions using complex number ω will find many applications in various fields of continuum mechanics, especially in the mechanics of granular materials.[3]

REFERENCES

1) Takagi, T., Daisugaku Kogi, 3rd ed. (Lecture on algebra), Kyoritsu (1965), pp. 171–180
2) Kanatani, K., Mechanical foundation of the plastic deformation of granular materials, in: *Deformation and Failure of Granular Materials* (P.A. Vermeer, H.J. Luger, eds.) A.A. Balkema (1982), pp. 119–127.
3) Satake, M., Consideration on the stress-dilatancy equation through the work increment tensor, in: *Micromechanics of Granular Materials* (M. Satake and J.T. Jenkins, eds.) Elsevier (1988), pp.61–70.

APPENDIX

To obtain M with the properties of Eqs. (32)–(34), we introduce a tensor defined by

$$X = L \cdot M = \begin{pmatrix} x_1 & & \\ & x_2 & \\ & & x_3 \end{pmatrix}. \tag{A1}$$

Substituting $M = L^{-1} \cdot X$ into Eqs. (33) and (34), we have

$$\left.\begin{array}{l} L^{-2} \cdot\cdot X^2 = \lambda_1^{-2} x_1^2 + \lambda_2^{-2} x_2^2 + \lambda_3^{-2} x_3^2 = 0 \\ |X| = \qquad\qquad\qquad x_1 x_2 x_3 = 1 \end{array}\right\}. \tag{A2}$$

On the other hand, as tr $X = 0$ from Eq. (32), we can write

$$X = u J + v \bar{J} \tag{A3}$$

using Cardano's formula[1], where u and v are two complex numbers to be dertermined as follows:

Substituting Eq. (A3) into Eq. (A2), we have simultaneous equations

$$\left.\begin{array}{l} a u^2 + 2 b u v + \bar{a} v^2 = 0 \\ u^3 + v^3 = 1 \end{array}\right\} \tag{A4}$$

where

$$\left.\begin{array}{l} a = \lambda_1^{-2} + \omega^2 \lambda_2^{-2} + \omega \lambda_3^{-2} \\ b = \lambda_1^{-2} + \lambda_2^{-2} + \lambda_3^{-2} \end{array}\right\}. \tag{A5}$$

Thus we obtain, when $a \neq 0$

$$u = \frac{1}{\sqrt[3]{1 + \xi^3}}, \quad v = \frac{\xi}{\sqrt[3]{1 + \xi^3}} \tag{A6}$$

where

$$\xi = \frac{1}{\bar{a}}(\sqrt{3}\, l - b), \tag{A7}$$

$$l = \sqrt{\lambda_1^2 + \lambda_2^2 + \lambda_3^2}, \tag{A8}$$

and when $a = 0$

$$u = 1, \quad v = 0. \tag{A9}$$

Stress Analysis of an Elastic Half-Space Having an Axisymmetric Cavity

A Solution by Green's Functions for Axisymmetric Body Force Problems of an Elastic Half-Space

Hisao HASEGAWA and Kiyohisa KONDOU

Faculty of Engineering, Meiji University, Kawasaki, Kanagawa

This paper deals with the stress concentration problem of an elastic half-space having an axisymmetric cavity under uniform biaxial tension. A method of solution is developed for the problem by using Green's functions for axisymmetric body force problems of an elastic half-space. The method of solution may be said to be a kind of the variously called body force method, charge simulation method, indirect boundary element method with fictitious boundaries, and so on. To confirm the validity of the method of solution proposed here, numerical results are compared with the known one for the case of a spherical cavity. As an example of application of the present method, we consider the stress concentration problem of an elastic half-space having a circular cylindrical cavity of arbitrary length.

I. INTRODUCTION

This paper deals with the stress concentration problem of an elastic half-space having a cavity with an axisymmetric form under uniform biaxial tension. A method of solution is proposed for the problem that is a method of solution by distributions of body forces. For this purpose, we apply Green's functions[1] for axisymmetric body force problems of an elastic half-space. Green's functions used here are defined as a solution for the problem of an elastic half-space subjected to axisymmetric body forces acting on a circle in the interior of the half-space.

In order to confirm the validity of the present method of solution, it is applied to the problem of an elastic half-space having a spherical cavity and the numerical results are compared with the known one.[2] As an example of application of the method of solution proposed in this paper, we consider the stress concentration problem of an elastic half-space having a circular cylindrical cavity of arbitrary length.

The problems of stress concentration due to cavities, defects, inclusions, and so on in a solid body have been investigated by several authors because these problems are of great importance in engineering design. Most of these investigations, however, have dealt with the case of spherical or ellipsoidal form of cavities or inclusions.[2,3] The aim of this paper is an extension of the problems of spherical or ellipsoidal cavities to the problem of an axisymmetric cavity with arbitrary form.

II. BOUNDARY CONDITIONS

We use cylindrical coordinates (r, θ, z) in an elastic half-space having an axisymmetric cavity, as shown in Fig. 1, where n is a normal at a point (r_i, z_i) on the surface of cavity, and ϕ is an angle between n and the z-axis. If we assume that the half-space is subjected to a uniform biaxial tension stress $\sigma_r = p$ at a sufficiently large distance from the cavity and there are no surface forces at the surface of cavity and the boundary of the half-space, then the boundary conditions of this problem are expressed as follows:

(i) the surface of cavity (r_i, z_i); $\sigma_n = \tau_{n\phi} = \tau_{n\theta} = 0$

(ii) $z = 0, 0 \leq r < \infty$; $\sigma_z = \tau_{zr} = \tau_{z\theta} = 0$

(iii) $R \to \infty (R = \sqrt{r^2 + z^2})$; $\sigma_r \to p$ (1)

where σ_n and $\tau_{n\phi}$, $\tau_{n\theta}$ are normal and shearing stresses acting on the surface of cavity, and are expressed by

$$\sigma_n = \sigma_r \sin^2 \phi + \sigma_z \cos^2 \phi + 2\tau_{rz} \sin \phi \cos \phi$$
$$\tau_{n\phi} = (\sigma_r - \sigma_z) \sin \phi \cos \phi + \tau_{rz}(\cos^2 \phi - \sin^2 \phi)$$
$$\tau_{n\theta} = \tau_{r\theta} \sin \phi + \tau_{z\theta} \cos \phi. \qquad (2)$$

III. GREEN'S FUNCTIONS

In this paper we apply Green's functions for axisymmetric body force problems of an elastic half-space. Green's functions used here are defined as a solution to the elastic problem of a half-space satisfying the boundary conditions

$$z = 0, 0 \leq r < \infty; \sigma_z = \tau_{zr} = \tau_{z\theta} = 0 \qquad (3)$$

and subjected to the axisymmetric body forces

$$F_j = \frac{1}{2\pi r} \delta(r - a)\delta(z - h), \quad (j = 1, 3) \qquad (4)$$

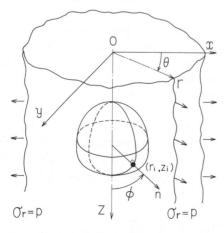

Fig. 1. An elastic half-space having an axisymmetric cavity under biaxial tention.

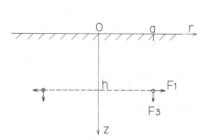

Fig. 2. An elastic half-space subjected to axisymmetric body forces on a circle ($r=a$, $z=h$).

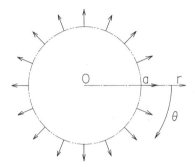
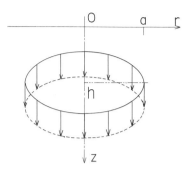

Fig. 3. Radial force.

Fig. 4. Axial force.

acting on a circle ($r = a$, $z = h$) in the interior of the half-space as shown in Fig. 2, where $\delta(\ )$ is a Dirac's delta function.

We shall express Green's functions defined above by

$$u_r^j = u_r^j(r, z, a, h), \qquad u_z^j = u_z^j(r, z, a, h)$$
$$\sigma_r^j = \sigma_r^j(r, z, a, h), \qquad \sigma_\theta^j = \sigma_\theta^j(r, z, a, h)$$
$$\sigma_z^j = \sigma_z^j(r, z, a, h), \qquad \tau_{rz}^j = \tau_{rz}^j(r, z, a, h) \qquad (5)$$

where $j = 1$ means the solution for the body force F_1 acting in the r-direction as shown in Fig. 3, and $j = 3$ is the solution for the body force F_3 acting in the z-direction as shown in Fig. 4. Concrete expressions for Green's functions of Eq. (5) have been shown by Hasegawa.[1]

IV. METHOD OF SOLUTION

IV-1. Integral Equation

The fundamental principle of the method of solution employed here is to distribute body forces in the interior of a half-space without cavities and to determine the intensities of the body forces distributed so as to satisfy the boundary conditions of an imaginary plane, which is to become a boundary of the cavity. Figure 5 shows an elastic half-space without cavities. The two-dot chain line in Fig. 5 represents a plane (real boundary L) which is to become a surface of an axisymmetric cavity, the broken line is a plane (fictitious boundary Γ) of distribution of body forces, and ε is the distance between these two planes.

When the body forces F_j of Eq. (4) are distributed on the broken line Γ, the stresses and displacements, due to the body forces distributed, at any point in the elastic half-space can be expressed by using Green's functions of Eq. (5) as follows:

$$\begin{Bmatrix} u_i \\ \sigma_{ij} \end{Bmatrix} = \int_\Gamma \left[f_1(\Gamma) \begin{Bmatrix} u_i^1 \\ \sigma_{ij}^1 \end{Bmatrix} + f_3(\Gamma) \begin{Bmatrix} u_i^3 \\ \sigma_{ij}^3 \end{Bmatrix} \right]_{\substack{a=a(\Gamma) \\ h=h(\Gamma)}} d\Gamma + \begin{Bmatrix} u_i^0 \\ \sigma_{ij}^0 \end{Bmatrix} \qquad (6)$$

where u_i^0 and σ_{ij}^0 are solutions of the problem of an elastic half-space without cavities under a uniform biaxial tension $\sigma_r = p$ and are expressed as follows:

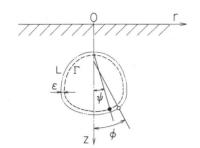

Fig. 5. An axisymmetric cavity in an elastic half-space.

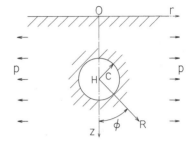

Fig. 6. An elastic half-space having a spherical cavity under biaxial tention.

$$u_r^0 = \frac{p(1-\nu)r}{2\mu(1+\nu)}, \quad u_\theta^0 = 0, \quad u_z^0 = \frac{-p\nu z}{\mu(1+\nu)}$$

$$\sigma_r^0 = \sigma_\theta^0 = p, \quad \sigma_z^0 = \tau_{rz}^0 = \tau_{r\theta}^0 = \tau_{z\theta}^0 = 0 \tag{7}$$

where μ is the modulus of elasticity in shear and ν is Poisson's ratio. The functions $f_1(\Gamma)$ and $f_3(\Gamma)$ in Eq. (6) are the intensities of the body forces distributed. Equation (6) satisfies always the boundary conditions (ii), (iii) of Eq. (1) regardless of the values of $f_1(\Gamma)$ and $f_3(\Gamma)$. Consequently, if we determine the values of $f_1(\Gamma)$ and $f_3(\Gamma)$ so as to satisfy the remaining boundary conditions (i) of Eq. (1), then all the boundary conditions of Eq. (1) are satisfied.

Application of the boundary conditions (i) of Eq. (1) to Eq. (6) yields the following integral equations with unknown functions $f_1(\Gamma)$ and $f_3(\Gamma)$:

$$\begin{Bmatrix} -p\sin^2\phi \\ -p\sin\phi\cos\phi \end{Bmatrix} = \int_\Gamma \left[f_1(\Gamma) \begin{Bmatrix} \sigma_n^1 \\ \tau_{n\phi}^1 \end{Bmatrix} + f_3(\Gamma) \begin{Bmatrix} \sigma_n^3 \\ \tau_{n\phi}^3 \end{Bmatrix} \right]_{\substack{a=a(\Gamma)\\h=h(\Gamma)}} d\Gamma. \tag{8}$$

If we can find $f_1(\Gamma)$ and $f_3(\Gamma)$ by solving the integral equations shown above, then the stresses and displacements at any point in the elastic half-space are obtained from Eq. (6).

IV-2. Linear Algebraic Equations

To solve the integral Equations (8) by reducing them to a set of linear algebraic equations, we perform an appropriate discretization of the body forces distributed. For this purpose, we shall assume that

$$\begin{Bmatrix} f_1(\Gamma) \\ f_3(\Gamma) \end{Bmatrix} = \sum_{n=1}^{N} \begin{Bmatrix} f_n^1 \\ f_n^3 \end{Bmatrix} \delta(\Gamma - \Gamma_n). \tag{9}$$

Equation (9) means that the body forces distributed are approximated by N concentrated ring forces, and f_n^k, ($k=1, 3$), stand for arbitrary constants corresponding to the intensities of the body forces distributed. Substitution of Eq. (9) into Eq. (8) yields

$$\begin{Bmatrix} -p\sin\phi^2 \\ -p\sin\phi\cos\phi \end{Bmatrix} = \sum_{n=1}^{N} \left[f_n^1 \begin{Bmatrix} \sigma_n^1 \\ \tau_{n\phi}^1 \end{Bmatrix} + f_n^3 \begin{Bmatrix} \sigma_n^3 \\ \tau_{n\phi}^3 \end{Bmatrix} \right]_{\substack{a=a_n\\h=h_n}}. \tag{10}$$

There are $2N$ arbitrary constants in Eq. (10). If we choose N points (r_i, z_i), $(i = 1 \sim N)$, on the surface of cavity, then we can construct a set of linear algebraic equations with unknown constants f_n^k. Solving the linear algebraic equations and substituting their solutions into the expressions,

$$\begin{Bmatrix} u_i \\ \sigma_{ij} \end{Bmatrix} = \sum_{n=1}^{N} \left[f_n^1 \begin{Bmatrix} u_i^1 \\ \sigma_{ij}^1 \end{Bmatrix} + f_n^3 \begin{Bmatrix} u_i^3 \\ \sigma_{ij}^3 \end{Bmatrix} \right]_{\substack{a=a_n \\ h=h_n}} + \begin{Bmatrix} u_i^0 \\ \sigma_{ij}^0 \end{Bmatrix} \quad (11)$$

then the stresses and displacements at any point (r, z) in the elastic half-space having an axisymmetric cavity are obtained.

V. NUMERICAL RESULTS

V-1. Convergency of Solution

In order to confirm the validity of the method of solution described above, we consider a cavity with a spherical form and compare the numerical results with the known one[2] for the problem. Figure 6 shows an elastic half-space with a spherical cavity, where C is the radius of the cavity and H is the distance between the boundary ($z = 0$) and the center of the cavity.

Figures 7 and 8 show the convergency of numerical results for the stresses σ_θ and σ_ϕ in the case of $C/H = 0.6$ under the assumptions of $\varepsilon/H = 0.06$ and Poisson's ratio $\nu = 0.25$. The two-dot chain lines show the results for $N = 5$, the one-dot chain lines are $N = 10$, the broken lines are $N = 20$, ○ marks are $N = 40$, and the solid lines represent the known results for the problem.[2] From these figures we see that the numerical results converge to the known results according to the increase in the number N of selected points.

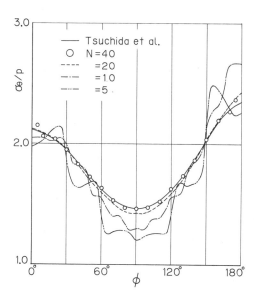

Fig. 7. Convergency of numerical solution.

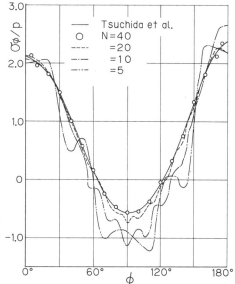

Fig. 8. Convergency of numerical solution.

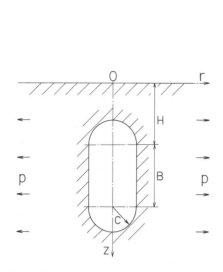

Fig. 9. An elastic half-space having a circular cylindrical cavity of arbitrary length.

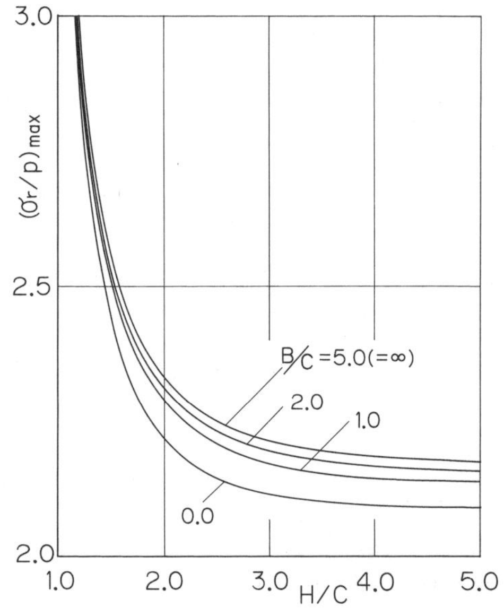

Fig. 10. Maximum stress σ_r at the top point ($r=0$, $z=H-C$) of the surface of the cylindrical cavity.

V-2. *Example of Application*

As an example of application of the method of solution proposed in this paper, we shall consider the stress concentration problem of an elastic half-space having a circular cylindrical cavity of arbitrary length, as shown in Fig. 9. It is assumed that the end faces of the cylindrical cavity are semispherical forms of radius C. The notations C, B, and H in Fig. 9 are the radius, length, and depth of the circular cylindrical cavity, respectively.

Numerical results for the present problem are shown in Fig. 10. These results represent the stress σ_r at the top point ($r=0$, $z=H-C$) on the surface of the cylindrical cavity, and the stress σ_r yields the maximum value. The results for $B/C = 0.0$ in Fig. 10 represent the value for the case of a spherical cavity, the results for $B/C = \infty$ are the value for the case of a semiinfinite cylindrical cavity, and the results for $H/C \gg 1.0$ are regarded as the value for the case of a cylindrical cavity in an infinite elastic body. From this figure, we see the following results:

(1) The value of stress σ_r becomes great when H/C approaches 1.0, regardless of the value of B/C.

(2) Under the condition $H/C =$ constant, the larger B/C, the larger σ_r. However, when B/C is more than 5.0, stress σ_r is not under the influence of B/C and is equal to the value for a cavity with semiinfinite length ($B/C = \infty$).

VI. CONCLUSION

Based on the results given in this paper, the following conclusions are justified.

1) A method of solution by distributions of body forces has been shown for the stress

concentration problem of an elastic half-space having an axisymmetric cavity under uniform biaxial tension.

2) The validity of the method of solution has been confirmed by comparison of the numerical results with the known one for the case of cavity with a spherical form.

3) As an example of application of the method of solution, we have investigated the stress concentration problem of an elastic half-space having a circular cylindrical cavity of arbitrary length. The influences of the depth and the length of the cylindrical cavity on the maximum stress were considered.

REFERENCES

1) Hasegawa, H., Green's functions for axisymmetric body force problems of an elastic half space and their application. *Bull. JSME*, Vol. **27**, No. 231 (1984), p. 1829.
2) Tsuchida, E. and Nakahara, I., Stress in a semi-infinite body having a spherical cavity with a pressure applied to the surface of the plane and cavity. *Trans. Japan Soc. Mech. Engrs.*, Vol. **37**, No. 297 (1971), p. 843 (in Japanese).
3) Tsutsui, S., Chikamatsu, S., and Saito, K., Stress fields of a semi-infinite body containing a spherical inclusion under uniform biaxial tension. *Trans. Japan Soc. Mech. Engrs.*, Vol. **40**, No. 337 (1974), p. 2284 (in Japanese).

Inplane Strength Analysis of Tapered Thin-Walled Steel Beam-Columns by Dynamic Relaxation Method under Displacement-Control Loading

Ichizou MIKAMI,* Yasuo MIURA,* Shigenori TANAKA,* and Yasuyoshi SHINNAI*

* Department of Civil Engineering, Kansai University, Suita, Osaka 564, 2* Toyo Information Systems Co., Ltd., Suita, Osaka 564, 3* Graduate School of Kansai University, Suita, Osaka 564

> Thin-walled steel tapered beam-columns subjected to end-bending and compression are analyzed to obtain the in-plane strength. The dynamic relaxation method (DRM) is used, where a new technique of loading by the compelled axial-displacement or slope is proposed. The technique is able to trace the relation between load and displacement over a peak. The numerical solutions by DRM are compared with experimental results for I-and box-sections, and it is found that the DRM can give very accurate solutions.

I. INTRODUCTION

Members of steel structures often have varying cross sections. Few studies have been done on the elasto-plastic buckling strength of the members: theoretically and experimentally by Prawel,[1] experimentally by Salter,[2] theoretically by Yoshida,[3] experimentally by Okumura,[4] and experimentally and theoretically by Shiomi.[5,6] There have been some investigations of practical design formulas,[6,7] but not of the analysis of elasto-plastic beam-columns with varying cross sections because it is difficult.

The dynamic relaxation method (DRM) is a method of repetition based on the finite difference method, and has some efficiency of memory size, computation time, and programming labor. The method can also be used advantageously on vector processors.[8,9] Authors effectively used DRM to analyze elasto-plastic buckling strength of uniform[10] and tapered[11] steel members. However, the load step must carefully be set, because the usual dynamic relaxation technique controls directly the acting load. Then it is difficult that a peak of load-displacement relation is correctly traced.

In the present paper, a displacement-control type of repetition technique is proposed to cause the load by compelling the slope or axial displacement at a edge. The displacement-control technique is able to trace closely the load-displacement relation after the peak and the ultimate strength of the beam-columns may be determined exactly. Therefore, the parametric analysis of the members becomes easy. The present solutions are compared with the test results by Okumura[4] and Shiomi.[5,6] Moreover, the present method is applied to the analysis of materially and geometrically nonlinear behavior.

II. ELASTO-PLASTIC ANALYSIS OF TAPERED BEAM-COLUMNS

A thin-walled steel tapered member of length L (cm) is simply supported at both ends,

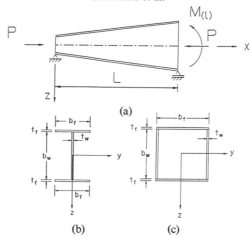

Fig. 1. Tapered beam-column and cross sections.

and is subjected to axial compressive force P at both ends and moment $M_{(l)}$ at the larger end, as shown in Fig. 1(a). The beam-column has an initial deflection in the z-direction and residual stress in the section. The following numerical treatment can be applied to any member whose monosymmetric cross section arbitrarily changes along the length. Hereafter, members with doubly symmetric cross section are treated; the I-section with linearly tapered webs and box-section with linearly tapered flanges and webs is shown in Fig. 1(b) and (c), respectively.

The following assumptions are used: 1) the cross section has its original shape after deformation; 2) the displacement is finite; 3) each cross section, originally plane, remains plane and normal to the longitudinal fibers after deformation; 4) the material is a perfect elasto-plastic body with modulus of elasticity E (kgf/cm²) and yield stress σ_Y (kgf/cm²); and 5) the shear strain can be disregarded.

II-1. Basic Equations

To apply the DRM, inertia and damping terms are added to the static equilibrium conditions in the x- and z-directions; i.e.,

$$m \frac{\partial \dot{u}}{\partial t} + k_u \dot{u} = \frac{\partial N}{\partial x} \quad (1)$$

$$m \frac{\partial \dot{w}}{\partial t} + k_w \dot{w} = \frac{\partial^2 M}{\partial x^2} + N \frac{\partial^2 (w + w_0)}{\partial x^2} \quad (2)$$

in which $u(x)$ and $w(x)$ are the additional displacements of the centroidal axis in the x- and z-directions (cm), respectively; $w_0(x)$ is the initial deflection in the z-direction (cm); $N(x)$ is the axial tension force (kgf); $M(x)$ is the bending moment (kgf-cm); m is the line mass density (kgf-s²/cm²); k_u and k_w are the damping coefficients in the x- and z-directions (kgf-s/cm²), respectively; and $\dot{u}(x)$ and $\dot{w}(x)$ are the displacement velocity (cm/s) which relates to the displacements as follows:

$$\dot{u} = \frac{\partial u}{\partial t} \quad (3)$$

$$\dot{w} = \frac{\partial w}{\partial t}. \tag{4}$$

The strain $\varepsilon(x, z)$ in the x-direction at the fiber of distance z from the centroidal axis may be represented in the terms of displacements u and w by:

$$\varepsilon = \frac{\partial u}{\partial x} - z \frac{\partial^2 w}{\partial x^2} \tag{5}$$

from which the elastic stress $\sigma_e(x, y, z)$ of the same fiber is calculated as follows:

$$\sigma_e = E \varepsilon + \sigma_{re} \tag{6}$$

in which $\sigma_{re}(x, y, z)$ is the residual stress. Thus, the stress $\sigma(x, y, z)$ (kgf/cm²) of the elasto-plastic member is

$$|\sigma_e| < \sigma_Y: \quad \sigma = \sigma_e \tag{7a}$$

$$|\sigma_e| \geq \sigma_Y: \quad \sigma = \pm \sigma_Y. \tag{7b}$$

The axial force $N(x)$ and the bending moment $M(x)$ at any cross section of area A are calculated from the following expressions using the value of stress σ:

$$N = \int_A \sigma \, dA \tag{8}$$

$$M = \int_A z \sigma \, dA. \tag{9}$$

II-2. Boundary Conditions

The boundary condition at the smaller end is given as follows:

$$x = 0: \quad u = 0, \quad w = 0, \quad M = 0. \tag{10a–c}$$

At the larger end, the loading is caused by the compelled slope $\theta_{(l)}$ or the compelled axial-displacement $u_{(l)}$ as shown in Fig. 2(a) and (b), respectively. When a large moment is applied, the load is controlled by the compelled slope $\theta_{(l)}$, while for large axial compression, the axial displacement $u_{(l)}$ is compelled to loading. In both cases, the following condition must be introduced to keep the ratio between moment and axial force at the larger end, $M_{(l)}/PL$, constant:

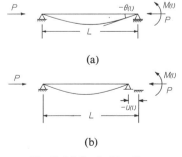

Fig. 2. Method of loading.

$$x = L: \quad \alpha\frac{M}{L} + N = 0 \text{ or } \frac{M}{L} + \beta N = 0. \tag{11a, b}$$

Loading by compelled solope. When the compelled slope $u_{(l)}$ at the larger end as shown in Fig. 2(a) causes the loading, the boundary conditions are

$$x = L: \quad w = 0, \quad m\frac{\partial \dot{w}}{\partial t} + k_b \dot{w} = \frac{\partial w}{\partial x} - \theta_{(l)} \tag{12a, b}$$

in which k_b is a damping coefficient (kgf-s/cm²), and Eq. (12b) is a dynamic representation of the slope[10] at the larger end that $\partial w/\partial x = \theta_{(l)}$.

The dynamic relaxation representation of Eq. (11a) is[10]

$$x = L: \quad m\frac{\partial \dot{u}}{\partial t} + k_u \dot{u} = \alpha\frac{M}{L^2} - \frac{N}{L} \tag{12c}$$

whose right side has been divided by L for numerical stability.

In the case of bending alone ($P = 0$), Eq. (12c) should be read as

$$x = L: \quad N = 0. \tag{12d}$$

Loading by compelled axial-displacement. The axial displacement $u_{(l)}$ is compelled at the larger end shown in Fig. 2(b). Therfore, the boundary condition is

$$x = L: \quad w = 0, \quad u = u_{(l)}. \tag{13a, b}$$

The condition of Eq. (11b) should be represented in the following dynamic relaxation form[10]:

$$x = L: \quad m\frac{\partial \dot{w}}{\partial t} + k_w \dot{w} = \frac{M}{L^2} + \beta\frac{N}{L}. \tag{13c}$$

In the case of axial compression alone ($M_{(l)} = 0$) Eq. (13c) should be read as

$$x = L: \quad M = 0. \tag{13d}$$

III. METHOD OF NUMERICAL COMPUTATION

The dynamic relaxation expressions of equilibrium conditions, Eq. (1) and Eq. (2), are the partial differential equations of field and time. Then they are discretized by using finite difference.

III-1. Initial Deflections
The initial shape in the z-direction $w_0(x)$ (cm) along the length is calculated as follows:

$$w_0 = f_0 \sin(\pi x/L) \tag{14}$$

in which f_0 is the midspan amplitude of the member's initial deflection.

III-2. Residual Stress
The residual stress distribution shown in Fig. 3(a) for the I-section[11] is determined to agree with[12] the results measured by Chubu University.[5] The distribution for the box-section shown in Fig. 3(b) is assumed.[11]

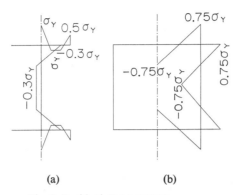

Fig. 3. Residual stress pattern.

III-3. Finite Difference Expressions

The beam-column is divided into n_x portions in the x-direction with mesh interval of Δx, and the mesh points are numbered from 0 to n_x. The flanges and webs are divided into a mesh of $n_{yf} \times n_{zf}$ and $n_{zw} \times 1$, respectively. The time is also divided by the interval of Δt.

Two techniques are adopted for the convenience of programming. First[13] is a technique by which the fictitious mesh points are not required, and the other[10] is a technique by which the finite difference may be expressed on the time interval of Δt, while being expressed on the time interval of $2\Delta t$ by the original dynamic relaxation technique.[14]

Equations (1) and (2) are expressed in the central difference form about the field at any mesh point i and in the forward difference form about the time at any time $j\Delta t$; thus the following expressions are obtained:

$$\dot{u}_{i(j)} = (1 - \Delta t\, k_u)\, \dot{u}_{i(j-1)} + \frac{\Delta t}{2\Delta x} \{N_{i+1} - N_{i-1}\}_{(j-1)} \tag{15}$$

$$\dot{w}_{i(j)} = (1 - \Delta t\, k_w)\dot{w}_{i(j-1)} + \frac{\Delta t}{(\Delta x)^2}\Big[\{M_{i+1} - 2M_i + M_{i-1}\}_{(j-1)}$$
$$+ N_{i(j-1)}\Big(\{w_{i+1} - 2w_i + w_{i-1}\}_{(j-1)} + \{w_{0,i+1} - 2w_{0,i} + w_{0,i-1}\}\Big)\Big] \tag{16}$$

in which $m = 1$ because the magnitude of m is independent of a static solution.

Equations (3) and (4) are expressed about the time in the backward difference as follows:

$$u_{i(j)} = u_{i(j-1)} + \Delta t\, \dot{u}_{i(j)} \tag{17}$$

$$w_{i(j)} = w_{i(j-1)} + \Delta t\, \dot{w}_{i(j)}. \tag{18}$$

The central difference expression of Eq. (5) about the field is

$$\varepsilon_{i(j)} = \frac{1}{2\Delta x}\{u_{i+1} - u_{i-1}\}_{(j)} - \frac{z}{(\Delta x)^2}\{w_{i+1} - 2w_i + w_{i-1}\}_{(j)} \tag{19a}$$

which can apply to the mesh points $i = 1$ to $n_x - 1$. For the mesh points $i = 0$ or n_x, the following higher-order expressions of forward or backward difference about the field are used.

$$i = 0: \quad \varepsilon_{i(j)} = \frac{1}{12\Delta x}\{-25u_0 + 48u_1 - 36u_2 + 16u_3 - 3u_4\}_{(j)}$$

$$- \frac{z}{12(\Delta x)^2}\{45w_0 - 154w_1 + 214w_2 - 156w_3 + 61w_4 - 10w_5\}_{(j)} \quad (19b)$$

$$i = n_x: \quad \varepsilon_{i(j)} = \frac{1}{12\Delta x}\{3u_{i-4} - 16u_{i-3} + 36u_{i-2} - 48u_{i-1} + 25u_i\}_{(j)}$$

$$- \frac{z}{12\Delta x^2}\{-10w_{i-5} + 61w_{i-4} - 156w_{i-3} + 214w_{i-2} - 154w_{i-1} + 45w_i\}_{(j)}. \quad (19c)$$

Expressions (19b) and (19c) contain five mesh points and have the truncation error of order (Δx).[4]

Equations (8) and (9) yield

$$N_{i(n)} = \iint_{top-flg} \sigma \, dy \, dz + \iint_{bottom-flg} \sigma \, dy \, dz + t_w \int_{web} \sigma \, dz \quad (20)$$

$$M_{i(n)} = \iint_{top-flg} \sigma z \, dy \, dz + \iint_{bottom-flg} \sigma z \, dy \, dz + t_w \iint_{web} \sigma z \, dz \quad (21)$$

from which the axial force N and the bending moment M at any cross section can be obtained by numerically integrating for the obtained values of stress $\sigma(y, z)_{i(j)}$.

III-4. Finite Difference Boundary Conditions

Equation (10) may be expressed in the finite difference form:

$$i = 0: \quad w_{i(j)} = u_{i(j)} = M_{i(j)} = \dot{w}_{i(j)} = \dot{u}_{i(j)} = 0. \quad (22a\text{-}e)$$

In the case of loading by the compelled slope $\theta_{(l)}$, Eq. (12a) becomes

$$i = n_x: \quad w_{i(j)} = \dot{w}_{i(j)} = 0. \quad (23a, b)$$

If Eq. (12b) is expressed about the time in the backward difference and about the field in the backward difference at $i = n_x$, the left side of Eq. (12b) vanishes from the condition of Eqs. (23a) and (12b). To avoid this inconvenience, the partial derivative with respect to time in Eq. (12b) is transformed into the backward difference at point $i = n_x - 1$. Thus the following equation is obtained:

$$i = n_x: \quad \dot{w}_{i-1(j)} = (1 - \Delta t\, k_b)\dot{w}_{i-1(j-1)} + \Delta t\, \theta_{(l)}$$

$$- \frac{\Delta t}{12\Delta x}\{3w_{i-4} - 16w_{i-3} + 36w_{i-2} - 48w_{i-1} + 25w_i\}_{(j-1)}. \quad (23c)$$

From Eq. (12c) the backward difference expression about the time is given as follows:

$$i = n_x: \quad \dot{u}_{i(j)} = (1 - \Delta t\, k_u)\dot{u}_{i(j-1)} - \Delta t\left(\alpha \frac{M_{i(j)}}{L^2} + \frac{N_{i(j)}}{L}\right). \quad (24)$$

In case of loading by compelled axial displacement $u_{(l)}$, the difference expressions of Eqs. (13a) and (13b) are

$$i = n_x: \quad w_{i(j)} = \dot{w}_{i(j)} = 0, \quad u_{i(j)} = u_{(l)}. \quad (25a\text{-}c)$$

The right side of Eq. (13c) collocates at point $i = n_x$, and the left side of Eq. (13c) is expressed in the backward difference with respect to time at point $i = n_x - 1$, because from the condition of Eqs. (25a) and (25b), the left side of Eq. (13c) vanishes at point $i = n_x$. Thus the required expression is obtained as follows:

$$i = n_x: \quad \dot{w}_{i-1(j)} = (1 - \Delta t\, k_w)\dot{w}_{i-1(j-1)} - \Delta t\left(\frac{M_{i(j)}}{L^2} + \beta\frac{N_{i(j)}}{L}\right). \quad (26)$$

In Eqs. (24) and (26), the second terms are changed into negative quantities to avoid divergence in the numerical repetition process.

III-5. Initial Values

At the first time step ($t = 0$), values of any parameter may be set at zero except the following values assumed to prevent the effect as impact load.

In the case of loading by compelled slope $\theta_{(l)}$,

$$\dot{u}_{i(0)} = u_{i(0)} = \dot{w}_{i(0)} = 0 \quad (27\text{a-c})$$

$$w_{i(0)} = -\theta_{(l)}\frac{L}{2}\left\{2\left(1 - \frac{i}{n_x}\right) - 3\left(1 - \frac{i}{n_x}\right)^2 + \left(1 - \frac{i}{n_x}\right)^3\right\}. \quad (27\text{d})$$

In the case of loading by compelled axial displacement $u_{(l)}$,

$$\dot{u}_{i(0)} = \dot{w}_{i(0)} = w_{i(0)} = 0, \quad u_{i(0)} = u_{(l)}\frac{i}{n_x}. \quad (28\text{a-d})$$

III-6. Numerical Computation

By progressing the time and by using the basic equations and the boundary conditions, the substitution is subsequently repeated and then the solution arrives at a static solution after the damped vibration. A flowchart of the computation process is shown in Fig. 4, whose detail is

Step 1: The initial values of $\dot{u}_{(0)}, u_{(0)}, \dot{w}_{(0)}, w_{(0)}, M_{(0)}, N_{(0)}$ are set by using Eq. (27) or Eq. (28).

Step 2: For each cross section $i = 0$ to n_x, the strains $\varepsilon_{(j)}$ on each mesh point are calculated from $u_{(j)}$ and $w_{(j)}$ by using Eqs. (19a) to (19c). The stresses in the elastic state are obtained from $\varepsilon_{(j)}$ and residual stress σ_{re} by Eq. (6), and judging the yielding and using Eq. (7) determine the stress $\sigma_{(j)}$ in the elasto-plastic state.

Step 3: For each cross section $i = 0$ to n_x, the axial force $N_{(j)}$ is computed from the stress $\sigma_{(j)}$ by using Eq. (20). At each cross section $i = 1$ to n_x, the bending moment $M_{(j)}$ is determined from the stress $\sigma_{(j)}$ using Eq. (23). At the edge section $i = 0$, it is found from Eq. (22c) that $M_{(j)} = 0$.

Step 4: Return to step 2 unless the displacements w and u converge.

Step 5: Progress the time with $j\Delta t$.

Step 6: The values of $\dot{u}_{(j)}$ at the section $i = 1$ to $n_x - 1$ are obtained from $\dot{u}_{(j-1)}$ and $N_{(j-1)}$ using Eq. (15). At $i = 0$, $\dot{u}_{(j)}$ is given by Eq. (22e). The remainder of $\dot{u}_{(j)}$ at $i = n_x$ is determined from Eq. (24) in the case of compelled slope control, while it is not required in the case of axial displacement control. In addition, $\dot{w}_{(j)}$ at $i = 1$ to $n_x - 2$ is computed from $\dot{w}_{j-1}, w_{j-1}, N_{j-1}, M_{j-1}, w_0$ using Eq. (16). The value of $\dot{w}_{(j)}$ at $i = n_x - 1$ is obtained from Eq. (23c) in the case of compelled slope control, and from Eq. (26) in the case of compelled axial displacement control. From Eq. (22d) and Eq. (23b) or Eq. (25b), it is found that $\dot{w}_{(j)} = 0$ for $i = 0$ and n_x.

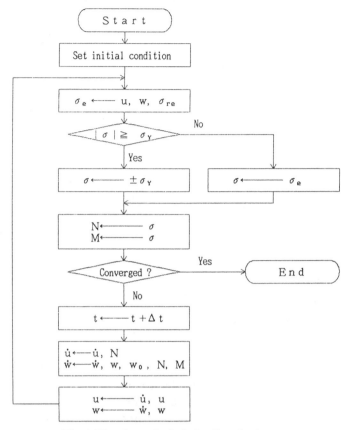

Fig. 4. Flowchart of calculation for a load step.

Step 7: The values of $u_{(j)}$ are calculated for $i = 1$ to n_x from $u_{(j-1)}$ and $\dot{u}_{(j)}$ using Eq. (17), except $u_{(j)}$ at $i = n_x$ given by Eq. (25c) in the case of compelled axial displacement control. From Eq. (22b) it is found that $u_{(j)} = 0$ for $i = 0$. Using Eq. (18), $w_{(j)}$ is determined for $i = 1$ to $n_x - 1$ from $w_{(j-1)}$ and $\dot{w}_{(j)}$. It is found from Eq. (22a) and Eq. (23a) or Eq. (25a) that $w_{(j)} = 0$ for $i = 0$ and n_x.

It is seen from the above computation process that analyzing the damped longitudinal vibration of Eq. (1) and the damped transverse vibration of Eq. (2) using the DRM yields the distribution of elasto-plastic stress and displacement under a single repetition cycle. By adopting the technique of Eq. (12c) or Eq. (13c), loading with a constant ratio between P and $M_{(1)}$ can be caused by compelled slope or axial displacement.

IV. NUMERICAL RESULTS

In this section, a mesh pattern of $n_x = 24$, $n_{yf} = 24$, $n_{zf} = 3$, $n_{zw} = 24$ is used based on the results of the numerical discussion.[11] The time interval of $\Delta t = 0.0005$ s is also assumed.

IV-1. Damping Coefficient

The amplitude of damping coefficients k_w, k_u, and k_b influences the velocity of conver-

Table 1. Optimal and critical damping factors for compelled slope.

$\theta_{(l)}$	$k_{w,cr}$	$\dfrac{k_{w,opt}}{k_{w,cr}}$	$k_{u,cr}$	$\dfrac{k_{u,opt}}{k_{u,cr}}$	$k_{b,cr}$	$\dfrac{k_{b,opt}}{k_{b,cr}}$
−0.005	35.0	2.0	40.0	1.3	0.5000	0.0001
−0.008	−	3.3	−	4.1	−	0.0001
−0.015	−	13.3	−	6.3	−	0.0001

gence and the accuracy of the solution in the elasto-plastic range. Then the optimum value of damping coefficients is examined for a tapered member of length $L = 200$ cm and with an I-section of $b_f = 10$ cm, $t_f = 1$ cm, $b_{w(s)} = 20$ cm, $b_{w(l)} = 25$ cm, $t_{w(l)} = 0.8$ cm, where the subscripts (s) and (l) denote the smaller end and the larger end, respectively. The initial deflection of $f_0 = L/1000$ is assumed and the material is $\sigma_y = 2400$ kgf/cm^2 and $E = 2.1 \times 10^6$ kgf/cm^2.

First, a free vibration of a beam-column can occur by putting $k_w = k_u = k_b = 0$ in the elastic range under both compression and bending. From the numerical response of w_{12}, u_{12}, and w_{23} the natural period T is examined and thus the critical value of the damping coefficient may be determined as $k_{cr} = 4\pi m/T$. The results shown in Table 1 and Table 2 are obtained, among which $k_{w,cr}$ and $k_{u,cr}$ are nearly equal to each other in the case of compelled slope control, while $k_{u,cr}$ becomes 10 times or more greater than $k_{w,cr}$ in the case of compelled axial displacement control. The values of k_b to be used are smaller than $k_{w,cr}$ and $k_{u,cr}$ under compelled slope control.

As the amplitude of k_w, k_u, or k_b increases, damped vibration occurs, and then the value of k for the over-damping determines the optimum damping coefficient k_{opt}. The numerical results are shown in Table 1 and Table 2 with the ratio to k_{cr}. It is seen from both tables that the optimum value of k_b is constant over the elastic and elasto-plastic states and equal to about $k_{b,cr}/1000$. The optimum value of k_w or k_u increases as the plastic state progresses, and $k_{w,opt}$ is larger than $k_{u,opt}$.

IV-2. Convergence of Solution

The tapered I-member mentioned above is analyzed using the optimum damping coefficient from Table 1 and Table 2, and the convergence to a static solution is examined.

In the case of large moment $M_{(l)}$ ($\alpha = 3.0562$), the solutions are obtained under loading by the compelled slope for the two load steps: $\theta_{(l)} = -0.008$ where a part of the member is in the elasto-plastic state and $\theta_{(l)} = -0.015$ where the applied load is close to the ultimate strength. Figure 5 shows the deflection w_{12} and axial displacement u_{12} at the midspan of the member, the deflection w_{23} at the mesh point 23 where Eq. (23c) is applied, in the case of $\theta_{(l)} = -0.008$. Figure 6 shows the bending moment M_{24} and the axial force N_{24} at the larger end. It is found from these figures that the displacements w and u converge smoothly and rapidly with the lapse of time, and Eq. (23c) may be satisfied at an extremely

Table 2. Optimal and critical damping factors for compelled axial displacement.

$u_{(l)}$	$k_{w,cr}$	$\dfrac{k_{w,opt}}{k_{w,cr}}$	$k_{u,cr}$	$\dfrac{k_{u,opt}}{k_{u,cr}}$
−0.10	15.0	1.7	160.0	2.0
−0.15	−	1.8	−	2.3
−0.25	−	8.6	−	2.4

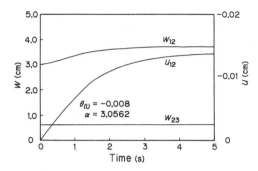

Fig. 5. Time variation of displacements.

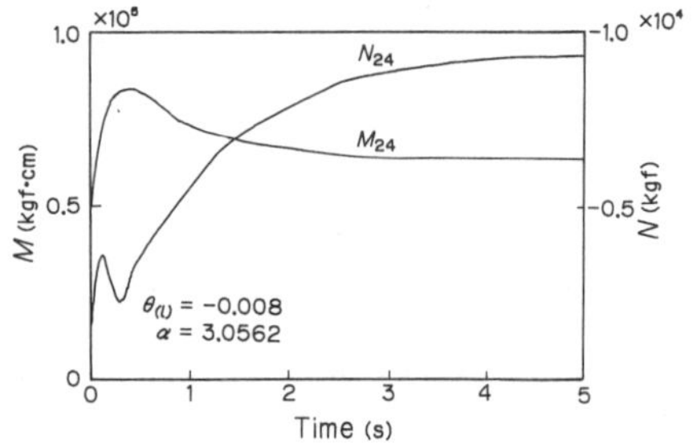

Fig. 6. Time variation of stress resultants.

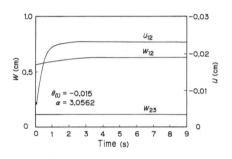
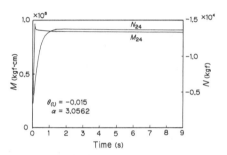

Fig. 7. Time variation of displacements.

Fig. 8. Time variation of stress resultants.

early step. The stress resultants N and M smoothly converge as time progresses after early vibration. The deflection w has an earlier convergence than the displacement u, and the axial force N converges earlier than the bending moment M because the slope at the larger end $\theta_{(l)}$ is given.

The convergence of displacements and stress resultants in the ultimate state is shown in Fig. 7 and Fig. 8. In this state, both the displacements and the stress resultants converge promptly.

In the other case of large compressive load P ($\beta = 0.015956$), two states are analyzed under loading by compelled axial displacements: $u_{(l)} = -0.15$ where a part of the member is elasto-plastic and $u_{(l)} = -0.25$ where the applied load is close to the ultimate strength.

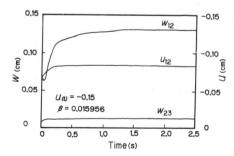

Fig. 9. Time variation of displacements.

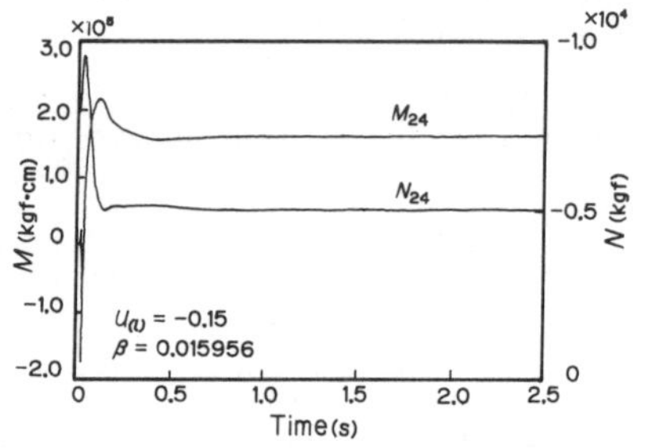

Fig. 10. Time variation of stress resultants.

In the case of $u_{(l)} = -0.15$, the deflection w_{12} and the axial displacement u_{12} at the midspan of the member, and the deflection w_{23} at the mesh point 23 where Eq. (26) is applied, are shown in Fig. 9. The bending moment M_{24} and axial force N_{24} at the larger end are as shown in Fig. 10. In this case, the displacements w and u and the stress resultants M and N are converging rapidly and smoothly. The axial displacement u has an earlier convergence than the deflection w, and the axial force N converges earlier than bending moment M, because the axial displacement at the larger end is compelled. It is also found that Eq. (26) is satisfied at an extremely early step.

Figures 11 and 12 show the convergence of displacements and stress resultants in the close ultimate state. In this state, the displacements and the stress resultants promptly converge.

IV-3. Control of Load or Displacement

Model No. IT-1.6-2 tested by Chubu University[5] is analyzed using both techniques of load control and slope control. The relation between load and deflection at the midspan of the member is shown in Fig. 13. The numerical results are not different from each other. The displacement control technique, however, is able to trace the relation between load and displacement after the peak.

IV-4. Comparison with Test Results

The maximum failure loads for the tapered steel beam-columns tested by Chubu Uni-

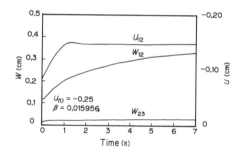

Fig. 11. Time variation of displacements.

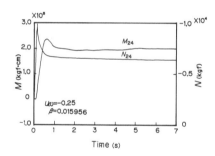

Fig. 12. Time variation of stress resultants.

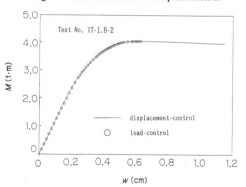

Fig. 13. Relation between load and deflection.

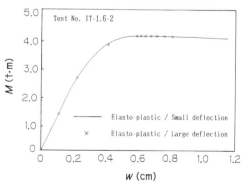

Fig. 14. Relation between load and deflection.

versity[4] and Tokyo University[5,6] are compared with the numerical solutions of ultimate strength to evaluate the present analysis method. The in-plane buckling of five welding tapered specimens with a doubly symmetrical I-section was experimented by Chubu University. The theoretical strength is calculated using the measured distribution of initial deflection[5] and the assumed residual stress distribution which agrees[12] with the measured value.[5] At Tokyo University, in-plane buckling was tested for 36 welding tapered specimens

Table 3. Comparison between test and theoretical results.

Ref.	Model	$\dfrac{M_{(1)}}{PL}$	$P/P_{Y(s)}$		$\dfrac{①}{②}$
			DRM ①	Test ②	
5)	IT-1.4–1	0.0379	0.587	0.554	1.061
	IT-1.6–2	0.0458	0.591	0.561	1.053
	IT-1.8–3	0.101	0.403	0.389	1.037
	IT-2.2–5	0.163	0.324	0.282	1.149
	IT-2.4–6	0.334	0.186	0.174	1.069
4)	30–1–1	0.0145	0.681	0.704	0.967
	30–1–2	0.0289	0.540	0.586	0.925
	30–1–3	0.0433	0.445	0.528	0.843
	30–8–1	0.0190	0.999	1.055	1.054
	30–8–2	0.0379	0.895	0.846	1.059
	30–8–3	0.0570	0.786	0.724	1.085
	30–15–1	0.0202	1.010	1.000	1.010
	30–15–2	0.0404	0.939	0.936	1.002
	30–15–3	0.0607	0.837	0.771	1.083
	50–1–1	0.0090	0.684	0.665	1.029
	50–1–2	0.0173	0.553	0.580	0.953
	50–1–3	0.0260	0.462	0.466	0.991
	50–8–1	0.0114	0.920	0.905	1.017
	50–8–2	0.0228	0.792	0.777	1.019
	50–8–3	0.0342	0.692	0.662	1.045
	50–15–1	0.0121	0.961	1.001	0.953
	50–15–2	0.0242	0.838	0.840	0.998
	50–15–3	0.0364	0.737	0.700	1.052
	70–1–1	0.0062	0.574	0.570	1.007
	70–1–2	0.0124	0.492	0.444	1.108
	70–1–3	0.0186	0.472	0.413	1.143
	70–8–1	0.0081	0.777	0.792	0.983
	70–8–2	0.0163	0.669	0.669	1.000
	70–8–3	0.0244	0.590	0.650	0.899
	70–15–1	0.0086	0.820	0.899	0.911
	70–15–2	0.0173	0.709	0.712	0.996
	70–15–3	0.0260	0.628	0.604	1.043
	90–1–1	0.0048	0.454	0.448	1.013
	90–1–2	0.0096	0.377	0.394	0.957
	90–1–3	0.0144	0.325	0.363	0.895
	90–8–1	0.0063	0.682	0.694	0.981
	90–8–2	0.0127	0.503	0.562	0.960
	90–8–3	0.0190	0.471	0.535	0.879
	90–15–1	0.0067	0.655	0.739	0.887
	90–15–2	0.0135	0.572	0.591	0.965
	90–15–3	0.0202	0.510	0.498	1.024

Table 4. Statistic values of ①/② in Table 3.

Ref.	Mean	Standard deviation
5)	1.0748	0.0551
4)	0.997	0.0664
All models	1.035	0.0760

with a doubly symmetrical box-section, where the four cases of slenderness ratio, the three kinds of taper, and the three cases of eccentricity are chosen. The initial deflection of $f_0 = L/2500$ and the residual stress distribution shown in Fig. 3(b) are assumed for numerical analysis.

The maximum loads of test and analysis are shown in Table 3, where $P_{Y(s)}$ is the full plastic axial force of the smaller end. Table 4 shows the mean value and the standard deviation calculated for the ratio between theoretical and experimental values of $P/P_{Y(s)}$. The difference between the theoretical and experimental values is about 10% except for some models, and it is found that the present analysis gives reasonable results.

IV-5. Materially and Geometrically Nonlinear Analysis

In the preceding sections, materially nonlinear and geometrically linear behavior was analyzed. The effect of geometrical nonlinearity is examined here.

The relation between strain and displacement, Eq. (5), may be rewritten as follows:

$$\varepsilon = \frac{\partial u}{\partial x} + \frac{1}{2}\left(\frac{\partial w}{\partial x}\right)^2 + \frac{\partial w}{\partial x}\frac{\partial w_0}{\partial x} - z\frac{\partial^2 w}{\partial x^2}. \tag{29}$$

Therefore, Eq. (19a) should be read as

$$\varepsilon_{i(j)} = \frac{1}{2\Delta x}\{u_{i+1} - u_{i-1}\}_{(j)} + \frac{1}{8\Delta x^2}\{w_{i+1} - w_{i-1}\}^2_{(j)}$$

$$+ \frac{1}{4\Delta x^2}\{w_{0,i+1} - w_{0,i-1}\}\{w_{i+1} - w_{i-1}\}_{(j)}$$

$$- \frac{z}{\Delta x^2}\{w_{j+1} - 2w_i + w_{i-1}\}_{(j)}. \tag{30}$$

Equations (19b) and (19c) can also be rewritten.

Equation (30), the expression rewritten from Eqs. (19b) and (19c), and the remainder are used to analyze the materially and geometrically nonlinear behavior of tapered steel beam-columns. As an example, the model of Fig. 13 is analyzed. Figure 14 shows the relation between load and midspan deflection. It is seen that the member has no effect on finite displacement.

V. CONCLUSIONS

An analysis method based on the DRM was presented. The method may be effectively applied to analyze the elasto-plastic in-plane buckling strength of steel beam-columns, having initial deflection and residual stress, with any taper, and subjected to axial compression and moment. By compelling a slope or an axial displacement at a member end, the loading is carried out at a constant ratio between both loads of moment and axial force, and thus the relation between displacement laod can trace over the peak. The numerical solutions are compared with the reported test results and it was found that the present analysis method gives reasonable maximum strength. Moreover, the present method was used to analyze the elasto-plastic and finite displacement behavior of tapered beam-columns for the purpose of examining the effect of geometrical nonlinearity. If the present method is used, it is possible to investigate numerically and widely the relation between various design factors and strength, and to obtain easily many information for the purpose of discussing design formula of tapered steel beam-columns.

Acknowledgments

The authors wish to thank Mr. Eiji Kuwahara and Mr. Kenji Kobori for their cooperation during graduate research at the Department of Civil Engineering, Kansai University. The numerical computations presented here were done on a FACOM M-380 computer at Kansai University.

REFERENCES

1) Prawel, S.P., Morrell, M.L., and Lee, G.C., Bending and buckling strength of tapered structural members. *Welding Res. Suppl., Welding J.*, Vol. **53**, No. 2 (1974), pp. 75s–84s.
2) Salter, J.B., Anderson, D., and May, I.M., Tests on tapered steel columns. *J. Struct. Engng., ASCE*, Vol. **58A**, No. 6 (1980), pp. 189–193.
3) Yoshida, H. and Nishida, S., Strength and deformation of tapered H-columns. *Proc. Jpn. Soc. Civil Engineers*, No. 220 (1973), pp. 17–27 (in Japanese).
4) Okumura, T., Abe, H., Takena, K., and Tomohiro, I., Ultimate strength of tapered box-shape steel columns. *Proc. Ann. Conf. JSCE* Sept. (1972), pp. 175–176 (in Japanese).
5) Shiomi, H., Nishikawa, T., and Kurata, M., Tests on tapered steel beam-columns. *Proc. Jpn. Soc. Civil Engineers*, No. 334 (1983), pp. 163–172 (in Japanese).
6) Shiomi, H. and Kurata, M., Strength formula for tapered beam-columns. *J. Struct. Engng., ASCE*, Vol. **110**, No. 7 (1984), pp. 1630–1643.
7) Lee, G.C., Morrell, M.L., and Ketter, R.L., Design of tapered members. *Welding Res. Council Bull.*, No. 173 (1972), pp. 1–32.
8) Mikami, I., Yamashina, J., and Tanaka, K., Finite deformation analysis of elastic cylindrical panels using vector processor. *J. Structural Engng., JSCE*, Vol. **32A** (1986), pp. 313–322 (in Japanese).
9) Mikami, I. and Tanaka, K., Finite deformation analysis of elastic-plastic cylindrical panels using vector processor. *J. Structural Engng., JSCE*, Vol. **33A** (1987), pp. 53–62 (in Japanese).
10) Mikami, I., Miura, Y., Matsumure, K., and Tanaka, K., Useful techniques for dynamic relaxation method. *Tech. Rep. Kansai Univ.*, No. 27 (1986), pp. 187–200.
11) Mikami, I., Miura, Y., Tsujimoto, A., and Tanaka, S., Inplane strength analysis of tapered thin-walled steel beam-columns by dynamic relaxation method. *J. Structural Engng., JSCE*, Vol. **33A** (1987), pp. 247–256 (in Japanese).
12) Nishida, S. and Hoshina, H., The estimation of residual stresses and its self equilibrium conditions in welded thin-walled members. *Proc. Jpn. Soc. Civil Engineers*, No. 334 (1983), pp. 163–172 (in Japanese).
13) Mikami, I., Dynamic and viscous relaxation methods using simulation language. *Theoret. Appl. Mech.*, Vol. **32** (University of Tokyo Press, 1984), pp. 507–515.
14) Day, A.S., An introduction to dynamic relaxation. *Engineer*, Vol. **219** (1965), pp. 218–221.
15) Rashton, K.R., Dynamic-relaxation solution of elastic-plate problems. *J. Strain Analysis*, Vol. **1**, No. 1 (1968), pp. 23–32.

Stability Analysis of Circular Cylindrical Shells Subjected to Tangential Follower Force

Kazuo MITSUI and Kohji SUMINO

Department of Mathematical Engineering, College of Industrial Technology, Nihon University, Chiba

In this paper, the dynamic method of stability analysis is applied to the nonconservative stability problem of circular cylindrical shells subjected to tangential follower forces which are directed along the generators of the deformed shells. Loading of this kind may be due to tangential frictional forces, for example, developed through the motion of granular material in a circular silo. In the first part of this paper, a numerical examination of the stability of circular cylindrical shells clamped at one edge is considered, and it is shown that loss of stability can take place only in the form of dynamic instability. In order to study the destabilizing effect of damping, in the second part, an examination of the stability of circular cylindrical shells clamped at one edge and supported viscoelastially at the other is performed.

Nomenclature

a:	radius of the cylinder	E:	Young's modulus
$l\lambda$:	length of the cylinder	D:	bending stiffness; $Eh^3/12(1-\nu^2)$
h:	thickness	F:	uniform tangential follower force
λ:	slenderness ratio	m:	circumferential buckling waves
ρ:	density	ω:	characteristic exponent
x, ϑ:	coordinate variables	K_1:	spring constant
t:	time variable	C_1:	damping coefficient
u, v, w:	displacement components	$N_x, N_\vartheta, N_{\vartheta x}$:	stress resultants
ν:	Poisson's ratio	$M_x, M_\vartheta, M_{\vartheta x}$:	stress couples

I. INTRODUCTION

Preliminary attempts to clarify the influence of tangential follower forces on the elastic stability of shells have already been described in several papers. For example, the stability of clamped spherical shells under combined hydrostatic pressure and tangential follower force has been studied by the dynamic method, and it has been shown that loss of stability can take place not only in the form of static instability but also in the form of dynamic instability, and the critical load of the system may be reduced by almost 50% through the influence of the tangential follower force in some cases.[1,2] The stability of circular cylindrical shells with fixed ends and subjected to tangential follower forces has been investigated by means of Liapunov's second method.[3] Loading of this kind may be due to tangential frictional forces, for example, developed through the motion of granular material in a circular silo. The effect of the axial friction force due to the grain on stability has been investigated.[4] However, none of the reports clarified whether dynamic instability may occur in circular cylindrical shells subjected to tangential follower forces or whether the presence of sufficiently small, velocity-dependent forces has a destabilizing effect. The purpose of the present study is to show theoretically that dynamic instability takes place in a cylindrical shell under a uniformly distributed tangential follower force and the magnitude of the critical load may be reduced through the influence of the velocity-dependent force due to the viscoelastic support at the edge.

II. NONLINEAR BASIC EQUATIONS

Assume that a circular cylindrical shell with radius a, length l, and uniform thickness h, shown in Fig. 1, is subjected to a tangential follower force F which is directed along the generator of the deformed shell. Nonelinear basic equations of the buckling problem are given as follows:

$$N_{x,x} + \frac{1}{a} N_{\vartheta x, \vartheta} - F = \rho h u_{,tt},$$

$$N_{\vartheta x, x} + \frac{1}{a} N_{\vartheta, \vartheta} = \rho h v_{,tt},$$

$$M_{x,xx} + \frac{2}{a} M_{\vartheta x, \vartheta x} + \frac{1}{a^2} M_{\vartheta, \vartheta\vartheta} + \frac{1}{a} N_\vartheta$$

$$+ \left[N_x w_{,x} + \frac{1}{a} N_{\vartheta x} w_{,\vartheta} \right]_{,x}$$

$$+ \left[\frac{1}{a} N_{\vartheta x} w_{,x} + \frac{1}{a^2} N_\vartheta w_{,\vartheta} \right]_{,\vartheta}$$

$$- F w_{,x} = \rho h w_{,tt}. \tag{1}$$

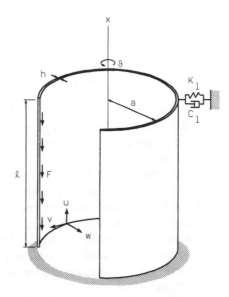

Fig. 1. Shell geometry and coordinate system.

Further, the boundary conditions become

$$u = 0,\ v = 0,\ w = 0,\ w_{,x} = 0 \text{ at } x = 0,$$

$$N_x = 0,\ N_{\vartheta x} = 0,\ M_x = 0,$$

$$M_{x,x} + \frac{2}{a} M_{\vartheta x,\vartheta} + \frac{1}{a} N_{\vartheta x} w_{,\vartheta} + N_x w_{,x}$$

$$+ K_1 w + C_1 w_{,t} = 0 \quad \text{at } x = l. \tag{2}$$

The stress resultants and stress couples are obtained as

$$N_x = \frac{Eh}{1-\nu^2}\left[u_{,x} + \frac{1}{2}w_{,x}^2 + \nu\left(\frac{1}{a}(v_{,\vartheta} - w) + \frac{1}{2}\frac{1}{a^2}w_{,\vartheta}^2\right)\right],$$

$$N_\vartheta = \frac{Eh}{1-\nu^2}\left[\frac{1}{a}(v_{,\vartheta} - w) + \frac{1}{2}\frac{1}{a^2}w_{,\vartheta}^2 + \nu\left(u_{,x} + \frac{1}{2}w_{,x}^2\right)\right],$$

$$N_{\vartheta x} = \frac{Eh}{1+\nu}\frac{1}{2}\left[\frac{1}{a}u_{,\vartheta} + v_{,x} + \frac{1}{a}w_{,x}w_{,\vartheta}\right],$$

$$M_x = -D\left(w_{,xx} + \frac{\nu}{a^2}w_{,\vartheta\vartheta}\right),$$

$$M_\vartheta = -D\left(\frac{1}{a^2}w_{,\vartheta\vartheta} + \nu w_{,xx}\right),$$

$$M_{\vartheta x} = -D(1-\nu)\frac{1}{a}w_{,\vartheta x}. \tag{3}$$

II-1. Equilibrium Conditions

In the prebuckling axisymmetric state, the displacements and stress resultants are a function of x only. Therefore the equilibrium equations are given by the following equations:

$$N_{x0,x} - F = 0,$$

$$M_{x0,xx} + \frac{1}{a} N_{\vartheta 0} + [N_{x0} w_{0,x}]_{,x} - F w_{0,x} = 0. \tag{4}$$

The corresponding boundary conditions become

$$u_0 = 0,\ w_0 = 0,\ w_{0,x} = 0 \text{ at } x = 0,$$

$$N_{x0} = 0,\ M_{x0} = 0,\ M_{x0,x} + K_1 w_0 = 0 \text{ at } x = l. \tag{5}$$

The corresponding stress resultants and stress couples become

$$N_{x0} = \frac{Eh}{1-\nu^2}\left(u_{0,x} + \frac{1}{2}w_{0,x}^2 - \frac{\nu}{a}w_0\right),$$

$$N_{\vartheta 0} = \frac{Eh}{1-\nu^2}\left(-\frac{1}{a}w_0 + \nu\left(u_{0,x} + \frac{1}{2}w_{0,x}^2\right)\right),$$

$$M_{x0} = -Dw_{0,xx}. \tag{6}$$

From Eqs. (4), (5), and (6), the equilibrium conditions expressed by the displacement component w are given by the ordinary differential Eq. (7) and the boundary conditions by Eq. (8).

$$-D w_{0,xxxx} - \frac{\nu}{a} \int_x^l F\,dx - \int_x^l F\,dx\, w_{0,xx} - \frac{Eh}{a^2} w_0 = 0. \tag{7}$$

$$w_0 = 0,\ w_{0,x} = 0 \quad \text{at } x = 0,$$

$$w_{0,xx} = 0,\ -D w_{0,xxx} + K_1 w_0 = 0 \quad \text{at } x = l. \tag{8}$$

II-2. Stability Conditions

The components of the characteristics of the disturbed motion will be of the form

$$(u, v, w) = (u_0, v_0, w_0) + (u_1, v_1, w_1),$$

$$(N_x, N_\vartheta, N_{\vartheta x}) = (N_{x0}, N_{\vartheta 0}, N_{\vartheta x0}) + (N_{x1}, N_{\vartheta 1}, N_{\vartheta x1}),$$

$$(M_x, M_\vartheta, M_{\vartheta x}) = (M_{x0}, M_{\vartheta 0}, M_{\vartheta x0}) + (M_{x1}, M_{\vartheta 1}, M_{\vartheta x1}),$$

where the quantities with subscript 0 correspond to the prebuckling axisymmetric state, and the quantities with subscript 1 correspond to the infinitesimal increments during buckling. Substituting these expressions into Eqs. (1), (2), and (3), considering the equilibrium Eqs. (4), (5), and (6) and retaining only the linear terms of the increments, the variational equations and the corresponding boundary conditions for stability analysis become

$$N_{x1,x} + \frac{1}{a} N_{\vartheta x1,\vartheta} = \rho h u_{1,tt},$$

$$N_{\vartheta x1,x} + \frac{1}{a} N_{\vartheta 1,\vartheta} = \rho h v_{1,tt},$$

$$M_{x1,xx} + \frac{2}{a} M_{\vartheta x1,\vartheta x} + \frac{1}{a^2} M_{\vartheta 1,\vartheta\vartheta} + \frac{1}{a} N_{\vartheta 1}$$

$$- \int_x^l F\,dx\, w_{1,xx} + N_{x1,x} w_{0,x} + N_{x1} w_{0,xx}$$

$$+ \frac{1}{a} N_{\vartheta x1,\vartheta} w_{0,x} - \frac{1}{a^2}\left(\nu \int_x^l F\,dx + \frac{Eh}{a} w_0\right) w_{1,\vartheta\vartheta} = \rho h w_{1,tt}, \tag{9}$$

$$u_1 = 0,\ v_1 = 0,\ w_1 = 0,\ w_{1,x} = 0 \text{ at } x = 0,$$

$$N_{x1} = 0,\ N_{\vartheta x1} = 0,\ M_{x1} = 0,$$

$$M_{x1,x} + \frac{2}{a} M_{\vartheta x,1,\vartheta} - \int_x^l F\,dx\, w_{1,x} + N_{x1} w_{0,x}$$

$$+ K_1 w_1 + C_1 w_{1,t} = 0 \quad \text{at } x = l. \tag{10}$$

The corresponding stress resultants and stress couples become

$$N_{x1} = \frac{Eh}{1-\nu^2}\left(u_{1,x} + w_{0,x}w_{1,x} + \frac{\nu}{a}(v_{1,\vartheta} - w_1)\right),$$

$$N_{\vartheta 1} = \frac{Eh}{1-\nu^2}\left(\frac{1}{a}(v_{1,\vartheta} - w_1) + \nu(u_{1,x} + w_{0,x}w_{1,x})\right),$$

$$N_{\vartheta x1} = \frac{Eh}{1+\nu}\frac{1}{2}\left(\frac{1}{a}u_{1,\vartheta} + v_{1,x} + \frac{1}{a}w_{0,x}w_{1,\vartheta}\right),$$

$$M_{x1} = -D\left(w_{1,xx} + \frac{\nu}{a^2}w_{1,\vartheta\vartheta}\right),$$

$$M_{\vartheta 1} = -D\left(\frac{1}{a^2}w_{1,\vartheta\vartheta} + \nu w_{1,xx}\right),$$

$$M_{\vartheta x1} = -D(1-\nu)\frac{1}{a}w_{1,\vartheta x}. \tag{11}$$

When the type of load is a dead load, the static boundary-value problem (9) is self-adjoint[5] by virtue of the boundary conditions (10).

III. NUMERICAL ANALYSIS

III-1. Equilibrium Analysis

Substituting the following differences equations into the equilibrium conditions (7) and (8), the differential equations will be replaced by the system of linear equations for calculating deflections $w_1, w_2, \ldots w_{n+4}$ and w_{n+5}.

$$w_{k,x}(x) = \frac{w_{k+1} - w_{k-1}}{2\Delta x}, \quad w_{k,xxx}(x) = \frac{w_{k+2} - 2w_{k+1} + 2w_{k-1} - w_{k-2}}{2\Delta x^2},$$

$$w_{k,xx}(x) = \frac{w_{k+1} - 2w_k + w_{k-1}}{\Delta x^2}, \quad w_{k,xxxx}(x) = \frac{w_{k+2} - 4w_{k+1} + 6w_k - 4w_{k-1} + w_{k-2}}{\Delta x^4}$$

Where n is the number of the divisions, the quantities with subscripts 1, 2, $n+4$, and $n+5$ correspond to the deflections of the fictive exterior points, and the others correspond to the deflections of the interior points. When the shell dimensions and the boundary conditions are given, the prebuckling axisymmetric deformation can be determined.

III-2. Stability Analysis

Assuming that the shell buckles in m waves in the circumferential direction, variations u_1, v_1, and w_1 can be expressed by the following series:

$$u_1(x, \vartheta, t) = u_m(x, t)\cos m\vartheta,$$

$$v_1(x, \vartheta, t) = v_m(x, t)\sin m\vartheta,$$

$$w_1(x, \vartheta, t) = w_m(x, t)\cos m\vartheta.$$

Applying the finite difference method to the variational Eqs. (9) and the boundary conditions (10), the following ordinary differential equations can be obtained.

$$\ddot{f}_i + b_{ik}\dot{f}_k + a_{ik}f_k = 0, \qquad i, k = 1, 2, 3, \ldots, 3n \tag{12}$$

The solution of the differential Eqs. (12) may be expressed as

$$f_i(t) = \varphi_i \exp(\omega t).$$

The stability problem is reduced to the following algebraic eigenvalue problem:

$$(A + \omega B + \omega^2 E)X = 0$$

where $A = \|a_{ik}\|$, $B = \|b_{ik}\|$, $E = \|\delta_{ik}\|$, $X = \|\varphi_i\|$, and ω is the frequency of small vibrations about its equilibrium configuration. The frequencies ω can be obtained from the following characteristic equation:

$$\det(A + \omega B + \omega^2 E) = 0.$$

Therefore $\text{Re}(\omega) < 0$ is a sufficient condition for stability.

IV. NUMERICAL RESULTS

Poisson's ratio ν is taken as 0.3, and the thickness ratio a/h is taken as 100 to simplify the calculation. The buckling load and corresponding mode are determined for the geometric parameter λ ranging from 0.5 to 1.5, considering both the axisymmetric and nonaxisymmetric buckling modes. The convergence of the solution was examined by taking successively larger numbers of divisions into consideration. To illustrate the rate of convergence of the solution, the number of divisions n taken in the calculations and the cor-

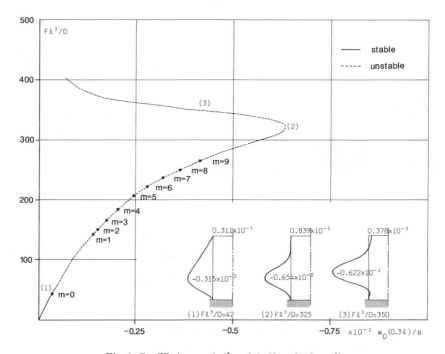

Fig. 2. Equilibrium path ($\lambda = 0.5$, $K_1 = 0$, $C_1 = 0$).

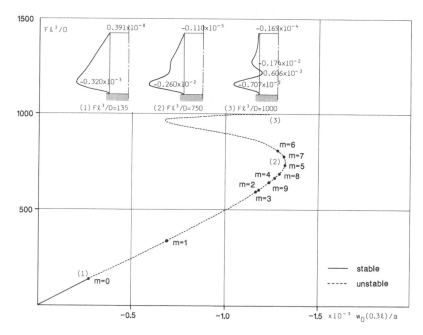

Fig. 3. Equilibrium path ($\lambda = 1.0$, $K_1 = 0$, $C_1 = 0$).

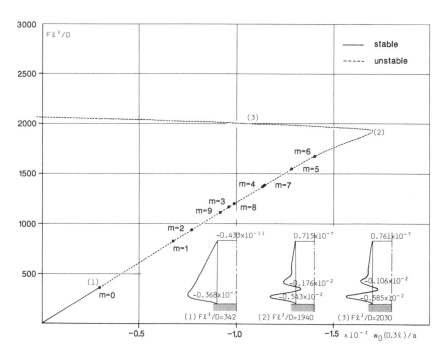

Fig. 4. Equilibrium path ($\lambda = 1.5$, $K_1 = 0$, $C_1 = 0$).

responding result are shown in Fig. 9 for typical values of the slenderness ratio λ. Equilibrium paths for the cylindrical shells with a slenderness ratio $\lambda = 0.5, 1.0$ and 1.5 are shown in Figs. 2, 3, and 4, respectively, together with the wave numbers m and typical prebuckling configurations. The stability of the respective equilibrium states on the equilibrium paths is examined by the dynamic method, namely the characteristic exponents of the variational equations considering the effect of the prebuckling deformations.

The behavior of the characteristic exponents of the undamped cylinder with a slenderness ratio $\lambda = 1.0$, as the loading parameter Fl^3/D varies, is graphically illustrated in Fig. 5. The illustrations in Fig. 5 include a perspective of the characteristic exponent curves, and also the orthographic projections on the real, imaginary, and complex planes. It is found that all the characteristic exponents ω are distinct and purely imaginary if $Fl^3/D < 135$. For $Fl^3/D = 135$ there exist two pairs of equal roots whose real parts are all zero. Thus the system is stable for $Fl^3/D \leqq 135$. For $Fl^3/D > 135$ there are two characteristic exponents with a positive real part and thus the system is unstable for $Fl^3/D > 135$. The vibration modes corresponding to the second and third characteristic exponents are illustrated in Fig. 6 and 7, respectively. In order to assess the influence of the slenderness ratio λ and the circumferential buckling wave m on the stability behavior of the cylindrical shell, the previous calculations have been repeated for various values of λ and various numbers of m. The critical values $F_{cr}l^3/D$ of the follower force for various numbers of circumferential

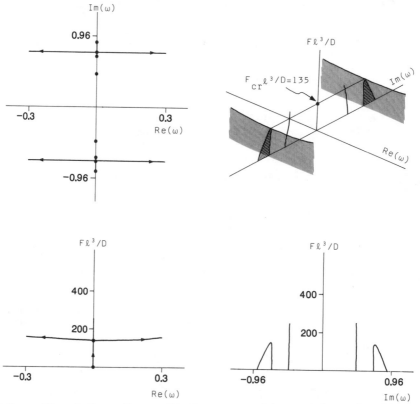

Fig. 5. Orthographic projections and perspective of root curves of characteristic equation with no damping ($\lambda = 1.0$, $K_1 = 0$, $C_1 = 0$).

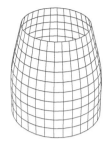
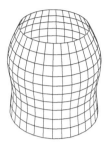

Fig. 6. Vibration mode ($\lambda = 1.0$, $K_1 = 0$, $C_1 = 0$).　　Fig. 7. Vibration mode ($\lambda = 1.0$, $K_1 = 0$, $C_1 = 0$).

Table 1. Variations of the critical load $F_{cr}l^3/D$ with circumferential wave number m ($K_1 = 0$, $C_1 = 0$).

λ	$m=0$	$m=1$	$m=2$	$m=3$	$m=4$	$m=5$	$m=6$	$m=7$	$m=8$	$m=9$
0.5	42	144	150	168	188	208	225	239	253	267
0.8	183	211	322	446	659	529	548	574	790	500
1.0	135	338	610	594	669	739	823	774	689	643
1.2	243	549	682	856	1018	1223	1151	1002	880	822
1.5	342	814	947	1171	1380	1560	1667	1385	1203	1153

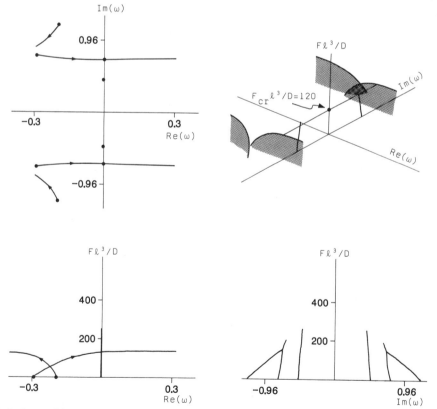

Fig. 8. Orthographic projections and perspective of root curves of characteristic equation with damping ($\lambda = 1.0$, $K_1 = 0$, $C_1 l^3/D = 0.1$).

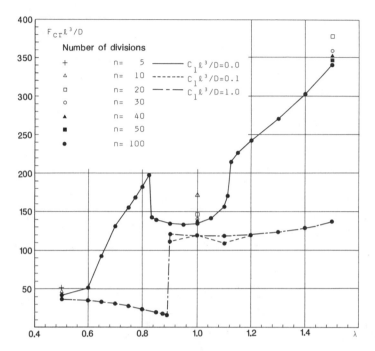

Fig. 9. Variations of the critical load $F_{cr}l^3/D$ with slenderness ratio λ ($K_1 = 0$).

Table 2. Variations of the critical load $F_{cr}l^3/D$ with circumferential wave number m ($K_1 = 0$, $C_1 l^3/D = 0.1$).

λ	$m=0$	$m=1$	$m=2$	$m=3$	$m=4$	$m=5$	$m=6$	$m=7$	$m=8$	$m=9$
0.5	37	135	143	161	183	204	223	238	253	266
0.8	25	195	314	382	438	484	522	563	511	474
1.0	120	329	471	590	667	729	788	762	672	623
1.2	122	468	677	798	908	1002	1097	991	861	799
1.5	139	813	936	1167	1355	1510	1664	1365	1174	1103

waves are shown in Table 1 at typical values of the slenderness ratio λ. In the range of present calculations, it is found that the smallest values of critical load parameters correspond to the circumferential wave number $m = 0$, namely the axisymmetric buckling modes. In Fig. 9, variation of the critical value $F_{cr}l^3/D$ of the undamped cylinder with slenderness ratio λ are illustrated by the solid line.

The numerical results for the damping coefficient $c_1 l^3/D = 0.1$ are illustrated in Fig. 8. For $Fl^3/D < 120$ all the characteristic exponents have a negative real part, and two characteristic exponents will have a positive real part for $Fl^3/D > 120$ and thus the critical value of the follower force is $F_{cr}l^3/D = 120$. The critical values $F_{cr}l^3/D$ of the follower force for various numbers of circumferential waves are shown in Table 2. Variations of the critical values $F_{cr}l^3/D$ of the follower force, taking into account the damping, are shown in Fig. 9, in which the dashed and chain lines correspond to the case of the damping coefficient $c_1 l^3/D = 0.1$ and 1.0, respectively. The critical values of the follower force are reduced through the influence of damping. It is found that critical values of the follower force here obtained for the cases of $c_1 l^3/D = 0.1$ and 1.0 are almost coincident with each other.

V. CONCLUSIONS

It has been shown that loss of stability can take place only in the form of dynamic instability and the critical values of the follower force are reduced under the influence of dissipative forces due to the viscoelastic support. In this paper, we have considered the circular cylindrical shell subjected only to the tangential follower force. The nonconservative stability problem of circular cylindrical shells subjected to combined normal and tangential follower forces is important as a problem to be solved in the future.

Acknowledgments

Part of this research was supported by a grant for scientific research from the Ministry of Education, Japan.

REFERENCES

1) Sumino, K. and Mitsui, K., An elastic stability analysis for spherical shells. *Bull. JSME*, Vol. **27**, No. 224 (1984), pp. 153–158.
2) Sumino, K. and Mitsui, K., Stability analysis of spherical shells under nonconservative forces. *Proc. IASS* (1985), pp. 261–275.
3) Leipholz, H.H.E., Stability of elastic cylindrical shells via Liapunov's second method. *Ingenieur Archiv*, Vol. **49** (1980), pp. 7–14.
4) Briassoulis, D. and Curtis, J., Design and analysis of silo for friction forces. *J. Struct. Engng.*, Vol. **111**, No. 6 (1985), pp. 1377–1398.
5) Sumino, K., A consistent theory for thin elastic shell stability. *Proc. IASS* (1986), pp. 25–32.

Buckled Configuration of Laterally Supported Columns in the Inelastic Range

Jiro Suzuya

Department of Architecture, Tohoku Institute of Technology, Sendai

The deflection configurations of columns, laterally supported by bracing, with one half-wave mode initial imperfection are investigated theoretically.

The numerical analysis is carried out on columns elastically supported at the midspan using the collocation method, and in each case of column analyzed different values of the slenderness ratio of columns and the stiffness of the bracing are chosen.

As the result of analysis, the process of forming a nonsymmetric and localized buckling pattern of the column laterally supported at midspan is clarified, and it is summarized as follows.

The deformed configurations of the columns in the elastic range is a combination of one half-wave and three half-waves mode patterns. After yielding of the column, the amplitude of the three half-waves mode component becomes larger, and the bifurcation of two half-wave modes occurs. When the stiffness of the bracing is large, the two half-waves mode component becomes remarkable in amplitude, compared to the one or three half-waves mode components, and thus the deflection configuration of the colum becomes nonsymmetric and localized.

I. INTRODUCTION

The most remarkable feature of the inelastic buckling of structures, such as bars, plates, and shells, is that the buckling patterns of the structures are not periodic but localized. Tvergaard and Needleman first analyzed this problem by solving a linear differential equation of a column on a nonlinear elastic foundation with an initial imperfection in the shape of the critical mode.[1,2] The results of the analysis showed that the yielding of the foundation is accompanied by the occurrence of several bifurcation modes different from the initial deflection mode, and a combination of the initial deflection pattern and the bifurcation patterns becomes a localized pattern.

The present author analyzed a geometrically nonlinear equation of a column on a nonlinear elastic foundation to investigate the mode components in the deflection of the column, and the process of forming a localized buckling pattern is summarized as follows.[3] The development of the yielding zones in the foundation is accompanied by several bifurcation modes, and as the yielding zone is expanding, the bifurcation modes with even smaller number of half-waves occur in succession, and thus the buckling patterns of the columns on a nonlinear elastic foundation become localized.

In the present paper, an elastic-plastic equilibrium equation of a laterally supported column is formulated and analyzed to investigate the deflection configuration of the laterally supported columns in the inelastic range.

Theoretical or experimental research on columns laterally supported by bracing have been carried out to investigate the sufficient bracing strength and stiffness in order for the strength of the columns to reach the expected value in the inelastic range.[4,5] In the experimental results of those studies, it was found that the deflection configurations of the columns, the strength of which reached the expected column strength, were not periodic but localized.

In this study, taking into consideration the results above, the deflection configuration of columns laterally supported at midspan is examined to investigate the process of formation of a localized buckling pattern.

II. ANALYSIS

II-1. Basic Equation

An analytical model of a column laterally supported by bracing at s points is shown in Fig. 1. The out-of-plane equilibrium equation of the column is written as follows,

$$EI\frac{d^2W/dx^2}{\{1-(dW/dx)^2\}^{3/2}} + P(W+W_0)$$
$$-\sum_{k=1}^{s} KW_k(1-x_k/L) + \sum_{k=1} KW_k(x-x_k) = 0 \qquad (1)$$

where EI is the flexural rigidity of the column, $W(x)$ is the lateral displacement of the column from initial configuration $W_0(x)$, and P is the axial force. K is the stiffness of the bracing, and the suffix k defines the values at the k-th support. The last term of the Eq. (1) defines the sum of the term within the limits of $x - x_k > 0$.

We assume that the displacement and the initial imperfection are represented by

$$W = \sum_{j=1}^{n} A_j \sin\frac{j\pi x}{L}$$
$$W = \sum_{j=1}^{n} C_j \sin\frac{j\pi x}{L} \qquad (2)$$

where A_j is the unknown displacement coefficient, and C_j is the initial configuration coefficient. Their nondimensional forms are introduced as

$$w = W/L, \qquad a_j = A_j/L$$
$$w = W_0/L, \qquad c_j = C_j/L \qquad (3)$$

where the suffix j corresponds to the number of half-waves and defines the mode.

Introducing the following nondimensional quantities, and substituting them into Eq. (1):

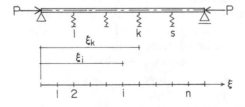

Fig. 1. Laterally supported column and coordinate system.

$$\xi = x/L$$
$$p = PL^2/\pi^2 EI = P/P_E$$
$$\alpha = KL^3/\pi^2 EI \qquad (4)$$

where L is the column length, and P_E is Euler's buckling load of the column.

After yielding of the column, we define the flexural rigidity as γEI, using stiffness reducing factor γ. Then we determine the location of n collocation points at equal intervals, and represent the value of ξ coordinate at the i-th collocation point as ξ_i. Then the equilibrium equation by the collocation method in the inelastic range is formulated at the i-th collocation point and written as

$$\sum_{j=1}^{n}(P - \gamma j^2) a_j \sin j \pi \xi_i + p \sum_{j=1}^{m} c_j \sin j \pi \xi_i$$
$$+ \frac{3}{2}\sum_{j=1}^{n} j^2 a_j \sin j \pi \xi_i (\sum_{j=1}^{n} j a_j \cos j \pi \xi_i)^2$$
$$- \sum_{j=1}^{n}\sum_{k=1}^{s} a_j \sin j \pi \xi_k (1 - \xi_k) \xi_i$$
$$\sum_{j=1}^{n}\sum_{k=1}^{} a_j \sin j \pi \xi_k (\xi_i - \xi_k) = 0. \qquad (5)$$

The stiffness reducing factor γ is the ratio of the bending moment M and the curvature Φ, and $\gamma = 1$ in the elastic range.

It is assumed that the cross section of the column is rectangular, and that the material of the column has the elastic-perfectly plastic stress-strain relationship shown in Fig. 2(a). Then the stiffness reducing factor is represented by

$$\gamma = \frac{2(1+n)}{|\phi|}\left(\frac{3}{2} - \frac{1+n}{|\phi|}\right) \qquad (6\text{-}1)$$

$$\gamma = \frac{3}{|\phi|}(1 - n^2) + \frac{1}{|\phi|} \qquad (6\text{-}2)$$

where n is the nondimensionalized axial thrust, and is the ratio of the axial thrust N to yield thrust N_0. ϕ is the nondimensionalized curvature, and is the ratio of the curvature Φ to the initial yielding curvature Φ_y of the rectangular section subjected to pure bending. After the extreme fiber starts to yield, the state of the section is classified into two regimes:

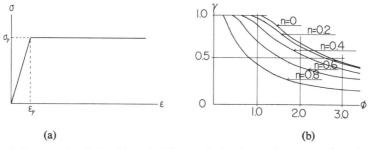

Fig. 2. Stress-strain relationship and stiffness reducing factor of a rectangular column.

Primary plastic regime: $|\phi| > 1 + n$
Secondary plastic regime: $|\phi| > 1/(1 + n)$.

Equation (6–1) is derived in the primary plastic regime, and Eq. (6–2) is derived in the secondary plastic regime. The $\gamma - \phi$ relationship of the rectangular cross section is shown in Fig. 2(b), and the corresponding strain distribution is also shown in Fig. 3. As shown in Fig. 3, the displacement is measured positively downward from the neutral axis, then the bending moment m is defined as positive, and the curvature is defined as negative when the upper fiber is in compression. Here the nondimensional bending moment m is defined as the ratio of the bending moment M to the initial yielding moment M_y of the rectangular cross section subjected to pure bending.

II-2. Iterative Procedure

The nonlinear equilibrium Eq. (5) is formulated on n collocation points ($i = 1, 2, \ldots, n$), and to analyze the set of the equation the iterative method is employed. The displacement factors a_j and the load factor p are incremented from the equilibrium state (p, a) to the next equilibrium state ($p + \Delta p, a + \delta$). In the iterative procedure, the value of the stiffness reducing factor γ is varied with the development of the deflection of the column and the axial force. We derive the increments of the stiffness reducing facto γ as

$$d\gamma_\phi = (d\gamma/d\phi) \cdot \Delta\phi$$
$$d\gamma_p = (d\gamma/dp) \cdot \Delta p. \qquad (7)$$

The curvature ϕ and the axial thrust n are represented by

$$n = -(\pi^2/12\,\varepsilon_y)(D/L)^2 \cdot p$$
$$\phi = -(\pi^2/2\,\varepsilon_y)(D/L)\sum_{j=1}^{n} j^2 a_j \sin j\pi\,\xi_i \qquad (8)$$

Where ε_y is the yield strain of the material of the column, and D is the depth of the column section.

From Eqs. (7) and (8), the incremental forms of the stiffness reducing factor are expressed in terms of the incremental axial thrust Δn and the incremental displacement factor δ.

Increasing the displacement factors and the load factor by the amount ($\delta, \Delta p$), substituting $a_j + \delta_j$, $p + \Delta p$ into the equilibrium Eq. (5), and introducing above derived incremental stiffness reducing factor, the linear incremental equation of the column in the inelastic range is formulated as

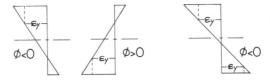

Primary plastic   Secondary plastic
Fig. 3. Strain distribution in the inelastic regime.

$$\sum_{j=1}^{n}\left\{p - \gamma\left(1 - \frac{3}{2}R1^2\right)j^2 - d\gamma_\phi R2\left(1 - \frac{3}{2}R1^2\right)j^2\right\}\delta_j \sin j\pi\xi_i$$

$$+ \sum_{j=1}^{n} 3\, R2 \cdot R1 \cdot j\delta_j \cos j\pi\xi_i$$

$$- \alpha \sum_{k=1}^{s}\sum_{j=1}^{n}\delta_j \sin j\pi\xi_k (1-\xi_k)\xi_i + \alpha \sum_{k=1}^{s}\sum_{j=1}^{n}\delta_j \sin j\pi\xi_k(\xi_i - \xi_k)$$

$$+ \Delta p\left\{R1 + \sum_{j=1}^{m} c_j \sin j\pi\xi_i - d\gamma_p\left(1 - \frac{3}{2}R1^2\right)R2 = 0 \right. \tag{9}$$

where

$$R1 = \sum_{j=1}^{n} ja_j \cos j\pi\xi_i$$

$$R2 = \sum_{j=1}^{j} j^2 a_j \sin j\pi\xi_i. \tag{10}$$

When reversion of the curvature occurs at any collocation point, in the process of changing the deflection pattern, the finite difference of the stiffness reducing factor is derived as

$$d\gamma_\phi = (1-\gamma)d\phi/\phi. \tag{11}$$

III. RESULTS

Analysis is carried out for some columns laterally supported by bracing at midspan, with one half-wave mode initial deflection. In each case different values of the bracing stiffness α and slenderness ratio λ of the columns are chosen. The following quantities are common to all columns: the amplitude of the initial deflection is $10^{-3}L$, the yield strain ε_y of the columns is 10^{-3}, and the relation between the slenderness ratio λ and the depth to length ratio D/L of the columns are represented by

$$\lambda = \pi^2(D/L)/12\,\varepsilon_y. \tag{12}$$

In the following, we discuss the mode components in the deflection of columns with $\lambda = 120$, 200, and 300, and the bracing stiffness $\alpha = 12$, 16, 20 and 24.

The maximum buckling strength of columns in the elastic range, laterally supported at midspan, is evaluated as $4P_E$, that is, the buckling strength of a simply supported column with length $L/2$. The value of the bracing stiffness required for the buckling strength of the column to reach $4P_E$ is evaluated as $\alpha = 16$.

The initial deflection patterns of columns laterally supported at midspan is a combination of symmetric mode components. In the initial deflection patterns, the mode components of $m = 1$ and $m = 3$ (m denotes the number of half-waves in the mode components) are dominative, and the amplitude of the other mode components are infinitesimally small. The amplitude ratio of the mode components of $m = 1$ to $m = 3$ is large for a column with weak bracing.

Initial yielding of the columns occurs at the point about one quarter of the column length from the both ends. For a column with weak bracing of $\alpha = 12$, 16, the yielding zones

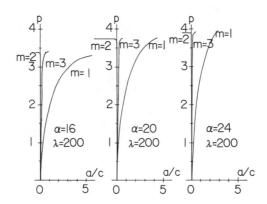

Fig. 4. Development of the mode component of deflection in the column $\lambda = 200$, $\alpha = 16, 20, 24$.

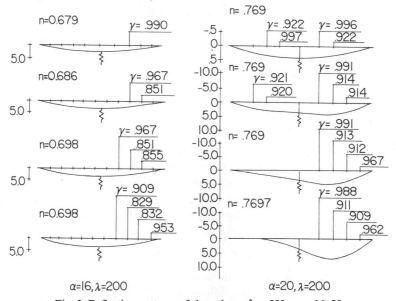

Fig. 5. Deflection patterns of the column $\lambda = 200$, $\alpha = 16, 20$.

expand to midspan, whereas for a column with stiff bracing of $\alpha = 20, 24$, the yielding zones expand to both side of the initial yielding point.

After the initial yielding of the columns, the bifurcation mode of $m = 2$ occurs in a column with stiff bracing of $\alpha \geq 16$, and no unsymmetric bifurcation modes are observed in a column with weak bracing of $\alpha = 12$.

Figure 4 shows the development of mode components in the deflection of columns with $\lambda = 200$. In Fig. 4 the abscissa a/c denotes the amplitude ratio of the mode components to the initial deflection.

As seen in Fig. 4, in the vicinity of the bifurcation point, the mode component $m = 1$ develops rapidly in the column with bracing stiffness $\alpha = 16$, and the deflection pattern of the column is almost symmetric before the maximum load is achieved. The bifurcation mode $m = 2$ develops rapidly in the column with bracing stiffness $\alpha = 20$ or $\alpha = 24$, and exceeds the mode component of $m = 1$, and then the deflection patterns of the columns become nonsymmetric in the vicinity of the maximum load point.

Figure 5 shows the deflection patterns of columns with $\lambda = 200$, and bracing stiffness $\alpha = 16$, $\alpha = 20$. As shown in the figure, the mode components of $m = 1$, the mode of the initial deflection, is a principal mode in the deflection of the column with bracing stiffness $\alpha = 16$, and then the deflection pattern of the column is symmetric. In the deflection of the column with bracing stiffness $\alpha = 20$, the mode $m = 2$ associated with bifurcation becomes remarkable in amplitude in the vicinity of the maximum load point, and the deflection pattern of the column becomes nonsymmetric and localized.

IV. CONCLUSIONS

As a result of analysis of columns laterally supported by bracing at midspan, the following information on the behavior of columns in the inelastic range is obtained. The initial deflection pattern of laterally supported columns with the one half-wave mode initial imperfection is symmetric. After the initial yielding of the columns, the yielding zone is expanded, and in case of a column with bracing stiffness of $\alpha > 16$, a two half-waves bifurcation mode occurs. In the vicinity of the maximum load point, the bifurcation mode becomes remarkable in amplitude, and then the deflection patterns of the columns of these columns become nonsymmetric and localized.

The bracing force of a column with bracing stiffness less than 16 ($\alpha < 16$) becomes exceedingly large in the vicinity of the maximum load point. In order for the strength of the column to reach the expected value, which is ordinarily the buckling strength of a simply supported column of length $L/2$, sufficient bracing stiffness is more than 16 ($\alpha > 16$).

REFERENCES

1) Tvergaad, V. and Needleman, A., On the localization of buckling pattens. *J. Appl. Mech.*, Vol. **47** (1980), pp. 613–619.
2) Tvergaard, V. and Needleman, A., On the development of localized buckling patterns. *Proc. IUTAM Symp. COLLAPSE*, Cambridge University Press (1982), pp. 1–17.
3) Suzuya, J., Buckled configuration of columns on nonlinear elastic foundation and its localization. *Proc. 36th Japan NCTAM* (1988).
4) Saisho, M., Tanaka, H., Takanashi, K., and Udagawa, K., Lateral bracing of compression members. *Trans. AIJ*, No. 184 (1971), pp. 73–79.
5) Matsui, C. and Matsumura, H., Study on lateral bracing of axially compressed members (Part 1). *Trans. AIJ*, No. 205 (1973), pp. 23–29.
6) Matsui, C. and Matsumura, H., Study on lateral bracing of axially compressed members (Part 2). *Trans. AIJ*, No. 208 (1973), pp. 15–21.
7) Chen, W.F. and Atsuta, T., *Theory of Beam-Columns*, Vol. **1** (McGraw-Hill, 1976).

Buckling of an Annular Sector Plate Subjected to In-Plane Moment

Kazuo TAKAHASHI,* Yasunori KONISHI,* Yoshihiro NATSUAKI,[2*] and Michiaki HIRAKAWA*

* Department of Civil Engineering, Nagasaki University, Nagasaki, [2]* Katayama Iron Works, Co. Ltd., Osaka

> Buckling of an annular sector plate subjected to equal and opposite moments at the radial edges is examined. The governing differential equation of the plate is solved by a Galerkin method. Buckling moments and buckling shapes are obtained for the annular sector plate with radial edges simply supported and arbitrary boundary conditions along the circumferential edges.
> Numerical results are shown for various boundary conditions along the circumferential edges and geometrical parameters of the annular sector plate. Buckling properties of an annular sector plate are compared with those of a rectangular plate and a circular beam.

I. INTRODUCTION

Although many researchers have studied the bending and vibration of an annular sector plate, investigations on the stability appear to have been rather scanty.[1-4] Rubin[1] has considered an annular sector plate subjected to constant in-plane forces along the radial and circumferential directions. Srinivasan and Thiruvenkatachari[2] have presented buckling of an annular sector plate subjected to uniform forces along the circumferential edges. Chu[3] and Mikami et al.[4] have studied local buckling of a web plate of the circular curved beam with I-section subjected to in-plane moment and axial and shearing forces. However, buckling properties of an annular sector plate are not clear yet.

In this paper, buckling of an annular sector plate subjected to in-plane moment along the radial edges is analyzed. The governing differential equation is solved by a Galerkin method with vibration mode of the corresponding annular sector plate subjected to no in-plane force. Buckling moment and buckling shape are obtained for the annular sector plate with the radial edges simply supported and three different boundary conditions along the circumferential edges.

Numerical results are shown for various boundary conditions along the circumferential edges and aspect ratios of the annular sector plate and compared with a rectangular plate and a circular beam where comparisons are possible.

II. DIFFERENTIAL EQUATION AND BOUNDARY CONDITIONS

Figure 1 shows an annular sector plate of opening angle α. With the outer and inner radii denoted by a and b, respectively, the polar coordinates (r, θ) are taken in the neutral surface of the plate. Equal and opposite moments act along the radial edges. In-plane forces N_r, N_θ, and $N_{r\theta}$ due to the moment M are given as follows[5]

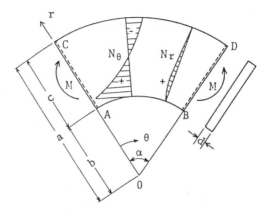

Fig. 1. Geometry and coordinates.

$$N_r = -\frac{4M}{N}\left(\frac{a^2b^2}{r^2}\ln\frac{a}{b} + a^2\ln\frac{r}{a} + b^2\ln\frac{b}{r}\right)$$

$$N_\theta = -\frac{4M}{N}\left(-\frac{a^2b^2}{r^2}\ln\frac{a}{b} + a^2\ln\frac{r}{a} + b^2\ln\frac{b}{r} + a^2 - b^2\right)$$

$$N_{r\theta} = 0 \tag{1}$$

where $N = (a^2 - b^2)^2 - 4a^2b^2(\ln(a/b))^2$ and N_r, N_θ, and $N_{r\theta}$ are functions of independent variable r. The governing differential equation in the polar coordinates of an annular sector plate with moments acting along the radial edges is written

$$D\nabla^4 w - \frac{1}{r}\frac{\partial}{\partial r}\left(r N_r \frac{\partial w}{\partial r}\right) - \frac{1}{r^2} N_\theta \frac{\partial^2 w}{\partial \theta^2} = 0 \tag{2}$$

where $D = Eh^3/12(1-\nu^2)$ is the bending stiffness, E is Young's modulus, ν is Poisson's ratio, w is the plate deflection, and $\nabla^2 = (\partial^2/\partial r^2 + 1/r\,\partial/\partial r + 1/r^2\,\partial^2/\partial\theta^2)$ is the Laplacian operator in the polar coordinates.

For simply supported radial edges, the boundary conditions along $\theta = 0$ and α are as follows:

$$w = M_\theta = 0 \tag{3}$$

where M_θ is the bending moment in the radial direction.

The following three boundary conditions along the circumferential edges ($r = b$ and a) are considered in the present analysis.

Case I: simply supported edges

$$w = M_r = 0;$$

Case II: clamped edges

$$w = \partial w/\partial\theta = 0; \tag{4}$$

Case III: free edges

$$M_r = V_r = 0$$

where M_r is the bending moment and V_r is the equivalent shearing force. By denoting

$$\xi = r/a$$

Eq. (2) can be expressed in the nondimensional form

$$L(w) = \nabla^4 w + \frac{4M}{\bar{N}D}\left[\frac{1}{\xi}\frac{\partial}{\partial \xi}\left\{\xi f_1(\xi)\frac{\partial w}{\partial \xi}\right\} + \frac{1}{\xi^2}f_2(\xi)\frac{\partial^2 w}{\partial \theta^2}\right] = 0 \tag{5}$$

where $\bar{N} = (1 - \beta^2)^2 - 4\beta^2(\ln(1/\beta))^2$, $f_1(\xi) = (\beta/\xi)^2 \ln(1/\beta) + \ln \xi + \beta^2 \ln(\beta/\xi)$, $f_2(\xi) = -(\beta/\xi)^2 \ln(1/\beta) + \ln \xi + \beta^2 \ln(\beta/\xi) + 1 - \beta^2$, and $\beta = b/a$ is the ratio of inner to outer radii.

III. METHOD OF SOLUTION

Taking these boundary conditions into account, we assume the solution of Eq. (5) by

$$w = \sum a_{sn} W_{sn}(\xi, \theta) \tag{6}$$

in which a_{sn} is an unknown constant and W_{sn} is an eigenfunction associated with free vibrations satisfying the boundary conditions of the plate, defined as[6]

$$W_{sn} = R_{sn}(\xi) \sin \alpha_n \theta \tag{7}$$

where

$$R_{sn} = A_{sn} J_{\alpha_n}(k_{sn}\xi) + B_{sn} Y_{\alpha_n}(k_{sn}\xi) + C_{sn} I_{\alpha_n}(k_{sn}\xi) + D_{sn} K_{\alpha_n}(k_{sn}\xi),$$

in which A_{sn}, B_{sn}, C_{sn} and D_{sn} are constants of integration dependent on the boundary conditions, $J\alpha_n$ and $Y\alpha_n$ are the Bessel function, $I\alpha_n$ and $K\alpha_n$ are the modified Bessel function, k_{sn} is the eigenvalue of vibration, $\alpha_n = n\pi/\alpha$, and $n = 1, 2, \ldots$ is an integer.

Substituting Eq. (6) into Eq. (5) and applying a Galerkin method, one has

$$\int_\beta^1 L W_{pn} \xi\, d\xi = 0 \tag{8}$$

where $p = 1, 2, \cdots$

By performing integrations of Eq. (8), the following equations for the unknown constant a_{pn} are obtained:

$$k_{pn}^4 I_{pn} a_{pn} - (M/D) \sum_{s=1} a_{pn} I_{spn} = 0 \tag{9}$$

where

$$I_{pn} = \int_\beta^1 R_{pn}^2 \xi\, d\xi,$$

$$I_{spn} = \int_\beta^1 \left\{\xi f_1(\xi)\frac{dR_{sn}}{d\xi}\frac{dR_{pn}}{d\xi} + \frac{\alpha_n^2}{\xi}f_2(\xi) R_{sn} R_{pn}\right\} d\xi.$$

Equation (9) may be written in the following matrix form

$$[I]\{X\} = (M/D)[G]\{X\} \tag{10}$$

where $\{X\} = \{a_{1n} a_{2n} \cdots a_{Nn}\}^T$ is the column vector, $[I]$ is the unit matrix, and $[G]$ is the square matrix.

Upon introducing $\lambda_{cr} = M/D$, Eq. (10) becomes an eigenvalue problem of the matrix $[G]$:

$$[G]\{X\} = \lambda\{X\} \tag{11}$$

where $\lambda = 1/\lambda_{cr}$.

The geometrical parameters in the present analysis are the opening angle α and the radius ratio β of an annular sector plate. To compare the buckling properties of an annular sector plate with those of a rectangular plate, it will be better to define the aspect ratio of an annular sector plate which denotes the ratio of the mean arc length $l = (a+b)/2$ to the loaded length c as

$$\mu = \frac{l}{c} = \frac{\alpha(1+\beta)}{2(1-\beta)}. \tag{12}$$

IV. NUMERICAL RESULTS

IV-1. Convergence Study

Figure 2 shows the convergence of the present solution for the annular sector plate with the opening angle $\alpha = 60°$ and the aspect ratio $\mu = 1.0$. The minimum buckling eigenvalue λ_{cr} vs. the number of term n is plotted for three different boundary conditions. The first term solution for case III and five terms solution for cases I and II are accurate enough in the calculation. The present solution for case I exactly coincides with the results obtained by Mikami et al.[4]

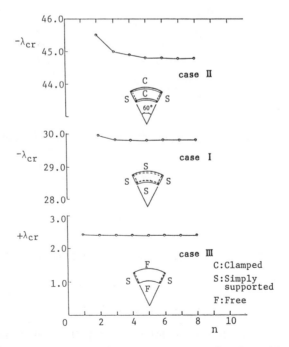

Fig. 2. Convergence of buckling eigenvalue λ_{cr} for the annualr sector plate with $\mu = 1.0$ and $\alpha = 60°$.

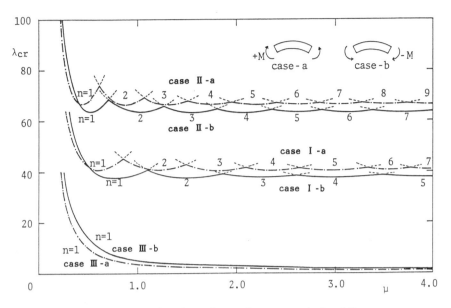

Fig. 3. Buckling eigenvalue vs. the aspect ratio $\beta = 0.8$.

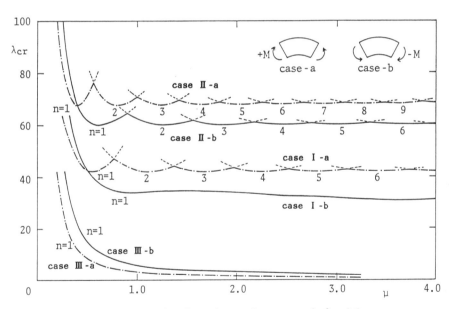

Fig. 4. Buckling eigenvalue vs. the aspect ratio $\beta = 0.6$.

IV-2. Buckling Analysis

Figures 3 through 6 show the variation of the buckling eigenvalue λ_{cr} with the aspect ratio μ for annular sector plates with radius ratios $\beta = 0.8, 0.6, 0.4$, and 0.2. In these figures, notations case-a and case-b show the buckling eigenvalue curves caused by the positive

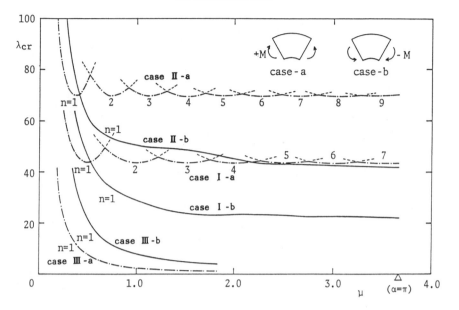

Fig. 5. Buckling eigenvalue vs. the aspect ratio $\beta = 0.4$.

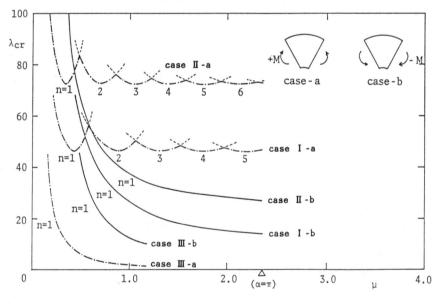

Fig. 6. Buckling eigenvalue vs. the aspect ratio $\beta = 0.2$.

and negative moments, respectively. Notation n denotes the half-wave numbers in the circumferential direction.

The buckling eigenvalue vs. the radius ratio $\beta = 0.8$ shown in Fig. 3 is similar to that of a rectangular plate.[7] That is, buckling eigenvalue curves of cases I and II have the same local minimum independent of the half-wave number n and the buckling eigenvalue curve of case III decreases monotonously with the increase in μ. This fact is not true for eigen-

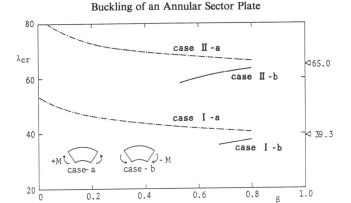

Fig. 7. Buckling eigenvalue vs. radius ratio, cases I and II.

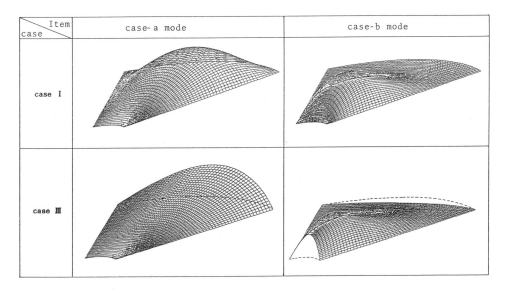

Fig. 8. Buckling modes subjected to positive and negative moments $\beta = 0.2$ and $\alpha = 60°$.

value curves of case I-b and case II-b when the radius ratio β is less than 0.6 (case I-b) and 0.4 (case II-b) as can be seen in Figs. 4 and 5.

Figure 7 summarizes the relation between the buckling eigenvalue and the radius ratio for annular sector plates whose buckling pattern is the same as that of a rectangular plate. The buckling coefficients of a rectangular plate for cases I and II are also shown in Fig. 7. It is impossible to estimate the buckling moment by using a rectangular plate analogy when radius ratio β is less than 0.8.

Figure 8 shows buckling modes of the annular sector plate with $\alpha = 60°$ and $\beta = 0.2$ subjected to positive and negative moments. Deflection of the compressive force region is large and that of the tensile force region is small independent of signs of the moment.

Figure 9 shows the comparisons of the buckling eigenvalue between the present solution for case III with $\beta = 0.6$ and the bending-torsional buckling eigenvalue of the circular beam based upon the beam theory which neglects deformation of the cross section.[7] The

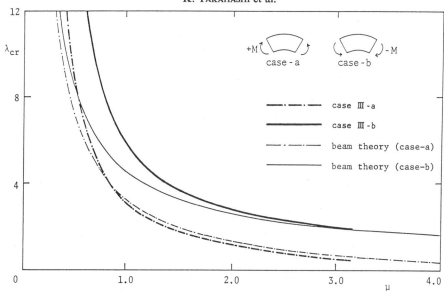

Fig. 9. Comparisons between results obtained by the annular sector plate and beam theory, $\beta = 0.6$.

buckling eigenvalue of the positive moment of the annular sector plate with considerably small aspect ratio can be estimated by the buckling problem of the circular beam as shown in Fig. 9. On the other hand, the results obtained by the beam theory agree well with those obtained by the thin plate theory when the aspect ratio is greater than 2 when the negative moment acts. This may be explained because the buckling mode subjected to the positive moment consists of rigid body deformation as shown in Fig. 8.

V. CONCLUSIONS

The buckling problem of an annular sector plate subjected to in-plane moments is reported in this paper. The buckling moment vs. the aspect ratio curves are obtained for various radii ratios and three different boundary conditions. These results will be valuable in the design of an annular sector plate subjected to in-plane moments.

REFERENCES

1) Rubin, C., Stability of polar-orthotropic sector plates. *J. Appl. Mech.*, Vol. **45**, No. 2 (1978), pp. 448–450.
2) Srinivasan, R.S. and Thiruvenkatachari, V., Stability of annular sector plates with variable thickness. *Am. Inst. Aeronaut. J.*, Vol. **22**, No. 2 (1984), pp. 315–317.
3) Chu, K.Y., Beuluntersuchung von ebenen Stegblechen kressformig gekrumter Trager mit I-Querschnitt. *Sthalbau*, Heft **5** (1966), pp. 129–142.
4) Mikami, I., Akamatsu, Y., and Takeda, H., Elastic local and coupled buckling of vertically curved I-girders under pure bending. *Proc. JSCE*, Vol. **230** (1974), pp. 45–54.
5) Timoshenko, S.P. and Gere, J.M., *Theory of Elastic Stability*, 2nd ed., New York: McGraw-Hill Book Co. (1961).
6) Yamasaki, T., Chisyaki, T., and Kaneko, T., Free transverse vibration of circular ring sector plates. *Tech. Rep. Kyushu Univ.*, Vol. **42**, No. 4 (1969), pp. 576–583.
7) Column Research Committee of Japan, *Handbook of Structural Stability*, Corona Publishing Co. Ltd. (1971).

Development of Rail/Wheel High-Speed Contact Fatigue Testing Machine and Test Results

Makoto Ishida,* Yukio Satoh,[2]* Yoshihiko Sato,[3]* and Shinsaku Matsuyama[4]*

Track and Structure Laboratory, Railway Technical Research Institute, Tokyo, [2] Material Laboratory, Railway Technical Research Institute, Tokyo, [3]* Japan Mechanized Works & Maintenance of Way Company, Ltd., Tokyo, [4]* Toyo Electric MFG Co. Ltd., Kanagawa

> Currently, rail shellings that are a kind of rolling contact fatigue failure cost a considerable amount for rail renewal on the Shinkansen lines. To cope with this situation we developed a rail/wheel high-speed contact fatigue testing machine based on investigations of conventional testing machines. This machine features additional functions to simulate dynamic conditions between rail and wheel and to clarify the high-speed rolling contact fatigue mechanism of rails including the occurrence of rail shellings. This report presents technical details of this machine and results of traction coefficient tests and rolling contact fatigue tests using it.

I. INTRODUCTION

Currently, rail shellings that are a kind of rolling contact fatigue failure cost a considerable amount for rail renewal on the Shinkansen lines. To cope with this situation some testing machines have been developed in our institute or elsewhere and simulation of rail shellings has been tried using these machines in order to reveal the occurrence mechanism of rail shellings.

We have broadly identified two types, i.e., running surface shelling and gauge corner shelling. In Shikansen rails the running surface shelling poses a problem. Thus we have tried a simulation of running surface shelling. But the occurrence mechanism has not yet been revealed and no effective countermeasures have been available.

We have designed and manufactured a rail/wheel high-speed contact fatigue testing machine. This machine features additional functions to simulate dynamic conditions between rail and wheel and to clarify high-speed rolling contact fatigue mechanism of rails including the occurrence of rail shellings. This report presents technical details of this machine and results of traction coefficient tests and rolling contact fatigue tests using it.

II. TECHNICAL DETAILS

We have studied and compared the conventional testing machines, resulting in development of this contact type machine which consists of the cylindrical test wheels (rail piece and wheel piece). We added to this machine the functions to set the contact angle and the attack angle and to simulate the dynamic conditions between rail and wheel. Table 1 gives technical details of this machine. Figure 1 shows the machine layout.

(1) Test wheels

The rail piece and wheel piece are made of the same materials as the real things. The

Table 1. Technical details of the testing machine.

Items	Technical details
Test wheels	Rail piece ϕ 250–ϕ 550 mm
	Wheel piece ϕ 500 mm
Test speed	310 km/h (maximum)
Radial load	90 kN (maximum)
Thrust load	50 kN (maximum)
Slip ratio	$-10 - +100\%$
Torque	$0 - 1950$ N·m
Attack angle	$-3° - +3$
Contact angle	$0° - 3°$
Test atmosphere	Dry, water lubrication, and oil lubrication

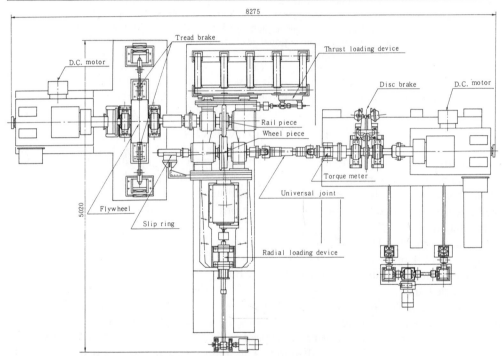

Fig. 1. Layout of rail/wheel high-speed contact fatigue testing machine.

profiles of their contact parts are full scale. Both the rail piece and the wheel piece are manufactured by forging and shaving.

(2) Test speed (circumferential speed of test wheels)

Test speed can be set at a maximum of 310 km/h (86.1 m/s) in consideration of the present maximum train speed on Shinkansen lines and future speed-ups so that the speed effect with regard to occurrence of rail shellings can be examined.

(3) Radial load

Radial load can be set at a maximum of 90 kN under which the Hertzian contact pressure can be equal to that caused between rail and wheel under a wheel load (radial load between rail and wheel) of 400 kN for the purpose of studying the relation between occurrence of rail shellings and wheel load including dynamic load variation, usually called a variation of the wheel load,[1] which happens when a vehicle runs over the track. Moreover, a load control mechanism that keeps the load constant with a feedback system is provided.

(4) Thrust load

Thrust load can be set up to 50 kN in order to study the relation between occurrence of rail shellings, especially gauge corner shellings, and lateral force (thrust load between rail and wheel).

(5) Slip ratio control system (revolution control system)

Revolutions of rail piece and wheel piece are independently set by two d.c. motors to produce a slip ratio that is a circumferential speed of rail piece minus that of wheel piece, all divided by the average of circumferential speeds of rail piece and wheel piece from -10.0% to $+100\%$ at will. In addition a flywheel is provided to stabilize the circumferential speed of the rail piece.

The precision of the slip ratio control system is considered to be maximum $\pm 0.25\%$, being influenced by revolution change of the d.c. motor in use which is due to variations of temperature and voltage.

(6) Torque control system

On the slip ratio control system tests of the traction coefficient (coefficient of tangential force between rail and wheel) cannot be done at the creep region for reasons of precision.

The motor on the rail side equipped with the flywheel to decrease its revolution change is controlled with a revolution feedback system and the other motor on the wheel side is controlled with a torque feedback system, so that testing can be done in creep region.

The precision of the torque meter provided in this testing machine is $\pm 0.3\%$ of maximum torque, but the precision of slip ratio under the conditions shown in Figure 2 is considered to be about $\pm 0.05\%$ at the creep region except at the boundary zone between creep and slip. The relation between traction coefficient and slip ratio shown in Figure 2 is inferred from recent studies.[2] The torque can be set at a maximum of 1950 N·m.

(7) Attack angle

Attack angle is the angle between the tangential direction in curvature and the running direction of the wheel. It is possible to set this angle at up to $\pm 3°$. This maximum value is decided from the attack angle on the curve with radius of 100 m (minimum curvature radius in turnout) at which a two-axle wagon runs.

(8) Contact angle

Contact angle is the angle at which the rail is tiltingly laid on a tie plate or concrete tie. It is possible to set this angle up to a maximum of 3°. That is why it is 1/40 on Shinkansen lines, but it is maximum 1/20 (about 3°) on the narrow gauge lines.

(9) Test atmosphere

Tests under dry, water lubrication, and oil lubrication can be done. For the first time we planned a test under water lubrication to compare with the test results so far obtained, based on the possibility that rail shellings may be influenced by water.

(10) Program operation system

This machine can be operated with the program with which we can set the radial load,

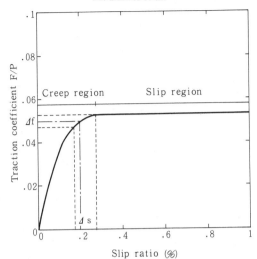

Δf : Error sphere on traction coefficient estimated from pricision of torque meter ($\pm 0.3\%$) under radial load 15kN (crresponding to wheel load 85kN)

Δs : Error sphere on slip ratio corresponding to traction coefficient

Fig. 2. Supposed relation between traction coefficient and slip ratio.

test speed, slip ratio (or torque), and so on in up to five steps at will through slip ratio control (or torque control). Thus we can take account of various running conditions of any vehicle in rolling contact fatigue tests using this machine.

III. TESTS FOR TRACTION COEFFICIENTS

First we carried out tests to search for the relation between traction coefficient and slip ratio in order to determine the test conditions for the rolling contact fatigue test.

We calculate the traction coefficient from the following formula:

$$f = \frac{F}{P} = \frac{T}{P \cdot r}$$

where f is traction coefficient; F, traction; T, torque of wheel piece side; P, radial load; and r, radius of wheel piece.

III-1. Test Conditions

The test conditions for rolling contact fatigue tests aimed at simulation of running surface shelling are considered to be: no thrust load, no attack angle, 1/40 of contact angle, and under water lubrication, from fact-finding studies on the Shinkansen.

We carried out this test under these conditions using the slip ratio control system. Figure 3 shows the state of rail/wheel interface in this test.

(1) Test speed

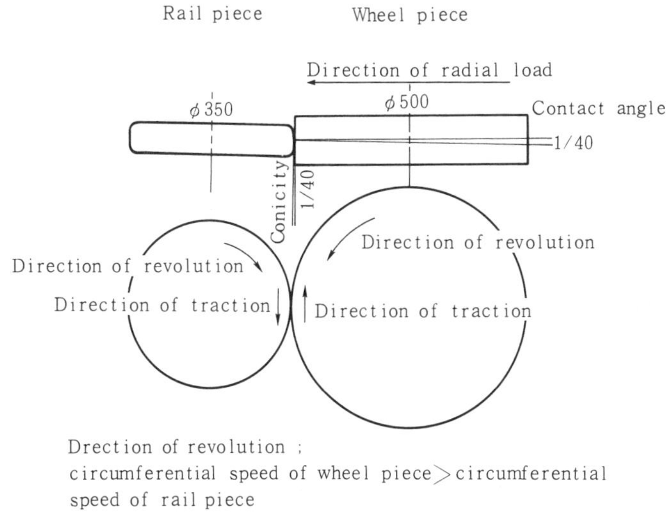

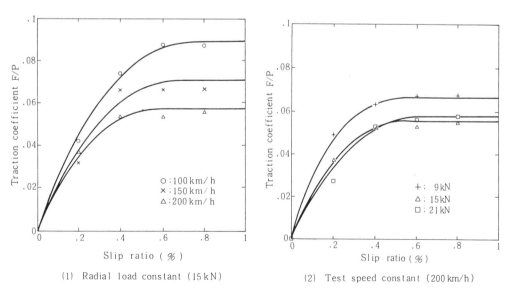

Fig. 3. Condition of rail/wheel interface.

100 km/h (27.8 m/s), 150 km/s (41.7 m/s), and 200 km/h (55.6 m/s) were adopted in consideration of the present operation speed of the Shinkansen.

(2) Radial load

9 kN (corresponding to wheel load 50 kN), 15 kN (corresponding to wheel load 85 kN), and 21 kN (corresponding to wheel load 120 kN) were adopted in consideration of variation of wheel load on the Shinkansen.

(3) Contact angle

1/40 (1.4°) was adopted to simulate the conditions on the Shinkansen.

Fig. 4. Results of traction coefficient tests.

III-2. Test Results and Discussion

Figure 4 shows test results. From this figure we confirmed the speed effect under the condition of radial load being constant but could not confirm the radial load effect under the condition of test speed being constant with respect to the relation between traction coefficient and slip ratio under water lubrication. As the radial load effect is considered to be more influenced than the speed effect by surface roughness and contact pressure, we must judge from tests under defined conditions.

IV. ROLLING CONTACT FATIGUE TESTS

The object of this test is to confirm a plastic flow of rail piece similar to that of the rails on the Shinkansen and to cause a rolling contact fatigue failure.

IV-1. Test Conditions

In view of Figure 4, we examined the test under 0.1 % slip ratio with the torque control system. We had to change the test conditions depending on the occurrence and progress of cracks on the rolling surface of the rail piece. Thrust load, attack angle, and contact angle were similar to those in tests for traction coefficient.

(1) Test speed

At first the test speed was set at 200 km/h (55.6 m/s), but it was gradually decreased to less than 100 km/h (27.8 m/s) so that the axle box vibrating acceleration which increases with the progress of the cracks would not exceed 29.4 m/s^2 (3.0 G, G; gravity acceleration).

(2) Radial load

Radial load as well as the test speed was decreased to 13 kN (corresponding to wheel load 75 kN) from 18 kN (corresponding to wheel load 100 kN) so that the axle box vibrating acceleration would not increase.

(3) Torque

Torque was decreased to 50 N·m under the condition of less than 100 km/h of test speed and 13 kN of radial load from 80 N·m under the condition of 200 km/h of test speed and 18 kN of radial load.

IV-2. Test Results and Discussion

Figure 5 shows the appearance of cracks on the rolling surface of the rail piece at accumulated revolutions of 1.7×10^7. These cracks occurred in the center of the rolling

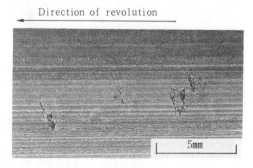

Fig. 5. Cracks occurring on rolling surface of rail piece.

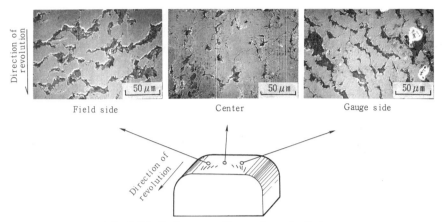

Fig. 6. Plastic flow on rolling surface of rail piece.

surface and were similar to the rolling contact fatigue failures[3] which occurred in the test with a high-speed rolling fatigue testing machine.

Figure 6 shows the state of plastic flow of the rail piece. On the rolling surface a plastic flow which was convex in counter direction of revolution (corresponding to the direction of a train) was observed and looked much like the one[4] observed on the running surface of a rail in use. That is considered to be the reason why the profile of the contact part of test wheels is the same as the one of a real rail and real wheel, and the contact angle is set so as to cause a differential slip based on the diameter difference of the wheel piece.

On the part of the rail piece that contacted the side of the larger wheel piece some pits[5] appeared due to stripping of the rolling surface. These are considered to be similar to the ones developed more remarkably under higher speed in the test with the high-speed rolling fatigue testing machine.

V. CONCLUSIONS

To study the rolling contact fatigue mechanism including occurrence of rail shelling we developed a rail/wheel high-speed contact fatigue testing machine which is considered to be able to simulate the dynamic condition between rail and wheel from comparison and study with conventional testing machines. We confirmed the following from tests of traction coefficient and rolling contact fatigue:

(1) Test speed influences the relation between traction coefficient and slip ratio under water lubrication in terms of constant radial load.

(2) Failure on the rolling surface of a rail piece occurred and developed in a rolling contact fatigue test continued until accumulated revolutions of 1.7×10^7 under 0.1% slip ratio was reached, which is considered to be similar to the one that occurred in the test with the high-speed rolling fatigue testing machine. Plastic flow observed on the rolling surface looks much like the one observed on the running surface of a rail in use.

The rolling contact fatigue failures occurring in this test are considered to be similar to rail shellings occurring on the running surface of a rail in use, but we cannot judge these cracks as rail shellings because their development is slow. Nevertheless we can say that this test result shows the possibility of making clear the relation between traction acting between rail and wheel and rolling contact fatigue failure including rail shelling because it seems that this test can simulate plastic flow occurring on the running surface of a rail in use.

REFERENCES

1) Sato, Y. and Satoh, Y., Cause and effects of wheel load variation on the high speed operating line. *Railroad Track Mechanics and Technology* (Pergamon Press, 1978), pp. 63–78.
2) Ohyama, T., Nakano, S., Natsui, Y., and Ohya, M., Influence of surface characteristics on adhesion force between wheel and rail. *Trans. Jpn. Soc. Mech. Eng.*, No. 475. c (1986), p. 941.
3) Matsuyama, S., Sato, Y., Kashiwaya, K., and Inoue, Y., Plastic behavior of the rolling contact surface of rails. *Tetsu-to-Hagane*, Vol. **69**, No. 13 (1983), S1259.
4) Sugino, K. and Kageyama, H., The wheel/rail contact condition as viewed in terms of plastic flow on the running surface of rail. *Prep. Symp. Wheel/Rail Contact Problems* (1979), p. 9.
5) Satoh, Y. and Matsuyama, S., Influence of contact speed on rolling fatigue under water lubrication. *Proc. Jpn. Soc. Strength and Fracture of Materials* (1987), p. 9.

INDEX OF AUTHORS

Adachi, T., 23

Cheng, X.M., 73

Datta, S.K., 207

Fujimoto, K., 137
Fukusawa, Y., 305

Gomi, Y., 305

Hasegawa, H., 339
Hasimoto, H., 3
Hatano, M., 51
Hatta, K., 63
Hayashikawa, T., 263
Hirakawa, M., 381
Hirano, Y., 159
Horii, K., 73
Horikawa, K., 9

Ishida, M., 389
Ishii, K., 105
Itabashi, M., 273

Kasano, H., 195
Kawamura, T., 289
Kawata, K., 273
Kawazoe, Y., 281
Kimpara, I., 149
Kobayashi, S., 23
Kondou, K., 339
Kondou, T., 289

Konishi, Y., 381
Kunoo, K., 219

Ledbetter, H.M., 207

Matsumae, Y., 73
Matsumoto, H., 195
Matsumoto, R., 83
Matsuno, K., 95
Matsuyama, S., 389
Mikami, I., 347
Mitsui, K., 361
Miura, K., 227
Miura, Y., 347
Miya, K., 235
Moon, D.H., 289
Mori, T., 63
Morishita, S., 253

Nakagawa, M., 183
Nakashima, M., 63
Natsuaki, Y., 381
Natsume, Y., 137
Noto, K., 83
Nozaki, T., 63

Ohira, H., 219
Okuzono, K., 253
Ono, K., 219
Oshima, N., 171

Saida, N., 31
Satake, M., 329

Sato, N., 219
Sato, Y., 389
Satoh, Y., 389
Shimoyama, K., 195
Shindo, Y., 207
Shinnai, Y., 347
Sogabe, H., 315
Sueoka, A., 289
Sugiyama, T., 305
Sumino, K., 315, 361
Suzuki, T., 23
Suzuya, J., 373

Takagi, T., 235
Takahashi, D., 105
Takahashi, K., 381
Takami, H., 105
Takami, H., 115
Takeda, Y., 105
Takei, M., 73
Tamura, H., 289
Tanaka, S., 347

Watanabe, N., 263
Watanabe, Y., 315
Watari, N., 171

Yamaguchi, Y., 39
Yamamoto, K., 125
Yoshihara, T., 137

Zhu, K., 115